T0250382

SHOTCRETE: ENGINEERING DEVELOPMENTS

PROCEEDINGS OF THE INTERNATIONAL CONFERENCE ON ENGINEERING DEVELOPMENTS
IN SHOTCRETE / HOBART / TASMANIA / AUSTRALIA / 2 - 4 APRIL 2001

Shotcrete:
Engineering Developments

Edited by
E.Stefan Bernard
University of Western Sydney, Australia

A.A.BALKEMA PUBLISHERS   LISSE / ABINGDON / EXTON (PA) / TOKYO

Cover photograph courtesy of Baulderstone Hornibrook Bilfinger & Berger Joint Venture

The texts of the various papers in this volume were set individually by typists under the supervision of either each of the authors concerned or the editor.

Published by: A.A.Balkema, a member of Swets & Zeitlinger Publishers
www.balkema.nl and www.szp.swets.nl

ISBN: 978-90-5809-176-5

Shotcrete: Engineering Developments, Bernard (ed.) © 2001 Swets & Zeitlinger, Lisse, ISBN 90 5809 176 7

Table of contents

Shotcrete: Engineering Developments, Bernard (ed.) © 2001 Swets & Zeitlinger, Lisse, ISBN 90 5809 176 7

Preface

This International Conference on Engineering Developments in Shotcrete is the first international conference organized by the Australian Shotcrete Society to foster the dissemination of information about developments in shotcrete technology to the Australian and international communities. It has been organized with the co-operation of the American Shotcrete Association and the Norwegian Concrete Association, both of whom have provided technical and logistic support, for which the society is indebted. The organizing committee has included individuals from the Australian shotcrete industry representing contractors, material suppliers, consultants, and academia. All the members of the committee contributed to making the conference function smoothly, but I am particularly grateful to John Mitchell and Tony Finn for their efforts in seeking sponsors and managing financial issues, and John Gelson and Warren Mahoney for assistance in compiling the all-important mailing list.

A number of conferences on shotcrete have taken place throughout the world in recent years, most of which have covered a broad range of topics within this field. The scope of this conference was restricted to a much narrower range, focusing on recent technical developments that may lead to significant improvements in the way shotcrete is used. While there are still some review papers included, the majority present original material recently developed in laboratories and on sites throughout the world. There are also a significant number of papers addressing the design of shotcrete linings, an issue that has not been examined in sufficient detail in the past. Case studies have been kept to a minimum.

Each paper presented at the conference has been independently reviewed by up to four peers from within the industry. The process of review has been lengthy and rigorous, and I would like to extend my appreciation and thanks to the reviewers for their assistance in making this conference possible.

At the heart of a vibrant and energetic industry are ideas, and central to the generation of new ideas is innovative talent. To encourage continued innovation, new people with an original perspective on shotcrete and experience in other fields must be brought into the industry. An international publishing house was therefore selected to print and distribute these proceedings so that the potential readership could be as wide as possible. The highly international distribution of authors has also helped ensure these proceedings represent a highly original selection of topics on engineering developments in shotcrete.

E. S. Bernard
Conference Chairman and Editor

Shotcrete: Engineering Developments, Bernard (ed.) © 2001 Swets & Zeitlinger, Lisse, ISBN 90 5809 176 7

Acknowledgements

The successful organization of a conference requires the efforts of a large number of people, as well as the financial support of sponsors. The organizing committee for this conference has been fortunate for the support of the following corporations and individuals. Administrative support was provided by Mures Convention Management, through their representative Mike Annand. Secretarial support was provided by Joan Oxford of the University of Western Sydney. Assistance in coordinating the conference was provided by Marilyn Netter and Peter Tatnall of the American Shotcrete Association, and Knut Berg of the Norwegian Concrete Association.

SPONSORS

Jetcrete (Australia) P/L
Master Builder's Technologies (Australia) P/L
Scancem Materials (Australia) P/L
Bekaert OneSteel Fibres Australia P/L
Sika (Australia) P/L
Hagihara Industries Inc.

REVIEWERS

S.Austin	Loughborough University, Britain
N.Banthia	University of British Columbia, Canada
D.Beaupré	Université Laval, Canada
E.S.Bernard	University of Western Sydney, Australia
H.Bleuler	Connell Wagner, Australia
M.Clements	Jetcrete Australia, Australia
K.-I.Davik.	Statens Vegvesen, Norway
S.Duffield	Northparkes, Australia
K.Garshol	MBT International, Norway
P.Groves	Coffey Geosciences, Australia
J.Holmgren	KTH, Sweden
H.Kirsten	SRK, South Africa
T.Melbye	MBT, Switzerland
D.R.Morgan	AGRA, Canada
T.Naaman	University of Michigan, USA
K.-A.Rieder	Grace Construction Products, USA
P.Tatnall,	Synthetic Industries, USA
J.-F.Trottier	Dalhousie University, Canada

Keynote papers

Shotcrete: Engineering Developments, Bernard (ed.) © 2001 Swets & Zeitlinger, Lisse, ISBN 90 5809 176 7

Fracture toughness of fiber reinforced shotcrete: issues and challenges

N.Banthia
University of British Columbia, Canada

ABSTRACT: All cement-based materials are brittle and show a poor response to loads applied in tension. These micro-fracturing materials, therefore, also portray a highly strain-rate sensitive behavior and demonstrate an inferior resistance to fatigue and impulsively applied loads. In shotcrete applications, these limitations are particularly worrisome since inadequate sub-base rigidity may often require a high level of material deformability, toughness and capability to absorb energy. One highly effective solution is that of fiber reinforcement that can dramatically enhance material toughness in shotcrete. This keynote will demonstrates how shotcrete is different from conventional concrete, highlight the leading trends in fiber-reinforced shotcrete, discuss the challenges and elaborate on the opportunities that lay ahead.

1 ISSUES AND CHALLENGES

1.1 Understanding the matrix in fiber reinforced shotcrete

The matrix in fiber-reinforced shotcrete is different from that in cast fiber reinforced concrete in rheology, internal structure, strength gain mechanisms and transport properties (Banthia 1999a, 1999b). Placement using pneumatic compaction and lack of forms in shotcrete requires that the material be more cohesive, adhere well to the surface and resist sloughing off. Smaller rounded aggregates are preferred over large angular ones and aggregate content is less. Shotcrete mixes have higher than normal cement contents, and supplementary cementing materials are more commonly used.

Once in place, shotcrete also has different spatial distribution of various mix components. Water in dry-mix shotcrete, given that it is introduced only at or near the nozzle, is far less uniformly distributed through the placement than in cast concrete. In the wet-process, although the water is expected to be uniformly distributed, the pneumatic compaction results in internal voids that are far different in size-ranges and spatial distribution than in cast concrete. Bleed channels are also absent in shotcrete.

One consequence, interestingly, is that in shotcrete the conditions of complete consolidation are never met, which means that the relationship between water/cement ratio and strength is not unique (Banthia 1999b). In fact, in dry-process shotcrete, an exact determination of the in-situ water/cement ratio itself is virtually impossible. For wet-mix shotcrete, while the water/cement ratio is the same as the one in the initial mix, compressive strengths can be sometimes as much as 30% lower, and flexural strength can be up to 20% higher than the cast counterpart (Banthia et al. 1994a). For a given compressive strength, the elastic moduli may also be much lower than cast concrete.

1.2 Fiber-matrix interaction

In any fiber-reinforced composite, the properties of the matrix such as the compressive strength, tensile strength, elastic modulus, shear modulus, and morphology of the fiber-matrix interface all contribute towards the development of composite properties. Fiber reinforced shotcrete, with the distinct properties of its matrix, develops mechanical and physical properties differently from its cast counterpart. Furthermore, in shotcrete, due to the pneumatic placement, fibers are distributed more or less in a two-dimensional random fashion as opposed to the usual 3-D distribution seen in cast fiber reinforced concrete (Banthia 1999b). This introduces anisotropy, and has a clear influence on the reinforcing efficiency of fibers.

1.3 The rebound issue

One primary concern with the dry-process shotcrete is the high rebound; nearly 20-40% of material and up to 75% of fiber may be lost through rebound (Armelin 1997, Armelin et al. 1997). Unfortunately, rebound material consists primarily of aggregates, and a lack of sufficient aggregate volume makes shotcrete susceptible to volumetric changes, creep and cracking due to plastic shrinkage (Banthia et al. 1998). A loss of fiber through rebound translates into a major loss of fracture toughness, deformability and the post-crack load carrying capacity in shotcrete. The use of various mineral admixtures such a fly-ash, silica fume and metakaolin reduces rebound, but the exact influence of shape, size and gradation of admixture particles on rebound is not well understood (Bindiganavile et al. 2000a). Equally fruitful have been the modeling attempts based on kinematic studies of particles and fibers, but much still remains to be done (Armelin et al. 1998a, 1998b 1998c and 1999; Banthia et al. 1997).

1.4 Fracture toughness of fiber reinforced shotcrete

Increased susceptibility to shrinkage induced cracking (especially in the dry-process shotcrete) has been one of the major causes of an inferior durability in service (Heere et al. 1996). In the early ages, fiber reinforcement is highly effective in terms of enhancing the resistance to plastic shrinkage cracking, but very limited research has been conducted to date to recognize the various mechanisms and to identify the most significant fiber or matrix parameters (Banthia et al. 1998).

Once hardened, shotcrete in most instances has greater deformability, toughness and energy absorption requirements than cast concrete. Predominantly, loads are applied through slow quasi-static ground movements resulting in strain-rates approximately 10^{-5} /sec to 10^{-6} /sec. However, in hard-rock mines and tunnels, rock-bursts may occur due to high in-situ, mining and seismically induced stresses (Kirsten 1997, Tannant et al. 1995). These rapid or dynamic ground deformation impose very severe toughness requirement on shotcrete and it is not uncommon to see rocks almost 1 m size ejected with typical ejection velocities of about 6 m/s and as high as 50 m/s. Proper instrumented impact tests alone can generate appropriate information for designing shotcrete containment structures to withstand such impulses (Banthia et al. 1999b, 1999c, 1999d, 1999e). Particularly lacking is a thorough understanding of the constitutive response of shotcrete materials under variable strain-rates, interaction between the retaining elements and the containment elements and its influence on the structural response of the entire assembly. In this regard, polymeric fibers on account of their visco-elastic nature are seen as particularly resistant to impact loads (Bindiganavile et al., 2000b and 2000c)

One perpetually discussed issue is that of standardized testing. The existing toughness tests such as ASTM C1018 have proven to be highly inadequate (Banthia et al. 1992, 1994b, 1995) and even contentious. Some newer tests such as the ASTM C1399 (Banthia et al., 1999a, 2000) have shown promise, but the range of their applicability is limited. More recently, plate tests have been introduced such as the EFNARC test and the round determinate panel test (Bernard, 1999; Bernard et al. 2000), but it will be some time before these are fully accepted by the international community.

2 CONCLUDING REMARKS

Fiber reinforcement is the most effective way of enhancing the mechanical performance and durability of shotcrete, and not surprisingly, the use of fibers is growing at a much faster rate in shotcrete than in any other application. This paper attempts to demonstrates how fiber reinforced shotcrete is different from conventional fiber reinforced concrete, draws attention to the leading trends in fiber reinforced shotcrete, and discuss the challenges and opportunities that lay ahead.

3 REFERENCES

Armelin, H.S. 1997. Rebound and toughening mechanisms in steel fiber reinforced dry-mix shotcrete, Ph.D. Thesis, University of British Columbia.

Armelin, H.S., Banthia, N., Morgan, D.R. and Steeves, C. 1997. Rebound in Dry-Mix Shotcrete, ACI-Concrete International, 19(9), pp. 54-60.

Armelin, H.S. and Banthia, N. 1998a. Development of a General Model of Aggregate Rebound in Dry-Mix Shotcrete (Part 2), RILEM Materials and Structures (Paris), 31, pp. 195-202.

Armelin, H.S. and Banthia, N. 1998b. Mechanics of Aggregate Rebound in Shotcrete (Part 1), RILEM Materials and Structures (Paris), 31, pp. 91-98.

Armelin, H.S. and Banthia, N. 1998c. Steel Fiber Rebound in Dry Mix Shotcrete: Influence of Fiber Geometry, ACI Conc. Int., 20(9), pp. 74-79.

Armelin, H.S., Banthia, N. and Mindess, S. 1999. Kinematics of Dry-Mix Shotcrete, ACI Materials J. 96(3), pp. 1-8.

Banthia, N., Trottier, J.-F., Wood, D. and Beaupré, D. 1992. Steel Fiber Dry-Mix Shotcrete: Influence of Fiber Geometry, Concrete Int.: Design and Construction, American Concrete Inst., 14 (5): 24-28.

Banthia, N., Trottier, J.-F. and Beaupré, D. 1994. Technical Note: Steel Fiber Reinforced Shotcrete: Comparisons with Cast Concrete, ASCE, J. of Materials in Civil Eng., 6 (3): 430-437.

Banthia, N., Trottier, J.-F., Beaupré, D. and Wood, D. 1994. Influence of Fiber Geometry in Wet-Mix Steel Fiber Reinforced Shotcrete, Concrete Int.-Design and Construction, 16(6), pp. 27-32.

Banthia, N. and Trottier, J.-F. 1995. Test Methods of Flexural Toughness Characterization: Some Concerns and a Proposition, Concrete Int.: Design & Construction, American Concrete Institute, Materials Journal, 92(1), pp. 48-57.

Banthia, N. and Armelin, H.S. 1997. Understanding Rebound in Fiber-Reinforced Dry-Mix Shotcrete, Fifth International Conference on Structural Failure, Durability and Retrofit, Singapore, pp. 471-481.

Banthia, N. and Campbell, K. 1998. Restrained Shrinkage Cracking in Bonded Fiber Reinforced Shotcrete, RILEM-Proc. 35, The Interfacial Transition Zone in Cementitious Composites, Eds. Katz, Bentur, Alexander and Arligui, E and F N. Spon, pp. 216-223.

Banthia, N. 1999a. Fiber Reinforced Shotcrete: Challenges, Proc. of Int. Workshop on High Performance Fiber Reinforced Cement-Based Composites (HPFRCC-3), Eds. Reinhardt and Naaman, Mainz, May 16-19, RILEM Publications-6, pp. 161-170.

Banthia, N. 1999b. Shotcrete: Is it Just another Concrete? Proc. Int. Conf. on Infrastructure Regeneration and Rehabilitation, Sheffield.

Banthia, N., Dubey, A. 1999a. Measurement of Flexural Toughness of Fiber Reinforced Concrete using a Novel Technique, Part 1: Assessment and Calibration, ACI Materials Journal, 96(6), pp. 651-656.

Banthia, N., Gupta, P. and Yan, C. 1999b. Impact Resistance of Fiber Reinforced Wet-Mix Shotcrete, Part 1: Beam Tests, Materials and Structures, RILEM (Paris), 32, pp. 563-570

Banthia, N., Gupta, P. and Yan, C. 1999c. Impact Resistance of Fiber Reinforced Wet-Mix Shotcrete, Part 2: Plate Tests, Materials and Structures, RILEM (Paris), 32, pp. 643-650.

Banthia, N., Gupta, P., Yan, C. and Morgan, D.R. 1999d. Toughness of Fiber Reinforced Shotcrete, Part 1: Beam Tests, Concrete International: Design & Construction, American Concrete Institute, pp. 59-62.

Banthia, N., Gupta, P., Yan, C. and Morgan, D.R. 1999e. Toughness of Fiber Reinforced Shotcrete, Part 2: Plate Tests, Concrete International: Design & Construction, American Concrete Institute, pp. 62-66.

Banthia, N., Dubey, A. 2000. Measurement of Flexural Toughness of Fiber Reinforced Concrete using a Novel Technique, Part 2: Performance of Various Composites, ACI Materials Journal, 97(1), pp. 3-11.

Bernard, E.S. 1999. Correlations in the Performance of Fiber Reinforced Shotcrete Beams and Panels, Civil Engineering Report CE9, School of Civil Engineering and Environment, UWS Nepean, July.

Bernard, E.S. and Pircher, M., 2000. The Use of Round Determinate Panels for the Assessment of Flexural Performance of Fiber Reinforced Concrete and Shotcrete, Cement, Concrete and Aggregates, ASTM, Submitted.

Bindiganavile, V and Banthia, N. 2000a. Rebound in Dry-Mix Shotcrete: Influence of Type of Mineral Admixture, ACI Materials Journal, 97(2), pp. 1-5.

Bindiganavile, V and Banthia, N. 2000b. Polymer and Steel Fiber Reinforced Cementitious Composites under Impact Loading, Part 1: Bond-Slip Response, American Concrete Institute, Materials Journal, In Press.

Bindiganavile, V and Banthia, N. 2000c. Polymer and Steel Fiber Reinforced Cementitious Composites under Impact Loading, Part 2: Flexural Toughness, American Concrete Institute, Materials Journal, In Press.

Heere, R., Morgan, D.R., Banthia, N. and Yogendran, Y. 1996. Evaluation of Shotcrete Repaired Concrete Dams in British Columbia, ACI, Conc. Int., 18(3), pp. 24-33.

Kirsten, H.A.D., Fiber Reinforced Shotcrete, World Tunneling, Nov. 1997, pp. 411-414.

Tannant, D.D., Kaiser, P.K. and McCreath, D.R., 1995. Large Scale Impact Tests on Shotcrete, Laurentian University, March 1995, 45 pp.

Shotcrete: Engineering Developments, Bernard (ed.) © 2001 Swets & Zeitlinger, Lisse, ISBN 90 5809 176 7

Modern advances and applications of sprayed concrete

T.A.Melbye
Director, MBT International Underground Construction Group, Switzerland

R.H.Dimmock
Project Manager, MBT International Underground Construction Group, United Kingdom

ABSTRACT: The purpose of this paper is to give an overview of modern sprayed concrete systems, but also to highlight that sprayed concrete should be regarded as concrete. To achieve this, the paper describes the rapid evolution of sprayed concrete and the international trends that have developed in the last ten years. Recent developments in application systems are explored, particularly pertaining to the use of modern, high performance admixtures and accelerators to ensure high quality, environmentally safe sprayed concrete, produced in line with recognised conventional concrete practices. Coupled with the mix design technology, new, more automated sprayed concrete equipment are reviewed that enable high production rates whilst reducing the risk of reduced sprayed concrete quality from human fatigue and errors. Finally, the paper discusses new developments in sprayed concrete and additional applications outside the underground construction industry that demonstrate sprayed concrete is a versatile permanent civil engineering material.

1 INTRODUCTION

Concrete can be considered as the most cost-effective, versatile building material, and when used with steel reinforcement, virtually all structural elements, even complex shapes can be formed. As conventional concrete is placed in its fluid state, there are often significant costs associated with the necessary shutters and formwork to hold the concrete in position whilst it sets and hardens. Sprayed concrete addresses this drawback by being able to be placed by spraying to a required structural geometry facilitated by the fast setting characteristics of the concrete mix, negating the need for shutters.

Sprayed concrete, or gunite, is not a new invention. Sprayed concrete has been known for more than 90 years. The first sprayed concrete jobs were done in the United States by the Cement Gun Company Allentown as early as 1907, with the sprayed concrete product named "gunite" using simple to operate, compressed air driven pumps, similar to that in Figure 1. Gunite was, and in some circumstances today, continues to refer to a fine-grained mortar concrete. Today, the terms "Sprayed concrete" and "Shotcrete" are commonly used, and are normally based on larger aggregate mixes (maximum 8 to 16 mm). The term "Shotcrete" is used widely in the USA, and "Sprayed concrete" is more typically used in Europe, but gaining rapid acceptance in other regions of the world.

For the purposes of this paper, and in order to avoid confusion, the EFNARC European Specifica-

Figure 1. Traditional dry-mix gunite pump.

tion for Sprayed Concrete define the expression "Sprayed Concrete" as a mixture of cement, aggregate and water projected pneumatically from a nozzle into place to produce a dense homogeneous mass. Later, this term will also be employed in the new European Norms and Standard for Sprayed Concrete (CEN) which is currently under preparation.

In recent years the use of sprayed concrete has equipped the modern underground construction industry in particular with a fast, cost effective lining system. Recent technological developments with the mix design and application equipment described in this paper have occurred as a result of the demands of this industry. Regional differences in the adoption of new sprayed concrete technology vary considerably in both the field of application and geographical location.

2 SPRAYED CONCRETE PROCESSES AND INTERNATIONAL TRENDS

Approximately 30 years ago, the only method available for the application of sprayed concrete was via the dry-mix process. This remained the dominant method until the mid nineties, when eventually the wet-mix process began to be used more widely. Currently the wet-mix method is being adopted on a worldwide basis as the preferred choice, particularly for ground support works. The following sections briefly describe the two processes, and offers some explanations to the geographical and application spread of the two systems.

2.1 Dry-mix process

The dry-mix process is a technique in which the cement and aggregate are batched either at a site-based plant or pre-batched and kiln dried into silos or bags. This material is fed into a dry-mix sprayed concrete pump, similar to that shown in Figure 2, and delivered to the nozzle by compressed air. At the nozzle, the water necessary for hydration is added to the discretion of the nozzleman.

Accelerators may be added in powder form at the pump hopper, or as a liquid with the water at the nozzle. In recent years, in countries such as Austria,

Figure 2. Typical dry-mix sprayed concrete pump.

the use of highly reactive cement in a pre-blended mix has replaced the need for accelerators, however these products must be stored in extremely dry conditions prior to use.

The dry-mix system's virtues are that it is a simple system, with few mechanical and mix design issues that can go wrong. It is often these two factors that have been the reason for selection of the dry-mix process in the past.

However, every process has its drawbacks. With the dry-mix method these can be described as follows:

- High costs of wear on gaskets and friction discs on rotor machines.
- The environmental and safety impact of the high concentrations of dust generated from the system must be taken into account.
- Another important problem in dry-spraying is rebound. Depending on the application, surface rebound can be expected to exceed 15% and be as high as 35 or 40%. Typically the range is 20 to 25%. Rebound can be considerably reduced by the use of new kinds of additives and admixtures such as microsilica or a hydration control system. The losses can be restricted to an average of 15% in these cases.
- Low equipment performance is another drawback often cited. Currently, machines are available which are capable of more than 10 m³/hour, but this is not possible by manual application methods: a spraying manipulator is required. However, due to the increase in equipment wear costs, outputs above 8 m³/hour become critical from an economical point of view. A typical output of between 4 and 6m³/hr should be considered.
- In quality control terms, the sprayed concrete material properties produced by the dry-mix process have a great degree of variance due to inadequate mixing between the nozzle and the substrate, unknown and variable water/cement ratios, and possibly pre-hydration of the mix delivered to the pump.
- Taking the above into account, it is understandable that sprayed concrete produced by the dry-mix method is considered "temporary support" in the tunnelling industry.

Due in no small part to the many years of experience with the dry spraying process there is now a great deal of know-how available. It is extremely important that the materials, equipment, and application techniques are selected to give the best possible results with regards to quality and economy.

Even though dry spraying is the older of the two technologies, because of on-going developments in machine and material technology it has been possible to continually extend the field of application. In the future it is expected that the dry spraying process will continue to play an important role. Main applications will be for projects with relatively small volumes and/or very flexible requirements such as concrete repair, which is considered the strongest market for the dry-mix process. Additionally, long conveying distances suit the dry-mix process, and it is consequently the method that remains in use in many mine workings worldwide.

2.2 *Wet-mix process*

The wet-mix sprayed concrete process uses concrete that is batched in a similar way to conventional concrete, making it possible to check and control the w/c ratio and thus the quality at any time. The consistency can be adjusted by means of admixtures.

With the wet-mix method, it is easier to produce a homogenous product with uniform quality throughout the spraying process. The ready-mix concrete is emptied into a pump and forwarded through a hose by pressure (thick stream transport). Today, piston pumps dominate the wet-mix sprayed concrete pump market. A typical wet-mix piston pump is illustrated in Figure 3.

At the nozzle, air is added to the concrete at a rate of 7 - 15 m^3/min. and at a pressure of 7 bar, depending on whether the spraying is performed manually or by robotic manipulator (see Figure 4). The air is added to increase the speed of the concrete so that good compaction is achieved as well as adherence to the substrate or surface. A mistake often made with wet spraying is that not enough air is used. In most cases only 4 - 8 m^3/min are added which gives a bad result in compressive strength, bond, and rebound. For robot spraying, a minimum of 12 m^3/min. is required. In addition to the air, liquid set accelerators are added at the nozzle to provide fast setting, high build characteristics.

The advantages of the wet-mix process are:
- Low aggregate rebound of about 5 - 10%. Additionally, if fibres are used, the rebound values are minimal compared to the dry-mix process.
- Better working environment, low dust levels.

- Thick layers because of effective use of admixing material.
- Controlled w/c ratio (down to 0.37) and subsequent durability benefits. This leads to the concrete being considered as a permanent element of the structure.
- Quality variance is minimised by virtue of the material being pre-mixed and a significantly reduced negative influence of the nozzleman.
- Higher capacity output from 4 to 25m^3/hr. Typically robotic manipulator spraying will deliver between 15 and 20 m^3/hr.
- Use of steel/synthetic fibres and new advanced admixtures/additives are made possible.
- Reduction in number of operatives.
- Better total economy for applied sprayed concrete.

The disadvantages of the wet-mix process are:
- Limited conveying distance (max. 300 m).
- Higher demands on mix design.
- Cleaning costs (can be solved with the use of hydration control admixtures).
- Limited open time/workability (which can be solved by using hydration control admixtures).
- If site batching is not used, reliance on local batching plants to dose specific types and volumes of admixtures, and provide a higher level of quality control during batching.

2.3 *Current and future trends*

Until the mid 1990's, dry-mix spraying had been the dominant method of applying sprayed concrete. However, in view of the benefits listed above for the wet-mix system, many contractors involved with projects requiring large volumes of sprayed concrete moved to the wet-mix process. In general, the majority of these projects were for tunnel construction and ground support works.

In view of the data presented in Table 1, the present situation worldwide is that 70 to 80% of sprayed concrete is being applied by the wet-mix method

Figure 3. Modern wet-mix sprayed concrete pump with integrated accelerator dosing system.

Figure 4. Robotic manipulator for safe, high output and quality wet-mix sprayed concrete application.

with a rapidly increasing tendency. In some areas, the wet-mix method has been dominant for 20 years (e.g. Scandinavia, Italy, and Switzerland with almost 100% wet-mix), and it is, in general, these countries that have exported the technology on an international basis. Within the next five years the wet-mix method is anticipated to represent more than 80 to 90% of all sprayed concrete works worldwide. Today, more than 8 million m^3 of sprayed concrete is being applied every year; this is also set to increase as the material is used in more applications.

A brief review of sprayed concrete development in Scandinavia and why it happened will illustrate why countries such as Italy, Switzerland, France, UK, Spain, Greece, Australia, Korea, Brazil and some other European countries have been moving in the same direction.

Between 1971 and 1980, the sprayed concrete process used in Norwegian tunnelling changed from dry-mix method only to 100% wet-mix. A similar change took place in Sweden and Finland. The next dramatic change was from mesh reinforcement to the incorporation of steel fibres. Again, the fastest adoption rate for this new technology happened in Norway. During the same period a similar change from manual hand spraying techniques to robotic manipulator arm application took place. Since 1976 steel fibres and microsilica have been added to wet-mix sprayed concrete in rapidly increasing volumes. It is not unfair to say that the Norwegians led the way toward widespread use of wet-mix sprayed concrete. They have the most experience and know most about wet spraying in large volumes.

Table 1. Recent world trends in sprayed concrete volumes and application processes (approximate).

Country/region	Dry %	Wet %	m^3/year	Tendency
Australasia	0	100	> 50,000	wet
Italy	0	100	700,000	wet
Scandinavia	0	100	250,000	wet
France	10	90	250,000	wet
Japan	10	90	2-3M	wet
Switzerland	10	90	300,000	wet
UK	10	90	> 50,000	wet
Asia/Pacific	20	80	> 1M	wet
Brazil	20	80	400,000	wet
Germany	20	80	500,000 to 1M	wet very recently
India/Nepal	20	80	300,000	wet (large projects)
Spain	20	80	300,000	wet
Greece	30	70	200,000	wet
Hong Kong	30	70	100,000	wet
Colombia	40	60	200,000	wet
Rest of Latin America	40	60	> 300,000	wet
China	60	40	> 1M	wet (large projects)
USA	70	30	500,000	wet
Austria	80	20	250,000	wet – slow change

The data presented in Table 1 is only an estimate, but is still believed to be reasonably correct, and can be used for the purpose of showing today's situation and tendencies. Some comments are given as a supplement.

When construction started on the London Underground for the Jubilee Line Extension, Heathrow Express Rail Link, and other well known projects in the area, the market was practically using 100% dry-mix sprayed concrete. Within slightly more than a year this had changed to wet-mix after one project demonstrated the significant programme and cost savings of using the wet-mix system. This is illustrated very well by the fact that MBT Intl. UGC had 17 MEYCO® Suprema wet-mix pumps in operation at any one time in London during 1994 to 1996.

Sometimes it is also claimed that in the "Developing World" countries only the simple, low cost dry-mix method can work properly. The reality is different since there is no relation between the choice of method, local development, and such popular regional simplifications. This is well illustrated by the figures in Table 1 for China, Brazil, Colombia, and India/Nepal compared to Austria. It is suggested that an explanation regarding the strength of the wet-mix method in such countries be based on the international contractors that are active on large projects, who consequently import the wet-mix technology.

Austria, and to a lesser degree Germany, have traditionally been very strong dry-mix areas, as Switzerland was some years ago. Switzerland converted to wet-mix in less than two years whilst Austria and Germany, in particular, have developed new systems within the dry-mix method, using extremely quick setting cement types (without gypsum). This allows spraying of thick layers in one go without the use of any admixtures or accelerators.

This new alternative with quick setting cement appears to be favorable at first sight because of lower material costs per m^3 mixed. But there are some serious negative aspects to be considered, such as high rebound (> 30%), dust (above all national limits), sensitivity to cement quality, high to very high energy cost, low capacity, complicated equipment and difficult handling of the material, as well as being labour intensive. Furthermore, there is no realistic chance of using steel fibres because of high fibre rebound (50 - 70%).

There are clear signs that a change in the direction of the wet-mix method will happen in Austria, as recently Germany has switched to the wet-mix method. Some pioneering large-scale projects have been completed successfully with the wet-mix method both in Germany and Austria (Königshainer Berge, Ditschhardt Tunnel, Irlahüll-Ingolstadt High-speed Link with 400000 m^3, Sieberg Tunnel, Blisadona Tunnel, Austria).

Another interesting area is Australia. Here mining led the civil construction industry in switching from dry to wet-mix, and also in part to steel fibre reinforcement. This has led to a substantially increased market share to almost 100% for wet-mix in both mining and civil construction works in Australia (e.g. Pasminco mine, Melbourne City Link).

It is clear that considering the rapid trend from dry to wet-mix sprayed concrete, future developments within the industry will be centred on the wet-mix process. This development will inevitably have its main focus on high performance admixtures and more mechanised spraying technologies, but perhaps will also need an improvement in human aspects, such as appropriate design and specification for permanent sprayed concrete structures and education of operatives about the new technologies.

The following sections deal with the state-of-the-art wet-mix technology available today, and considers potential future developments.

3 SPRAYED CONCRETE MIX DESIGN

As with traditional methods of casting, sprayed concrete also needs to comply with good concrete practice. All the usual demands on concrete technology (such as w/c ratio, amount of cement, correct consistency and curing) have to be complied with. The reason why so much poor quality concrete has been applied in many parts of the world is that it appears to be forgotten that spraying concrete is only a method of placing and that all concrete technological requirements still have to be fulfilled, such as a low w/c ratio and good workability.

Factors requiring attention to obtain good quality sprayed concrete with the wet-mix method are:
- cement
- microsilica/additives
- aggregates
- admixtures
- set accelerators
- curing
- fibres
- correct equipment
- correct execution

3.1 Cement type and quality

A wet-mix concrete needs a minimum of 200 - 210 litres of water per m^3 to maintain workability and to avoid slump loss. In order to keep the w/c ratio below 0.45, a minimum binder content of about 450kg is therefore needed (200 litres/0.45 = 444 kg binder).

It is possible to spray with a lower cement or binder content but this will result in a w/c ratio exceeding 0.5, reduced quality, and lower early and final strength. Furthermore, a lower binder content also increases the rebound dramatically, as indicated

in Figure 5. The plot demonstrates that a cement content of 450 kg dramatically reduces rebound compared to that achieved below 400 kg.

It is wrong to push the binder content as low as 400 kg, or even less, as is often attempted for economical reasons. This action is self-defeating, as the in-situ sprayed concrete is much more expensive due to the following reasons:
- Higher rebound (typically giving an additional 5 to 8% increase in concrete cost per m^3) compared to a mix with a higher binder content of 450kg/m^3 because the extra rebound costs being included as well as the costs for the applied concrete.
- Higher consumption of set accelerators because of a higher w/c ratio (a 1 to 2% increase in accelerator dosage will lead to an additional 3 to 5% increase in concrete cost per m^3, on top of the extra rebound costs identified above).
- Lower production rates because of more problems with consistency due to increased sensitivity of mix, and more volume to spray because of the higher rebound rate.

Considering these points, it is commercially prudent to add 50kg more cement/binder in order to get better economy, quality, and problem-free sprayed concrete. The additional 50kg of cement represents 4 to 5% of the concrete cost compared to extra rebound and accelerator costs alone which can cumulatively be between 8 and 13%.

Practical experience in Europe has demonstrated that 52.5 grade cements perform better than 42.5 grade cements. This is due to the superior reactivity with set accelerators, better early and final strength results, and cost savings with lower accelerator dosages.

Cement replacements are often considered for use with sprayed concrete mix designs, and their advised maximum quantities are given in Table 2. It should

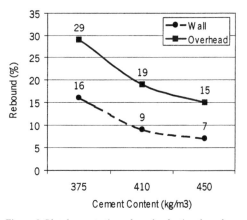

Figure 5. Plot demonstrating rebound reduction through cement increase.

Table 2. Maximum additions of cement replacement products (EFNARC 1996).

Cement replacement material	Maximum addition
Silica fume (microsilica)	15% of Portland cement
Fly ash	30% of Portland cement
	15% of Portland/Fly ash cement
	20% of Portland/ Blast furnace slag cement
GGBS	30% of Portland cement

be noted however, that testing of the mix in both fresh and hardened states should be undertaken to evaluate any detrimental effects of cement replacement.

3.1.1 Cement reactivity

One of the key factors that determine the successful application of sprayed concrete for tunnel linings is the cement-accelerator reactivity. It is essential that testing of cement-accelerator mortars be carried out with all locally available cements to establish the best setting characteristics and strength development prior to site trials.

Although complying with national cement specifications, not all cement types are suitable for sprayed concrete works as they can prove to be incompatible with certain accelerators, giving poor setting characteristics.

The fineness, chemical composition, and age of the cement essentially govern this variation in cement-accelerator reactivity. As a general guide, fineness should be above $350m^2/kg$ and preferably $400m^2/kg$ (fineness is measured in accordance with BS4550: Part 3: Section 3.3: 1978, or Blaine method ASTM 204-84). A C_3A content of not less than 5%, and preferably between 7 and 9%, is also recommended. It is always good practice with sprayed concrete applications to use fresh cement.

For guidance, typical accelerator types, dosages and setting times (Vicat needle) for sprayed concrete are listed in Table 3.

The recent arrival of high performance alkali-free accelerators with extremely fast gelling and setting times has effectively ruled out the use of the standard vicat test as often the 15 second mixing time destroys the gel. To address this, new laboratory testing methods are being examined that allow analysis of the new alkali-free accelerators and have

Table 3. Dosage and setting times for sprayed concrete accelerators.

Accelerator	Typical dosage range	Initial set	Final set
Alkali-free liquid	4 to 10%	< 4mins	< 8mins
Modified silicates	2 to 10%	Too fast to record	< 60mins (at 6% max.)

attempted to reflect more realistically the actual sprayed concrete process that occurs on site.

3.2 Microsilica

Silica fume, or microsilica, is considered to be a very reactive pozzolan. It has a high capacity to incorporate foreign ions, particularly alkalis. Microsilica has a definite filler effect in that it is believed to distribute the hydration products in a more homogeneous fashion in the available space. This leads to a concrete with reduced permeability, increased sulphate resistance, and improved freeze-thaw durability.

When considering the properties of microsilica concrete, it is important to keep in mind that microsilica can be used in two ways:

• as a cement replacement, in order to obtain a reduction in the cement content, usually for economic reasons,
• as an addition to improve concrete properties, both in the fresh and hardened states.

In sprayed concrete, the latter benefit is of paramount importance.

3.2.1 Special advantages of sprayed concrete with microsilica

Normal sprayed concrete qualities (i.e. 20 - 30 MPa cube strengths) can be produced without microsilica, whereas practical and economical production of higher strengths is more or less dependent on the use of microsilica. It seems favorable from a technical point of view to use 5 - 10% (by cement weight) of microsilica.

The correct use of microsilica can provide the sprayed concrete with the following properties:

• Better pumpability (lubricates and prevents bleeding and segregation)
• Reduced wear on the pumping equipment and hoses
• Increased cohesiveness of the fresh concrete and therefore reduced consumption of accelerator which is positive for the final compressive strength
• Increased bond strength to various substrates and between sprayed concrete layers
• Improved strengths
• Improved resistance to alkali aggregate reaction
• Reduced permeability
• Reduced rebound
• Improved sulphate resistance

In fibre reinforced sprayed concrete it also provides:

• Easier mixing and distributing of fibres
• Reduced fibre rebound
• Improved bonding between cement matrix and fibres

Because of these positive effects we wish to maintain that microsilica should always be added to the sprayed concrete in order to obtain the best possible quality.

Due to its fineness it is necessary to add a high rate of plasticisers/superplasticisers to disperse the microsilica when adding it to concrete. The dosage of admixture increases by approximately 20% compared to sprayed concrete without microsilica.

3.3 Aggregates

As for all special concrete, the aggregate quality is of major importance for the fresh concrete as well as for the hardened product. It is particularly important that the grain size distribution and other characteristics show only small variations. Of particular importance are the amount and characteristics of fines, i.e. the grain size distribution and grain size analysis. However, it is not relevant to talk about choice of aggregate, as normally the available material must be used and the prescription must be adapted to it.

Nevertheless, for wet-mix spraying, the following criteria have to be observed:
- Maximum diameter: 8-10 mm. This is because of limitations in the pumping equipment and in order to avoid too much rebound loss. From a technological point of view, one should wish for a larger maximum diameter.
- The granule distribution is also very important, particularly in the lower part. The fine material content in sieve size 0.125 mm should be min. 4-5 % and not higher than 8-9 %.
- Too little fine material gives segregation, bad lubrication and risk of clogging. However, in the case of fibre concrete a surplus of fine material is important, both for pumping and compaction. A high fine material content will give a viscous concrete.

As the margins in the sieve distribution are relatively small, it may often be convenient to combine two or more fractions, e.g. 0-2, 2-4 and 4-8 mm, by adjusting the proportion between them, to make a grading curve that fits within the ideal curve limits. Using more cement or microsilica will compensate for inadequate fine material. Increasing the dosage of water-reducing admixtures primarily compensates for too much fine material. The grain size distribution curve for the aggregate should fall within the envelope defined in Figure 6.

The quantity of 8mm particles should preferably not exceed 10%. The larger particles will rebound when spraying on a hard surface (when starting the application) or penetrate already placed concrete making craters difficult to fill. During screening, storing and handling of the aggregates, measures should be taken to prevent the presence of particles in excess of 8 mm. Coarse particles may block the nozzle and subsequent cleaning can be very time consuming.

An improvement in the grain size curve for a natural sand by the use of crushed materials often results in an increased water demand and poorer pumpability and compaction. Before crushed materials are employed as part of the aggregates, comparison tests should be done to establish whether the addition of crushed material gives an improved result.

More recently on major tunnel projects, the environmental need to recycle excavated rock into the aggregates for sprayed concrete has put higher demands on the admixture technology. On the Neat Tunnel, Switzerland, the recycled aggregates are less than ideal, with non-rounded particle geometry of schistose nature giving the mix high internal friction and high water demand. However, the use of microsilica, hyperplasticisers, and pumping aid admixtures has provided a workable solution.

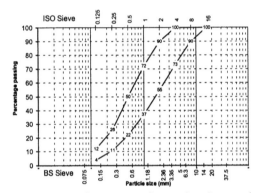

Figure 6. Proposed aggregate grading envelope for sprayed concrete (EFNARC 1996).

3.4 Admixtures (superplasticisers)

In order to obtain specific properties in the fresh and hardened concrete, concrete admixtures should always be used in the wet-mix spraying method. Concrete admixtures are not new inventions. The ancient Romans used different types of admixing material in their masonry, such as goat blood and pig fat in order to make it more moldable. The effect must have been good, since their structures are still standing.

The fact is that concrete admixtures are older than Portland cement, but it is only during the last 30 years that more stringent requirements for higher concrete quality and production have encouraged development, research and use of admixing materials. Water reducers are used to improve concrete workability and cohesiveness in the plastic state. The water reducer can give a significant increase in slump with the same w/c ratio, or the w/c ratio can be reduced to achieve the same slump as for a mix not containing the water reducer. The reduced w/c ratio relates to a direct increase in strength and durability. The higher slump adds to an increased pumpability.

The wet-mix method is attractive as the concrete is mixed and water is added under controlled and reproducible conditions, for instance, at a concrete plant. The w/c ratio, one of the fundamental factors in the concrete technology, is thereby controlled. One often forgets, however, that the equipment makes heavy demands on the fresh concrete first of all in terms of pumpability. Furthermore, the method requires a larger amount of fast setting admixing materials, which may lead to loss of strengths in the final product.

Today, combinations of naphthalene and melamine are often used. This is to obtain the best possible production-friendly concrete. Naphthalenes/melamines (superplasticisers) are chemically distinct from ligno-sulphonates (plasticisers/water reducers) which are not used for sprayed concrete applications. They are better known as high range water reducers since they can be used at high dosages without the problems of set retardation or excessive air entrainment often associated with high rates of addition of conventional water reducers. The action of superplasticisers to disperse "fines" makes them ideal and necessary admixtures for sprayed concrete. The slump increase achieved by adding conventional superplasticisers is time and temperature dependent. However, pumpability can only be maintained for a limited time (20 to 90 minutes) after mixing, and excessive dosages of admixtures can result in a total loss of cohesiveness and in segregation. Normal dosage is from 4 to 10 kg/m^3 depending on the quality requirements, w/c ratio, required consistency, as well as cement and aggregate type.

3.4.1 *Hyperplasticisers - A new generation of special superplasticisers*

A new generation of high performance superplasticisers has entered the market during the last two years. They are based on modified polycarboxylic ether and offer a much higher water reduction than traditional superplasticisers whilst maintaining workability at lower dose rates. These products are referred to as "Hyperplasticisers". As an example of their performance, Glenium™ (polycarboxylic) hyperplasticiser can effect a greater than 40% water reduction, while BNS/melamine superplasticisers can achieve at best a 30% water reduction.

This opens up new possibilities for sprayed concrete. With these types of admixtures you can produce sprayed concrete with a w/c ratio of 0.38 and with a slump of 150-200 mm. The lowering of the w/c ratio results in the following benefits:
- Greatly increased durability
- Longer / better slump retention
- Faster setting
- Higher early strength
- Higher final strength

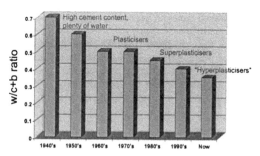

Figure 7. Development of lower w/c+b ratios through admixture technology.

- Reduced rebound
- Possibility to control water ingress or to lower the dosage of accelerators (2 - 3 %)

Glenium™ is already being used widely in Europe in combination with alkali-free accelerators. A chronology of lowered water/binder ratios (w/c+b) through admixture technology is given in Figure 7. This indicates the future direction of sprayed concrete admixtures.

3.5 *Hydration control system*

Perhaps the one technology that has made the wet-mix process so successful in recent years has been the use of hydration control admixtures.

Before hydration control admixtures were introduced, the wet-mix sprayed concrete system had no flexibility, in that batched concrete had to be used with 1 to 2 hours (and even in less time in ambient temperatures in excess of 20°C), otherwise it was discarded as the hydration process made the material unusable (see "Traditional sprayed concrete" in Figure 8). This caused several cost and programme problems:
- The supply and utilization of sprayed concrete mixes for infrastructure projects in congested environments created problems for both the contractor and ready-mixed concrete supplier.
- Long trucking distances from the batching plant to the site, delays in construction sequences as well as plant and equipment breakdowns ensured that much of the concrete actually sprayed was beyond its 'pot-life'.
- In addition to this, environmental regulations imposed restrictions upon the working hours of batching plants in urban areas, meaning that a contractor who required sprayed concrete mixes to be supplied 24 hours per day, was only able to obtain material for 12 hours a day.

To solve these logistical problems, hydration control admixtures, such as MBT's Delvocrete Hydration Control System, were introduced in the mid 1990's. This system comprises two liquid components. The first component is the stabiliser admixture that is added to the mix at the batching plant.

Traditional sprayed concrete

New flexibility with hydration control

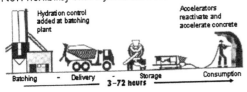

Figure 8. Hydration control for total flexibility in sprayed concrete.

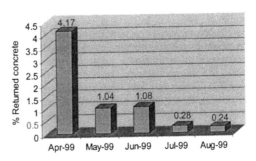

Figure 9. Reduction in returned concrete from site with use of hydration control system.

This admixture coats the cement grains and prevents hydration from occurring. Depending on dose, the stabiliser can control hydration from 3 to 72 hours, as indicate in Figure 8. However with most sites, a stabilised period of 6 to 8 hours is normally sufficient for sprayed concrete operations.

The second component to be added at the nozzle with the air is the activator that disperses the stabiliser from the cement grains and immediately activates the hydration process. It ensures that all concrete which is sprayed through the nozzle contains a 'fresh' cement that has undergone little or no hydration reactions allowing maximum setting characteristics and early strength development.

The system brings revolutionary benefits to sprayed concrete and is currently being used on a significant number of major projects in Europe, America, the Middle East, and Far East. By means of example, the currently under-construction 3.5km North Downs Tunnel which forms part of the new Channel Tunnel rail link in the UK, used the following mix design for the permanent sprayed concrete tunnel lining:

- 360 kg Rapid hardening cement 52.5N
- 90 kg PFA
- 1730 kg aggregates (60:40 - sand:crushed granite aggregates)
- 2.5 kg hyperplasticiser
- 4.5 kg stabiliser
- 7% alkali-free accelerator
- w/c+b ratio = 0.38
- Slump: target 200 mm

This mix design provided a stabilised concrete for up to 6 hours, and the alkali-free accelerator was used as the activator. Figure 9 represents data supplied by the contractor to demonstrate the reduction in the number of concrete trucks returned

with the introduction of Delvocrete Stabiliser hydration control in May 1999.

3.6 Set accelerators

The wet-mix method requires the addition of fast setting admixtures at the nozzle. The primary effect of the material is to reduce the slump (consistency) at the moment of spraying from liquid to earth moist while the concrete is still in the air, so that it will adhere to the surface when the layer thickness is increased. This slump reduction must take place in a matter of seconds.

With the use of set accelerators, spraying on vertical surfaces and overhead becomes possible. The setting effect allows application of sprayed concrete for initial support - an important function in the New Austrian Tunnelling Method (NATM). Water ingress (e.g. from the rock behind) usually calls for a higher proportion of admixtures to further accelerate the setting of the sprayed concrete.

Accelerators are added in liquid form via a pressure tank or a special dosage pump (piston or worm pumps). The dosage of accelerator will vary, depending on the ability of the operator, the surface, and the w/c ratio (high w/c ratios will increase the need for accelerators in order to reduce consistency).

A side effect of the accelerators is a decrease in the final strength. Compared to plain concrete (without accelerators), the 28-day strength can be reduced significantly. Therefore, the accelerator consumption should be kept at a minimum at all times (lower consumption on walls than in the roof).

Different types of accelerators are used in sprayed concrete:

- waterglass
- modified sodium silicate
- consistency activators
- aluminates (sodium or potassium or a mix)
- alkali-free

Water glass should normally not be used because high dosages (>10 - 12%, normally 20%) are needed consequently decreasing strengths and producing a

15

low quality concrete and a false security. Normal water glass is essentially banned in Europe.

Accelerators containing aluminates should not be used because of their strong negative influence on working conditions and environment. Due to their high pH > 13 they are very aggressive to skin and eyes causing severe burns. Additionally, these accelerators have a pronounced negative effect on the final strength and durability of the sprayed concrete.

3.6.1 *Modified sodium silicates*

Modified sodium silicates provide a momentarily gluing effect (less than 10 seconds) in the sprayed concrete mix (loss of slump) and take no part in the hydration process, unlike alkali-free accelerators. Modified sodium silicates bind the water in the mix and dosage is subsequently dependent on the w/c ratio: the higher the w/c ratio the more modified sodium silicate is required in order to glue the water in the concrete mix.

Modified sodium silicates do not give very high strength within the first 2 to 4 hours and are therefore not the ideal accelerator for soft ground tunnelling applications, where high early strength is required. Normal final setting occurs after 30 minutes, (depending on cement type and ambient temperature). The advantages of modified silicates include:
- Work with all types of cement
- Less decrease in final strength than with aluminate based accelerators at normal dosages (4 to 6%)
- Very good gluing effect
- Environmentally friendly, not so aggressive to skin. The pH is less than 11.5, but still direct skin contact must be avoided and gloves and goggles should always be used
- Much lower alkali content than aluminate based products (Na_2O < 8.5%).

Disadvantages of modified silicates:
- Temperature dependent (cannot be used at temperatures below + 5°C).
- Limited thickness: max. 80 - 150 mm

The European Sprayed Concrete Specification (EFNARC) only allows a maximum dosage of 8% by weight of the cementitious material for the use of liquid accelerators.

In most applications with a reasonable modified silicate dosage (3 - 6%) and good quality control, not more than 20% strength loss is acceptable. In practice the loss is between 10 - 15%. Note that an 18 year old wet-mix sprayed concrete recently tested in Norway had the same strength as after 28 days.

3.6.2 *Alkali-free sprayed concrete accelerators*

Of late, safety and ecological concerns have become dominant in the sprayed concrete accelerator market, and applicators have started to be reluctant to apply aggressive products. In addition, requirements for strength and durability of concrete structures are increasing. Strength loss or leaching effects suspected to be caused by strongly alkaline accelerators (aluminates) has forced our industry to provide answers and to develop products with better performances.

Due to their complex chemistry, alkali-free accelerators are legitimately much more expensive than traditional accelerators. However, accelerator prices have very little influence on the total cost of in-place sprayed concrete. Of much greater consequence are the time and rebound savings achieved, the enhancement of the quality, and the safe working

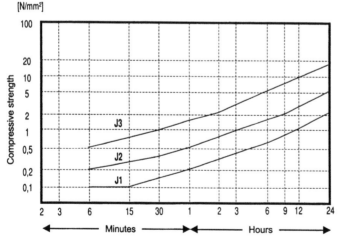

Sprayed Concrete Class	Application
J_1	Sprayed concrete suitable for the placing of thin layers on a dry base without special load bearing requirements to be met during the first hours after placing; it offers the advantages of low dust formation and rebound.
J_2	Sprayed concrete that is required to be placed as quickly as possible in thick layers (including overhead). Additionally, sprayed concrete can be applied to water bearing ground, and sections of lining that are immediately adjacent to construction operations involving immediate stress and strain changes, such as new excavations or spiling. In normal tunnel conditions J_2 should not be exceeded.
J_3	Sprayed concrete for support to highly friable rock or excessive ingress of water. Due to the high level of dust and rebound, this class should only be used in limited areas.

Figure 10. Early age sprayed concrete development (Austrian Sprayed Concrete Guidelines, 1999).

Figure 11. Meyco SA-170 alkali-free accelerated sprayed concrete to control active ground water, Blisadonna Tunnel, Austria, 2000.

environment.

The increasing demand for sprayed concrete accelerators termed *alkali-free* is always related to one or more of the following issues:

1 Reduction of risk of alkali-aggregate reaction, by removing the alkali content arising from the use of the common caustic aluminate-based accelerators.
2 Improvement of working safety by reduced aggressiveness of the accelerator in order to avoid skin burns, loss of eyesight, and respiratory health problems. The typical pH of alkali-free accelerators is between 2.5 and 4 (skin is pH 5.5).
3 Environmental protection by reducing the amount of aggressive leachates to the ground water, from both the in-situ sprayed concrete and rebound material deposited as landfill.
4 Reduced difference between the base mix and sprayed concrete final strength compared to older aluminate and waterglass accelerators that typically varied between 15 and 50%.

The focus within different markets, regarding the above points, is variable. Where most sprayed concrete is used for primary lining (in design considered temporary and not load bearing), points 2 and 3 are the most important. When sprayed concrete is used for permanent structures, items 1 and 4 become equally important.

As a result of the above demands, in excess of 25000 tonnes of alkali-free accelerator has been used worldwide since 1995.

3.6.3 A family of alkali-free accelerators

A significant R&D input into alkali-free accelerators has produced a family of products that are suited to most conditions, with performance equal to the old aggressive accelerators. Today alkali-free accelerators are successful because they perform both with the variation of cement quality and reactivity found around the world, but also meet demanding sprayed concrete performance requirements to stabilise poor

ground or running water, whilst maintaining good strengths and durability properties.

The most widely accepted sprayed concrete early age strength development specification is given in Figure 10 (Austrian Sprayed Concrete Guideline, 1999), where three classes are defined, J_1, J_2 and J_3. For most applications, achieving J_2 is sufficient to achieve good build rates with early strength development to stabilise tunnel excavations. The final strengths and durability of alkali-free accelerated sprayed concrete in this class will be excellent.

Should the ground be very poor, such as with alluvial deposits and with ground water flow, then J_3 class will be required. High performance alkali-free accelerators are available for these conditions, with very high early strength and moderate long-term strength, as demonstrated in Figure 11. Care must be taken to evaluate the long term durability of sprayed concrete in J_3 class, as often the sprayed concrete is not always compacted sufficiently, and may be cement rich due to high rebound rates.

3.6.4 Setting and early age strength development – vital considerations

Confusion exists between sprayed concrete setting characteristics and early age strength development. In relation to Figure 10, setting of the sprayed concrete should be considered from 0 to 0.5 MPa. This may occur between 6 minutes and 1 hour, depending on the dose and type of alkali-free accelerator used, and is measured with a Proctor penetrometer.

Early age strength development should be considered as measurements being above 0.5 MPa and those recorded up to 24 hours age. It is common practice to measure these strengths using the HILTI pull-out test method.

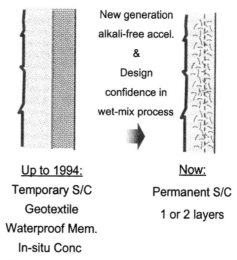

New generation alkali-free accel. & Design confidence in wet-mix process

Up to 1994:
Temporary S/C
Geotextile
Waterproof Mem.
In-situ Conc

Now:
Permanent S/C
1 or 2 layers

Figure 12. Wet-mix and alkali-free accelerators permit single shell linings for tunnels.

Particularly in tunnelling where the sprayed concrete lining will experience the greatest load directly behind the excavation face (refer to Figure 16), it is crucial that strength development is continuous after initial setting, and must not remain dormant until 6 to 10 hours age. In other words, traditional views of a "good" sprayed concrete purely based on fast setting characteristics is dangerous, as limited or no strength development afterwards may endanger the operatives at the tunnel face.

It is vital in soft ground tunnels that the choice of alkali-free accelerators for sprayed concrete should be based on providing setting and strength development characteristics that remain in the J2 and J3 class defined in Figure 10, particularly from 2 to 6 hours age.

3.6.5 Wet-mix and alkali-free accelerators enable economical permanent sprayed concrete structures

Wet-mix sprayed concrete coupled with new alkali-free accelerators has equipped the sprayed concrete industry with a method of placing high performance concrete. Of particular benefit is the tunnelling industry that has traditionally constructed tunnels with two shells. The first shell was typically made of temporary sprayed concrete that was not of sufficient quality to contribute to the long term design requirements. This was then followed by a permanent cast concrete shell on the inside.

With the use of alkali-free accelerators and the wet-mix process, primary tunnel linings can be considered as permanent support elements. As a permanent element of the structure, a lighter second lining may be required to improve watertightness and aesthetics, acting monolithically with the first layer to form a single shell. Where water-tightness and finish are not critical, a single layer may be adequate (see Figure 12).

By means of example, the North Downs Tunnel, UK, opted to use the primary sprayed concrete lining of the tunnel as a permanent structural element. This allowed the following savings to be achieved:
- Reduced excavation size
- Reduced overall lining thickness
- Reinforcement of second in-situ concrete layer removed
- Reduced landfill requirement
- Reduced programme length

By using alkali-free accelerators and the wet-mix process, the material and programme benefits allowed a cost saving of £10M for a tendered £80M project (13% saving). With this in mind, the tunnelling industry appears to have adopted sprayed concrete as a permanent lining material demonstrated by the ever increasing number of projects shown in Figure 13, particularly with the introduction of alkali-free accelerators in 1996. The apparent recent decrease is due to not all current data being provided to the ITA-AITES database.

3.7 Concrete improving (internal curing)

One would think that tunnels have ideal curing conditions with high humidity (water leakage), no wind and no sun exposure. However, this is not the case. Tunnels and other underground construction projects have some of the worst conditions for curing due to the ventilation that blows continuously dry (cold or hot) air into the tunnel. It can be compared with concrete exposed to a windy area.

3.7.1 Background

Curing is one of the most basic and important jobs in sprayed concrete because of the large cement and water content of the mix and the consequent high shrinkage and cracking potential of the applied concrete. Other reasons are the danger of rapid drying out due to the heavy ventilation as is usual in tunnels, and the fast hydration of accelerated sprayed concrete and the application in thin layers. Therefore, sprayed concrete should always be cured properly by means of an efficient curing agent. However, the use of curing agents involves several restrictions: they must be solvent-free (use in confined spaces), they must have no negative influence on the bonding between layers, and they must be applied immediately after placing of the sprayed concrete. Most of the in-place sprayed concrete around the world has no bonding and many cracks, due to the fact that no curing is applied.

With the use of sprayed concrete as a permanent final lining, long-term quality and performance requirements have become significant. These requirements are good bonding, high final density and compressive strengths to ensure freeze/thaw and chemical resistance, water-tightness, and a high degree of safety.

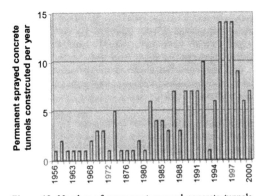

Figure 13. Number of permanent sprayed concrete tunnels constructed per year worldwide (source. ITA-AITES, 2000).

18

When curing sprayed concrete with a curing agent, one has to be very careful with the cleaning procedure of the substrate before applying a subsequent layer. Cleaning must be done with high-pressure air and a lot of water (use spraying pump and nozzle, adding air at the nozzle).

Another problem with curing agents is to be able to apply them quickly enough after finishing of spraying. To secure proper curing of sprayed concrete, the curing agents must be applied within 15 to 20 minutes after spraying. Due to the use of set accelerators, the hydration of sprayed concrete takes place a very short time after spraying (5 to 15 minutes). The hydration and temperature rise are most likely during the first minutes and hours after the application of the sprayed concrete and it is of great importance to protect the sprayed concrete at this critical stage.

Application of curing agents requires two time consuming working operations: application of curing agent, and cleaning/removal of the curing agent from the sprayed concrete surface between the layers in the case of multiple layers.

In many countries with experience in wet-mix sprayed concrete, like in Norway and Sweden, and in big projects worldwide, there is an obligation to cure the sprayed concrete with a curing agent.

Very good results have been achieved with the use of a special curing agent for sprayed concrete. It is used in many big projects and in different countries, everywhere with very good results. The use of specially designed curing agents for sprayed concrete improves bonding by 30 - 40% compared to no curing (air curing), reduces shrinkage and cracking and also gives a slightly higher density and compressive strength (at 28 days). These results are confirmed by several laboratory tests and field trials. However, in order to achieve these results, proper cleaning is required before subsequent layers of sprayed concrete can be applied. Even with easy-to-apply products, curing of sprayed concrete remains a time consuming job and is often regarded as a hindrance to other tunnelling operations. As a consequence the curing process is poorly achieved, if at all.

3.7.2 Concrete improving admixtures

Concrete improving (internal curing) means that a special admixture is added to the concrete/mortar during batching. This admixture produces an internal barrier in the concrete, which secures safer hydration than the application of conventional curing agents. The benefits resulting from this new technology are impressive:

- The time consuming application and, in the case of multiple sprayed concrete layers, removal of curing agents are no longer necessary.
- Curing is guaranteed from the onset of hydration.

- There is no negative influence on bonding between layers enabling structures to act monolithically without the risk of de-lamination.
- Acts on the whole thickness of the concrete lining, rather than just the exposed surface.

Figure 14 illustrates typical bond strength results for sprayed concrete samples that have no curing, a spray applied curing membrane, and finally, a sprayed concrete containing Meyco TCC735 Concrete Improving admixture. The externally cured sprayed concrete gives the worst performance as the membrane has a negative influence on the bond strength. The effectiveness of the concrete improver is also seen with age, as demonstrated by the increase between 10 and 28 days.

As a consequence of this optimum curing effect, all other sprayed concrete characteristics are improved: density, final strengths, freeze-thaw and chemical resistance, watertightness, reduced cracking and shrinkage. In addition, it also improves pumpability and workability of sprayed concrete, even with low-grade aggregates. It particularly improves the pumpability of steel fibre reinforced sprayed concrete mixes.

4 NEW FIBRE REINFORCED SPRAYED CONCRETE TECHNOLOGY

The advantages of fibre reinforcement in sprayed concrete have been demonstrated in numerous projects and applications around the world. When using state of the art technology, the technical performance of fibre reinforcement is generally equal to traditional mesh reinforcement. Additionally, it provides a number of other advantages:

- Overall productivity when applying shotcrete to drill and blast rock surfaces is often more than doubled.
- Substantially improved safety, while sprayed concrete and reinforcement can be placed by remote controlled manipulator (nobody venturing

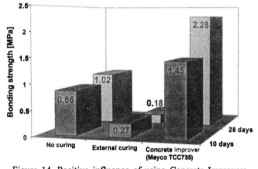

Figure 14. Positive influence of using Concrete Improvers to enable increased bond strength (e.g. Meyco TCC735).

below partly supported or unsupported ground to install the mesh).

- No areas of poor compaction behind overlap areas of 3 to 4 layers of mesh, causing very poor concrete quality and high risk of subsequent mesh corrosion and concrete cover spalling.
- The intended thickness and overall quantity of sprayed concrete can be achieved quite accurately and the problem of excess quantity to cover the mesh on rough substrates is avoided.
- One layer of mesh will typically be placed at varying depth in the sprayed concrete layer and cannot be placed in the cross section to target tension zones. The fibres will be present in the whole cross section, irrespective of where the tension will occur.
- Logistic advantage of avoiding handling and storage of reinforcement mesh under ground.

To be able to achieve the above advantages the wet-mix application method must be used. The reason for this is the lack of control and the large amount of fibre rebound in the dry process. Typically, steel fibre rebound in the dry process is above 50% and using synthetic fibres it is likely to be even higher. In comparison, steel fibre rebound is typically 10 to 15% in the wet mix process.

Until now, steel fibres have been the only alternative to mesh reinforcement. However, structural synthetic fibres have reached a performance level that is comparable to steel fibres, and are set to be a viable alternative in the future.

4.1 Structural synthetic fibres

Structural synthetic fibres made of polymers have recently become available that can equal the post-crack performance of steel fibres in many applications. These new plastic fibres are made of high quality materials and are delivered in lengths ranging from 30 to 60 mm with a geometry specifically designed to resist matrix pull-out (see Figure 15), thereby enhancing concrete performance even after it has developed stress cracks. Test results from Australia and Europe (Bernard 1999, Banthia & Yan) show that this type of fibre can achieve a level of toughness required in tunnels if dosed moderately (10 - 13 kg/m^3). The tests show that these fibres absorb at least 700 - 900 Joules according to the EFNARC plate test. This result is approximately equal to the result achieved with 30 kg/m^3 of high quality fibres.

Additional benefits of structural synthetic fibres over steel fibres are listed below:
- Properly designed high volume synthetic fibres are typically more user friendly in pumping, spraying and finishing compared to steel fibres.
- Structural synthetic fibres result in lower pump pressures and less wear and tear on the pumping equipment, hoses, and nozzles than steel fibres.

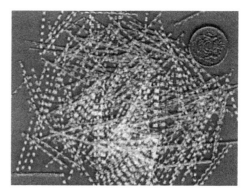

Figure 15. High Performance Polypropylene fibres (HPP) for sprayed concrete.

- Where finished surfaces are required, synthetic fibre reinforced sprayed concretes typically are easier to cut, trim and finish than steel fibre reinforced sprayed concretes.
- High volume structural synthetic fibre reinforced sprayed concrete typically displays better residual load carrying capacity at larger deformations compared to steel fibre reinforcement. This is particularly beneficial in squeezing ground and mining activities where rock bursts are likely.
- The performance of steel fibres is based on their ability to bridge cracks. If exposed they can corrode thereby reducing their reinforcing role. Synthetic fibres, being made of an inert polymer, can bridge cracks without a corrosion risk.
- Concrete structures in severe chemical exposure environments, such as sub-sea and coastal tunnels, will benefit from non-corrosive fibres.
- Protruding synthetic fibres do not cause skin lacerations, which is often considered a negative aspect of exposed steel fibres.

4.2 Future insight into fibre reinforced sprayed concrete for permanent structures

Figure 16 simplifies the current knowledge levels concerning fibre reinforced sprayed concrete tunnel linings as a permanent structural element. As the tunnel environment is probably the most demanding on sprayed concrete performance both during construction and throughout the operational life of the structure, the knowledge gained can be used in most other ground and civil engineering applications.

The following list suggests areas of study into fibre reinforced sprayed concrete that should receive attention of the next five years so that a more complete understanding of the material performance is acquired:
- As fibre reinforced sprayed concrete was introduced in large projects during the mid 1970s, long-term durability assessment has only been based on in-situ tunnel samples, particularly in

20

Norway, up to an age of 25 years. Early indications from these studies have demonstrated the sprayed concrete to be in good condition, with compressive strengths similar to those achieved at 28 days. Fibres have been shown to corrode near exposed surfaces, but show little signs of corrosion within the main section of the linings. Considering these sprayed concrete mixes had w/c ratios in excess of 0.45, the durability of more recent sprayed concrete linings will be further improved. However, studies on the long-term durability of existing sprayed concrete linings needs to be continued so that an understanding of structural decay and whole life costing models can be established.

- From Figure 16 it can be noted that significant loads on the tunnel lining normally occur within a tunnel diameter of the face excavation. In modern tunnelling, this represents sprayed concrete of less than 1 day age. It is acknowledged that the young sprayed concrete lining behaves plastically at this stage, allowing a re-distribution of stresses throughout the lining, thereby reducing adverse moments. However, the role achieved by fibre reinforcement during these vital early ages has not been evaluated to a large extent. When referring to durability of sprayed concrete linings, it is this early stage that may define the future performance of the lining, and it is conceivable that fibre reinforcement has a significant benefit in this regard.

- Significant amounts of test data on the post-crack performance of fibre reinforced sprayed concrete is available today. This is very beneficial to the hard rock tunnelling and mining market, where rock support relies on this material performance for both short-term construction safety, and long-term operational security. In respect to soft ground tunnels, this information is of minimal value apart from safety benefits during construction. The soft ground tunnelling industry requires further insight into the benefits of fibre-reinforced sprayed concrete, for example in terms of achievable flexural strengths relating to fibre type, and to the benefits and design recommendations to achieve durable, anti-crack sprayed concrete linings. Currently, only small-scale tests on concrete rings using unknown steel fibre types have guided the industry to a dose of 40 kg/m^3 of steel fibres to maintain crack widths of less than 0.2 mm. More recent studies have shown relationships between fibre diameter and length to calculate fibre dosages for anti-crack reinforcement, but no actual tests have been achieved to the authors' knowledge.

5 SPRAYED CONCRETE EQUIPMENT

The application of sprayed concrete using poor equipment may ruin all the efforts made through the development of a good concrete mix design and

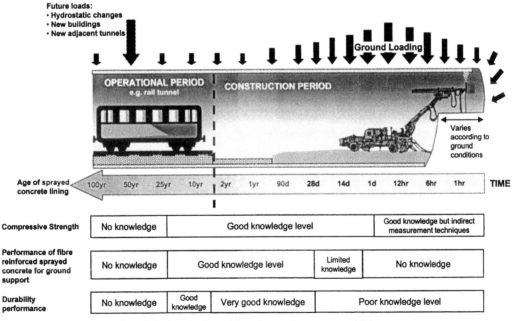

Figure 16. Short to long-term knowledge base for fibre reinforced sprayed concrete tunnel linings.

21

other actions taken to produce a high quality concrete. Only with a balanced system of reliable spraying equipment, high performance products, and competent service can the required quality and efficiency be achieved.

Parallel to the development in material technology there has been constant innovative development in the equipment sector to produce machines suited to the new products and that are adaptable to the ever-changing conditions in the construction business. The result is a wide range of systems that cover all sprayed concrete works: from major tunnelling contracts with large quantities of concrete to be sprayed, down to small volume repair works. Common to all developments in equipment is the tendency toward integrated and automated systems that ensure higher production output, consistent and controllable quality, as well as safer and more operator-friendly working conditions.

Development of dry-mix equipment is considered to be at its limit, and in view of the international trend towards wet-mix application methods, present development is almost entirely via wet-mix sprayed concrete equipment.

5.1 Developments

To ensure even spraying, the latest equipment developments aim at providing a pulsation-free conveyance of the wet-mix from the pump to the nozzle.

This is put into practice with the MEYCO® Suprema: the electronically controlled push-over system that is integrated into the output adjustment brings the pulsation of the material flow to a level that is hardly noticeable at the nozzle. An integrated memory programmable control system (PLC) supervises, coordinates and controls all functions of the machine. The PLC system allows checking and controlling of data which can also be printed out, e.g. dosing quantity of admixtures, output capacity, etc. A dosing unit for liquid admixtures is integrated into the drive system of the machine and connected to the PLC system. This synchronises the dosing of accelerator to the concrete output of the pump. This is of critical importance when constructing permanent sprayed concrete structures so as to reduce the risk of reduced strength and durability with overdosing.

5.2 Spraying manipulators

Spraying manipulators or robots are suitable for use wherever large quantities of sprayed concrete are applied, especially in tunnel and gallery constructions or for protection of building pits and slopes. Thanks to mechanised and automated equipment, even large volumes of sprayed concrete (dry-mix and wet-mix) can be applied under constantly optimum conditions and without fatigue for the nozzleman who also profits from higher safety and improved general working conditions.

The spraying robots typically consist of:
- Lance-mounting with nozzle
- Boom
- Remote control
- Drive unit
- Turntable or adapter-console (for different mounting versions).

5.3 Spraying mobiles

Many suppliers also offer complete mobile systems with integrated equipment for the complete spraying job, as shown in Figure 17. A Spraying Mobile typically consists of:
- Wet-mix spraying machine/pump
- Spraying manipulator
- Accelerator storage tank
- Dosing unit for accelerator
- Cable-reel with hydraulic drive
- Air compressor, capacity >12 m³/min
- Central connection and control system for external power
- High pressure water cleaner with water tank
- Working lights

The benefits of mechanised spraying can be summarised in the following:
- Reduced spraying cycles due to higher output capacity and the elimination of time-consuming installation and removal of the scaffolding, particularly in tunnels with variable profiles.
- Cost savings thanks to reduced rebound, and labour savings.
- Improved quality of the in-place sprayed concrete thanks to even spraying.
- Improved working conditions for the nozzleman thanks to protection from cave-ins, rebound, dust, and accelerators.

Essential to some specialist sprayed concrete contractors is the flexibility to take a spraymobile onto the road network and be able to travel between

Figure 17. Meyco Spraymobile – complete sprayed concrete system on board vehicle.

sprayed concrete project sites, without the need for unnecessary site set-up time associated with individual compressors and generators etc. To address this need the Meyco Roadrunner (Figure 18) was developed, and has found a strong market in Scandinavia.

For the mining market, a small spraymobile is required, such as the Meyco Cobra. Due to the restricted size of mine roadways and drifts, this type of rig must have a robust, articulated chassis. Again all equipment for spraying wet-mix sprayed concrete is included on the rig. An example of this type of machine is given in Figure 19.

5.4 *The latest generation of concrete spraying robots*

Spraying manipulators or robots are suitable for use wherever large quantities of sprayed concrete are applied. A new machine, the Meyco Logica, based on the worldwide well known principle of the MEYCO® Robojet, has now been developed in co-operation with industry and academia. A new automatic and human-oriented control system has been developed for this manipulator with 8 degrees of freedom. The new tool enables the operator to manipulate the spraying jet in various modes, from purely manual to semi automatic and fully automatic, within selected tunnel areas. In one of the modes the operator uses a 6-D joystick (see Figure 20). The calculation of the kinematics is done by the control system. A laser scanner sensor measures the tunnel geometry and this information is used to control automatically the distance and the angle of the spraying jet.

The aim of this control is not to automate the whole job of spraying but to simplify the task and enable the operator to use the robot as an intelligent tool and to work in an efficient way with a high level of quality. With the correct angle, and constant spraying distance, a remarkable reduction in rebound and therefore savings in material cost are achieved. Furthermore, if the tunnel profile is also measured after spraying the system will provide information about the thickness of the applied sprayed concrete layer which was, until recently, only possible with core drilling and measurement. If an exact final shape of a tunnel profile is required, the control system instructs the robot to spray to these defined limits automatically.

5.5 *Nozzle systems*

Nozzle systems are an important part of the spraying equipment. Through proper mixing of accelerators and air in the wet-mix spraying method, nozzles can contribute to:
- Lowering rebound
- Improving bond strengths
- Improving compaction
- Reducing dust levels

Figure 18. Meyco Roadrunner - road worthy spraymobile.

Figure 20. Meyco Logica - fully automatic spraymobile system.

Figure 19. Meyco Cobra - Mining industry spraymobile.

Figure 21. A wet-mix nozzle.

23

Only with the correct nozzle system, adapted to the type of application (wet-mix/dry-mix method, robot/hand application) and the accelerator/activator used, can low wear and outstanding quality of the in-place sprayed concrete be obtained. A typical wet-mix nozzle system is shown in Figure 21.

6 NEW DEVELOPMENTS

This section highlights some of the improvements that have been made relating to the sprayed concrete industry, other than those that have been reviewed in the previous sections.

6.1 *Achieving water-tight sprayable membranes*

With the advent of single shell permanent sprayed concrete linings, there has also been a request by the industry to provide watertight sprayed concrete. This is of particular importance with public access tunnels and highway tunnels that are exposed to freezing conditions during winter months, and also electrified rail tunnels. It has been shown that most permanent sprayed concrete exhibits an extremely low permeability (typically 1×10^{-14} m/s), however water ingress still tends to occur at construction joints, at locations of embedded steel, and rock bolts.

Traditionally, polymer sheet membranes have been used, but this system has been shown to be sensitive to the quality of heat sealed joints and tunnel geometry, particularly at junctions. Furthermore, when sheet membranes have been installed with an

Figure 22. Using a simple screw pump, two men can apply a sprayable membrane at up to 50m² per hour.

inner lining of sprayed concrete, the following adverse conditions can occur:
1 As the sheet membranes are point fixed, sprayed inner linings may not to be in intimate contact via the membrane to the substrate. This may lead to asymmetrical loading of the tunnel lining.
2 To aid the build of sprayed concrete onto sheet membranes, a layer of welded mesh is used. Due to the sheet membrane being point fixed, the quality of sprayed concrete between the mesh and the sheet membrane is often inferior, and may lead to durability concerns.
3 The bond strength between sprayed concrete inner lining and sheet membrane is inadequate and leads to potential de-bonding, particularly in the crown sections of the tunnel profile. This is a detrimental effect when constructing monolithic structures.
4 As there is little bond strength at the concrete - sheet membrane interface, any ground water will migrate in an unrestricted manner. Should the membrane be breached, the ground water will inevitably seep into the inside tunnel surface at any lining construction joint or crack over a considerable length of tunnel lining.

To combat these problems, a water-based polymer sprayable membrane called Masterseal® 340F has recently been developed. This sprayable membrane has excellent double-sided bond strength (0.8 to 1.3 MPa), allowing it to be used in composite structures, and thereby effectively preventing any potential ground water paths on both membrane–concrete interfaces being created. Masterseal® 340F also has a strain capacity of 80 to 140% over a wide range of temperatures allowing it to bridge most cracks that may occur in the concrete structure. Being a water-based dispersion with no hazardous components, it is safe to handle and apply in confined spaces. The product can be sprayed using a screw pump and requires two operatives to apply up to 50m²/hr, even in the most complex of tunnel geometries, where sheet membranes have always demonstrated their limitation, as shown in Figure 22.

As presented in Figure 23, in single shell lining applications, Masterseal® 340F is applied after the first layer of permanent fibre-reinforced sprayed concrete, where the sprayed surface should be as regular as possible to allow an economical application of membrane 5 to 8 mm thick (all fibres are also covered). A second layer of permanent steel fibre reinforced sprayed concrete can then be applied to the inside. As the bond strength between the Masterseal® 340F and the two layers of permanent sprayed concrete is about 1 MPa, the structure can act monolithically, with the sprayable membrane resisting up to 15 bar. As this application considers no water drainage, the second layer of sprayed concrete must be designed to resist any potential hydrostatic load over the life of the structure.

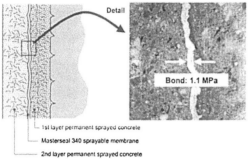

Figure 23. Composite, watertight structures using Master-seal 340F sprayable membrane.

6.2 *Modern admixtures for deep shaft construction*

The Sedrun Access Shaft constitutes one of three intermediate adits along the 57km Gotthard Base Tunnel, part of the New Alpine Rail Transit system in Switzerland. Because of the elevation of the terrain along the tunnel, the Sedrun intermediate adit is being constructed as an 800m deep vertical shaft with a diameter of 9 m.

Having readily available concrete at the bottom of a shaft during sinking involves several challenges for concrete technology and concrete logistics. Transport by means of a vertical free flow through a pipeline requires special mix design considerations, as well as special technical solutions regarding the actual transportation line.

Typical problems are blocking of the pipeline due to bridging inside the pipe and segregation of the mix, causing the coarser fragments of the aggregate to travel faster than the fines. This in turn causes blockages consisting of gravel. Critical fresh concrete properties to prevent these problems are high slump retention to avoid bridging and a high level of cohesiveness to avoid segregation and bleeding.

Figure 24. Screed and float finish permanent sprayed concrete tunnel.

To overcome these difficulties, the desired combination of the fresh concrete properties of high slump retention and high cohesiveness led to the following mix design features:
- High binder content
- Hyperplasticiser with extremely high water reducing effect
- Special aids for increased cohesiveness

Products are available to provide particularly good cohesiveness and pumpability with difficult and uneven gradation conditions in the aggregate.

The bottom of the pipeline was equipped with a kettle for remixing and removal of kinetic energy, allowing concrete to flow gently into a kibble for buffer storage and subsequent spraying.

6.3 *Surface finish*

Depending on the intended role of the permanent sprayed concrete structure, several surface finishes can be provided varying from a sprayed concrete finish to a float finished surface.

Sprayed concrete can be surface finished by screeding and hand-floating to produce a surface similar in quality to that of a cast in-situ concrete. This process is performed on a sprayed mortar layer applied to the final structural sprayed concrete layer, which is typically 25mm thick. Polypropylene fibres may be included in the mix design to control surface crazing produced by thermal and surface drying effects. In tunnels, the screeding process is relatively simple to perform using 25mm diameter screed profiles bent to the finished profile of the tunnel, and if required, further improvement to the surface finish can be attained by hand float work, as illustrated in Figure 24. Purpose made power floats are also available to aid this task. As these finishing layers tend to be relatively thin coats, concrete improving admixtures should be used to aid bond strengths and efficiently cure the concrete.

Following recent tunnel fires where considerable structural damage resulted, the use of 50mm thick sprayed cementitious based fire protective layers can be applied. These coatings provide fire protection up to 1350°C, and can be float or screed finished as described above.

If a high reflectance and colour are required, the application of a pigmented cementitious fairing coat or a coloured epoxy coating to the may be used.

6.4 *Electronic penetrometers*

As discussed earlier, the selection of the sprayed concrete system is based on setting characteristics and early age strength gain, determined essentially by cement-accelerator reactivity and low w/c ratios. In the laboratory, the setting characteristics are examined using the Vicat test, but during development of the mix during site trials, the testing has been

Figure 25. Traditional Proctor penetrometer where full body weight required (Jubilee Line Project, UK, 1994).

achieved using a Proctor penetrometer, as indicated in Figure 25.

There are numerous influences on the recorded result using this equipment, such as:

- Penetrometer needle hits large aggregate rather than sandy matrix, giving higher results
- Angle of penetration not always perpendicular to sprayed concrete surface
- Penetration often exceeds the required 15mm depth due to the excessive force applied to the penetrometer (see Figure 25, where full body weight needs to be applied)
- Use and results dependent on operator to high degree.

To combat these negative influences, the use of an electronic penetrometer is advocated (as illustrated in Figure 26) providing the following benefits:

- Shorter pocket sized instrument making it easier to handle and reduce errors
- More accurate, faster measurements producing

Figure 26. Pocket-sized electronic penetrometer for measuring set and early age strength development.

data with significantly less variance than traditional penetrometers

- Guidance ring on needle to help with correct penetration depth
- Calibration of instrument is simple
- All results can be downloaded to a PC
- Can also be used as a 50 kg balance which is useful during site trials

7 RAISING COMPETENCE

There are a few major sprayed concrete consumers who through practical experience, research, and development, have acquired significant know-how. Equipment and control methods have also gone through a development process that has led to a rational production as well as a more uniform quality of the final product. From an international point of view it is safe to say that we have come a long way from when sprayed concrete was used for securing rock, but it is also fair to say that we are lagging behind when using sprayed concrete for building and repair works. It is not easy to explain this. The know-how exists, however, it is not fully implemented.

Prevailing regulations make special concrete technological demands on the people doing the spraying work. Present requirements have led to better training of personnel involved. The result of this is an improved quality of work. The number of special contractors who are working with sprayed concrete has increased over the last few years, which has improved the quality of the application.

Sprayed concrete structures are heavily reliant on human competence during construction, and therefore the design should reflect this by considering the "buildability" of these structures using sprayed concrete. Designing "buildability" ensures that safety and durability critical elements are either designed out, or simplified for ease of construction on the job site. Furthermore, design teams should be aware of the limitations of construction processes, and be familiar with the likely material performance.

Likewise, the construction team should be made aware of the design elements that are key factors in determining the safety and durability of the structure. To ensure the quality of the concrete lining is achieved, quality review systems should be adequate to control the production. It is of paramount importance that the communication link between design and construction teams should be maintained from the pre-design stage to project completion so that the above processes are promoted.

Modern sprayed concrete specifications now address the issues of achieving a quality controlled modern mix design, providing guidance on promoting durability and effective execution of the spraying processes. As an example, the new European Specification for Sprayed Concrete (1996) produced by EFNARC, provides comprehensive systems to attain permanent sprayed concrete. This specification has been the basis for new project-specific specifications worldwide, and is the basis of the new European Norm Sprayed Concrete Specification. Furthermore, the EFNARC Sprayed Concrete Specification tackles issues such as nozzleman training and certification, and also sets out systems for contractors and specifiers to consider the structures they are building and to adapt the sprayed concrete system and mix design accordingly.

To address the international issue of sprayed concrete training, an innovative service provided by the International Centre for Geotechnics and Underground Construction, based in Switzerland, is providing courses in modern sprayed concrete technology to address the shortcomings in the industry. Specific courses are available for designers and contractors, with specialist nozzleman training for robotic spraying, for example.

8 OTHER APPLICATIONS

The emphasis of this paper has been on sprayed concrete for tunnel and ground support applications, as this market uses the highest volume of concrete. Nevertheless, the versatility of sprayed concrete through the benefits of free-form design without the need for formwork, high quality concrete, early strength development and high bond strength has proved highly effective as demonstrated by the following examples.

8.1 White water course, Sydney 2000 Olympic Games, Australia

In Sydney, Australia, this white water rafting course was constructed for the 2000 Olympic Games using polypropylene fibre reinforced sprayed concrete. The inside walls were designed to resist turbulent water and repeated battering from canoes. This continuous structure, which contains joints, is capable of generating water-flow rates of up to four meters per second along the 300 linear metre course.

8.2 Oro Valley channel, Arizona, USA

In Oro Valley, Arizona, 50mm long polypropylene fibers were used in the sprayed concrete mix design instead of traditional reinforcement for this channel lining due to concerns about post-crack integrity. The sprayed concrete thickness was 150 mm.

8.3 Seismic dam retrofit, Little Rock Dam, California, USA

Situated 2 km from the San Andreas Fault, the Little Rock Dam required a retrofit using a ready mix supplied, air-entrained, steel fibre reinforced, silica fume, wet-mix sprayed concrete.

Critical to the successful implementation of the design was the achievement of a minimum direct bond (pull-off) strength of 1.0 MPa and a crack-free material. The use of a high performance steel fibre reinforced sprayed concrete and the rigorous implementation of the specified curing regime were considered critical in ensuring successful completion of this project in a severe desert climate.

8.4 Second Dover Cruise Liner Pier, UK

In view of the aggressive environment, 400×2 m diameter steel piles that support the new cruise liner pier and side fenders required additional corrosion protection between low and high tide marks. After reviewing all coating systems, high performance sprayed concrete was selected due to fast application and strength gain, but also long-term durability.

The sprayed concrete mix contained monofilament polypropylene fibres, a high microsilica content, concrete improver admixture for bond, and was accelerated with alkali-free set accelerators. The concrete was applied during the outgoing tide from a pontoon, so that on the return tide the concrete had acquired adequate strength and could not be eroded. Sprayed concrete for coastal protection works has significant benefits over conventional methods.

8.5 Cement Storage Domes, Ontario, Canada

Located in Ontario, Canada, a 55 m diameter, 29 m high dome to store 66000 tonnes of cement powder was constructed using only high performance sprayed concrete applied to a pre-inflated plastic form. Such structures have found a considerable market in both North and South America, with one structure also being constructed in Germany.

9 CONCLUSIONS

Wet-mix sprayed concrete applied using modern environmentally safe admixtures and high performance equipment equips the industry with an economical tool to construct permanent, high strength, durable concrete structures. The application process has become highly automated thereby significantly reducing the degree of human influence that has, in the past, prevented clients from considering sprayed concrete as a permanent structural material.

The development of sprayed concrete has, and continues to be, centred on the wet-mix method which has been used in the underground support market in significant volumes since the 1970's. The larger international contractors that have had prior knowledge of the benefits have orchestrated the international spread of the wet system. Consequently, wet-mix sprayed concrete has been confined to countries with large-scale tunnel or mining projects. The transfer of technology to the other domestic markets that use the dry-mix method has been slow, but the tendency has been towards wet-mix in the last 10 years. The total volume in Europe alone is more than 3 million cubic metres per year. In the opinion of the authors, this increasing trend will continue for several years to come.

The time and cost saving potential in application of wet-mix fibre reinforced sprayed concrete as permanent support is in most cases substantial and sometimes dramatic. It must be ensured that the design method allows the use of such permanent support measures. Also, contract terms that are counterproductive for the utilization of modern support techniques must be avoided.

Other important advantages like totally flexible logistics, very good working safety, and good environmental conditions, are completing the range of reasons in favour of using the wet-mix sprayed concrete technology. Wet mix sprayed concrete is no longer an experiment, as the solutions are well proven.

Sprayed concrete as a building method can have much greater application outside tunnelling, however, this system is still extremely under-used. One of the advantages of sprayed concrete is its flexibility and speed. Concrete which is to be applied simply with a hose against formwork, rock surface or concrete surface, may architecturally and constructively be varied. The only limit is imagination and the desire for experimentation.

We therefore call upon all contractors, architects, authorities and consultants to recognize that the concrete technology, experience, equipment, and materials exist and may be mobilised to increase the range of our building activities as soon as someone plucks up courage to embrace the building method

Table 4. Sprayed concrete technology development and future work

What has been achieved	Further work remaining
Spraymobiles	
Reliable machines with few breakdowns and low running costs.	Develop accurate concrete flow-meters.
A nearly pulsation free concrete flow to nozzle allowing homogeneous concrete to be placed.	Simplify transfer of recorded data into a laptop.
Sufficient spray capacity, theoretically up to 30m³ (effective 20-25m³), 5 to 8 times that of dry-mix systems.	Enable spraying automatically to a defined contour.
Accelerator dosing linked to concrete pump to not cause strength and durability problems, and to control site costs.	Completely pulsation-free machines with maximum piston efficiency.
Store data recorded during spraying.	
Spray a defined concrete thickness automatically.	
Design, Health and Safety	
Sprayed concrete accepted as permanent in some countries.	Sprayed concrete in many countries still considered temporary.
Specifications for dust have been established.	Concrete strength requirements still at 25 Mpa.
Caustic accelerators banned in some countries.	Promote single shell lined tunnels.
	Dust specifications not enforced in most countries.
	Spraying operations often done in unsupported areas.
	Caustic accelerators still in use in some countries.
	Mechanized robotic spraying offering increased safety to operatives at face.
	Introduce design and safety advantages to other markets e.g. concrete repair.
Construction Chemicals and Fibres	
New generation of superplasticisers.	Improve superplasticisers to cope with even more difficult aggregates.
Hydration control for up to 72 hours.	Find improved test procedure for accelerator laboratory tests.
Alkali-free accelerators with excellent setting characteristics.	Improve accelerator storage and handling properties.
Accurate weight and volume based batching systems for admixtures and fibres.	Implement systems to facilitate admixture and fibre batching on non-dedicated sprayed concrete plants.
	To create a more "rubber-like" sprayed concrete that does not crack too early under load, whilst remaining stiff enough to control ground movements.
	Further reduction in aggregate and fibre rebound.
	Fully understand the benefits of fibre reinforcement and implement in designs.

of the future: sprayed concrete.

Developments within sprayed concrete technology are listed in Table 4, with summarised areas of work that are required to further sprayed concrete implementation.

REFERENCES

Aldrian, W., Melbye, T. & Dimmock, R. 2000. Wet sprayed concrete – Achievements and further work. Draft paper for Felsbau publication. November 2000. 11 pages.

Annett, M., Earnshaw, G. & Leggett, M. 1997. Permanent sprayed concrete tunnel linings at Heathrow Airport. Tunnelling '97, Institution of Mining and Metallurgy, London, September 1997. pp517-533.

Austrian Concrete Society. 1999. Shotcrete Specification. March 1999. Published by the Austrian Concrete Society.

Banthia, N. & Yan, C. Toughness of fibre reinforced shotcrete panels (EFNARC) with S-152 deformed polypropylene macro-fiber. Department of Civil Engineering, The University of British Columbia, Vancouver, BC, Canada, VGT IZ4.

Bernard, E. S. 1999. Correlations in the Performance of Fibre Reinforced Shotcrete Beams and Panels. University of Western Sydney, Nepean, Australia, *Engineering Report No. CE9*. July 1999.

Dimmock, R.H. 1998. Draft Advice Note - Single Pass Tunnel Linings. Unpublished report for the UK Highways Agency. Transport Research Laboratory, report ref PR/CE/199/98. September.

Dimmock, R.H. 1998. Final Report - Single Pass Tunnel Linings. Unpublished report for the UK Highways Agency. Transport Research Laboratory, report ref PR/CE/143/98. June 1998.

Dimmock, R.H. 1999. Permanent sprayed concrete for UK tunnels. Proceedings of the 3rd International symposium on sprayed concrete. Modern use of wet-mix sprayed concrete for underground support. Gol, Norway 26-29 September 1999. pp 186-195.

Dimmock, R.H. 2000. Practical solutions for permanent sprayed concrete tunnel linings. Proceedings of the ITC Conference. Major Tunnel and Infrastructure Projects, Taiwan, ROC, 22-24 May 2000. pp191-202.

EFNARC. 1996. The European Specification for sprayed concrete. Published by EFNARC, Hampshire, UK.

EFNARC. 1999. The European Specification for sprayed concrete – Guidelines and Execution of Spraying (Revised Section 8 of Specification) Published by EFNARC, Hampshire, UK.

Garshol, K.F. 1998. International practices and trends. School on Shotcrete and it's Application. SAIMM - Randburg, South Africa.

Garshol, K.F. 1999. Durability of wet-mix sprayed concrete. Proceedings of the 3rd International symposium on sprayed concrete. Modern use of wet-mix sprayed concrete for underground support. Gol, Norway 26-29 September 1999. pp 259-271.

Garshol, K.F. 2000. MBT Internal report on HPP fibre reinforced sprayed concrete performance. Unpublished.

Holter, K., Poggio, P., Nel, P. & Blindenbacher, B. 2000. Properties and mix design of concrete to facilitate pipeline transport with a vertical drop of up to 800m during the construction of the Sedrun Access Shaft, Switzerland. Draft paper. November 2000. 4 pages.

Krebbs, C. 1999. Automated shotcrete application – Meyco Robojet Logica State-of-the-art declaration. MBT Meyco internal published report. February 1999. 8 pages.

Melbye, T., Aldrian, W. & Dimmock, R. 2000. International practices and experiences with alkali-free, non caustic liquid accelerators for sprayed concrete. MBT UGC International publication. July 2000.

Melbye, T.A. 1997. Sprayed Concrete for Rock Support. 8th edition. Published by MBT UGC International. August 2000.

Melbye, T.A. 1999. International practices and trends in sprayed concrete. International sprayed concrete conference, Kalgoorlie, Australia, March 1999.

Morgan, D.R. 1995. Dam seismic retrofit with high performance shotcrete. Paper from website: http://www.usherb.ca/CENTRES/beton/bulletin/mai95/morgeng.html.

Norris, P. & Powell, D. 1999. Towards quantification of the engineering properties of steel fibre reinforced sprayed concrete. Modern use of wet-mix sprayed concrete for underground support. Gol, Norway 26-29 September 1999. pp 393-402.

Schubert, P. & Aldrian, W. 1999. Wet against dry systems – A dynamic development in German speaking countries. Proceedings of the 3rd International symposium on sprayed concrete. Modern use of wet-mix sprayed concrete for underground support. Gol, Norway 26-29 September 1999. pp 439-445.

Other papers

Shotcrete: Engineering Developments, Bernard (ed.) © 2001 Swets & Zeitlinger, Lisse, ISBN 90 5809 176 7

Dynamically loaded young shotcrete linings

A. Ansell & J. Holmgren
Royal Institute of Technology (KTH), Stockholm, Sweden

ABSTRACT: A criteria for how close, in time and distance, to young shotcrete blasting can take place would be an important tool in planning for safe and economical tunnelling projects. As a first step, field tests with shotcrete at ages of 1 to 25 hours, sprayed on tunnel walls, have been conducted in a Swedish mine. The shotcreted areas were subjected to vibrations from explosive charges detonated inside the rock. The acceleration measurements showed that the shotcrete had withstood high particle velocity vibrations without being seriously damaged. Drumminess of shotcrete appeared, indicating that the major failure mechanism was sudden loss of adhesion to the rock. Elastic wave propagation theory has been implemented numerically to further study the failure mechanism. Due to the changes in shotcrete-rock impedance ratio as the shotcrete cures, the stresses at the interface between shotcrete and rock experience a maximum during a short period within the first 24 hours of shotcrete age.

1 BACKGROUND

The ability to project shotcrete on a rock surface at an early stage after blasting is vital to safety during the construction and function of a tunnel. A complication arises when the need for further blasting affects curing of the newly applied shotcrete. If concrete, cast or sprayed, is exposed to vibrations during early age while it is still in the process of curing, damage that threatens the function of the cured concrete may occur.

In tunnelling, the search for a more time-efficient construction process naturally focuses on the possibility of reducing the periods of waiting between stages of construction. As an example, the driving of two parallel tunnels requires coordination between the two excavations so that blasting in one tunnel does not, through vibrations, damage temporary support systems in the other tunnel prior to placing a more sturdy, permanent support. There also arises similar problems in mining. To be able to excavate as much ore volume as possible, the grid of drifts in a modern mine is dense. This means that supporting systems in one drift are likely to be affected by vibrations in a neighbouring drift. A criteria for how close, in time and distance, to the young shotcrete blasting can take place would be an important tool in planning for safe and economical tunnelling projects.

In the past there has been little, or no, research carried out on the behaviour of young shotcrete subjected to vibrations. Some interesting work has however been carried out on fully matured shotcrete. Wood & Tannant (1994), Tannant & McDowell (1993) and McCreath et al. (1994) presented results from tests in a Canadian mine, where steel fibre-reinforced and steel mesh-reinforced shotcrete linings have been subjected to vibrations from explosions. During the tests, it was found that steel fibre-reinforced shotcrete can maintain its functionality even though exposed to vibration levels of 1500–2000 mm/s. It was also seen that mesh-reinforced shotcrete performs better than steel fibre-reinforced shotcrete under very severe dynamic loading conditions. This is due to its ability to retain broken rock even when extensively cracked, which is not the case with most fibre reinforced shotcrete.

As a first step towards a damage criterion for young shotcrete, field tests were conducted at the LKAB mine in Kiruna, Sweden (Ansell 1999). Shotcrete, without mesh reinforcement and fibers, was sprayed on tunnel walls. Unreinforced shotcrete was used as it was expected to have lower resistance to vibrations than reinforced shotcrete. At ages of 1 to 25 hours, the areas were subjected to vibrations from explosive charges detonated inside the rock. Accelerations were recorded and later numerically integrated to give particle velocities. The maturing shotcrete linings were also tested to determine the development of compressive strength.

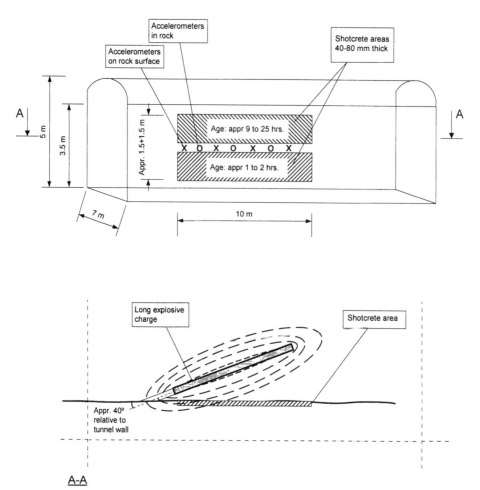

Figure 1. Schematic view of a test site. Explosive charge in rock behind shotcrete areas (in section AA).

2 FIELD TESTS

2.1 *Test set-up*

The performance of young shotcrete with respect to blast induced vibrations was tested during May 1998. The tests were carried out in cooperation with the research department of the mining company LKAB and the Swedish Rock Engineering Research (SveBeFo). A series of four tests, at four closely situated sites in a tunnel, were carried out. Each test was performed with a unique type of explosive charge, not repeated in any of the other tests. Two shotcrete areas of varying age, one fresh and one young, were subjected to vibrations in each test. A lining thickness of 50 mm was expected but variations between 40–80 mm were later measured. No shotcrete area was subjected to vibrations more than once. The geometry of each test site is described in

Figure 1. For each test site, the following procedure was repeated:

1. Projection of first shotcrete area, approximately 1.5×10 m².
2. Hardening of first shotcrete area, between 9 hrs. 35 min to 25 hrs.
3. Continuous testing of young shotcrete properties until detonation.
4. Projection of second shotcrete area, approximately 1.5×10 m².
5. Mounting and connection of accelerometers.
6. Charging of explosives.
7. Detonation of explosive charge, measurement of accelerations.
8. Time delay due to ventilation.
9. Damage detection on shotcrete and surrounding rock.

The testing of young shotcrete properties (Step 3, above) was intended to determine the compressive strength development of the young shotcrete. At each application of shotcrete (Steps 1 and 4, above), moulds were filled for 28 day compression tests on 150 mm cubes. At a later stage, adhesion tests were also carried out. The shotcrete used during the test was of the standard quality, without fibres, used by LKAB. The water-cement ratio was 0.45. On the four test sites, charge holes were drilled at an angle of approximately 40° relative to the tunnel surface. Within each of the test sites, measurement points for placement of accelerometers were located in the same horizontal plane as the charge holes. The shotcrete was sprayed on rock consisting of iron-ore, with a density of 4800 kg/m^3 and a compressive strength of 115 MPa. The rock contained a zone of partially crushed material close to the test sites.

2.2 Test results

The 28 day 150 mm compressive cube tests for the shotcrete from the test areas gave average compressive strengths of 31.3–36.9 MPa and an average density of 2161 kg/m^3. The four test sites were investigated to detect areas with drumminess, i.e. areas without adhesion at the shotcrete-rock interface. The investigations, using manual hammer soundings, were carried out at a shotcrete age of about 12–15 days. The planned adhesion tests, using a pull-out technique, had to be postponed due to resumed mining activities in the vicinity of the test sites. Although carried out almost six months after the application of the shotcrete, the results were judged to be of interest as a measure of the adhesion strength of the fully matured linings. A limited number of adhesion tests were performed. The results indicated an adhesion strength of $f_{ad} \geq 0.7$ MPa.

All four tests showed that where the rock walls were intact, so were the shotcrete linings. No cracks appeared anywhere on the intact shotcrete surfaces. Drumminess of shotcrete appeared at areas subjected to vibration levels above 500 mm/s. The major failure mechanism was sudden debonding at the rock-shotcrete interface. All tests resulted in ejection of large volumes of rock, creating 600–1000 mm deep craters with diameters of approximately 2.4–4.0 m. The results from Test 3 are shown in Figure 2. The areas of the lining found to have no adhesion are marked with shaded areas. The shape of the crater is shown by the dashed lines. The placement of accelerometers is also given, with "×" denoting surface mounted accelerometers and "o" denoting accelerometers inside fully grouted pipes, 500 mm inside the rock.

During the tests, accelerations were measured in two directions, orthogonal to the tunnel wall and parallel to the tunnel axis. The digitally recorded signals were later integrated to obtain velocities. The integration constants for each signal had to be chosen such that the velocities prior to the first arrival of the signals were set to zero. In the interpretation model used, it was assumed that only the first peak in each acceleration history provided a correct velocity, i.e. no reflections and super-positions have at such an early stage distorted the signal. This method of determining the maximum particle velocity v_{max}, or peak particle velocity (ppv), is sometimes referred to as the "first-arrival principle". The measured and numerically determined ppv values are also shown in Figure 2, close to the measurement points. Measurement points where accelerometers were lost, are shown within parenthesis. Additionally, theoretical ppv levels are denoted by labelled dashed lines. These levels, given for the rock surface, have been numerically calculated using the scaling relationship:

$$v_{max} = 832r^{-1.38} \qquad [mm/s] \qquad (1)$$

which was found to be valid at the test sites. The variable r is the scaled distance between explosive charge and point of observation, scaled with respect to the weight of the explosives used. Corrections for long charges and degree of coupling between charge hole and charge were also included.

3 NUMERICAL MODELLING

3.1 Stress wave theory

A numerical model, based on elastic wave propagation theory, has been used to study the failure mechanism. The main features of the elastic stress wave theory is briefly outlined below. For a more extensive discussion, see e.g. Meyers (1994). If a body (or structure) is exposed to a suddenly applied load, that acts for only a short time, the disturbance caused by the load will not be instantly transferred to all points within the body. The energy of a suddenly applied load, or impact, is transferred to remote, undisturbed parts of a body as stress waves. These waves, causing deformations and stresses, have a noticeable front called the wave front that travels through the material with constant velocity. This wave propagation is defined as a collective particle motion in a material around a state of equilibrium which is regained after passage of the wave.

The motion of the material particles enables transportation of energy through the material.

When an elastic medium is deformed locally, the propagation of two types of elastic stress waves are initiated. The first type of wave causes change in the volume, or dilatation, by forcing the material parti-

35

cles to move back and forth in a direction parallel to the direction of wave propagation. This wave type is therefore often called longitudinal wave or dilatational disturbance, or P-waves. The second type of wave causes distortions in the shape of a body, or shearing, by particle movement perpendicular to the direction of propagation. This wave type is often called a distortional wave, transverse wave, shear pulse, or S-wave (from *secondary*) since it reaches a point of observation following the P-wave. Thus, there are two different types of wave velocities: the particle velocity, v, and the velocity of propagation, c. In a one-dimensional elastic body, e.g. a bar, the propagation velocities of longitudinal and transverse waves depend only on the density of the material, ρ, and elastic constants. For longitudinal waves, the propagation velocity is:

$$c_1 = \sqrt{\frac{E}{\rho} \frac{(1-v)}{(1+v)(1-2v)}} \qquad (2)$$

and for distortional waves:

$$c_d = \sqrt{\frac{G}{\rho}} = \sqrt{\frac{E}{\rho} \frac{1}{(1+v)(1-2v)}} \qquad (3)$$

respectively, where E is the elastic modulus, G the shear modulus and v is Poisson's ratio.

When a travelling elastic wave reaches an interface, e.g. a free surface or a border to a medium of other material properties, it will be reflected back and/or transmitted to the other side of the interface. The simplest case occurs when the wave enters per-

pendicularly to the interface. An incident stress wave, corresponding to a particle velocity or acceleration, will normally give rise to a transmitted and a reflected wave. These relations are illustrated in Figure 3. As the figure demonstrates, when a compressive wave encounters a free surface, it reflects back as a tensile wave and vice versa, changing the stress sign, while the particle velocity is maintained. Also, when encountering a rigid boundary, the opposite occurs in that the stress sign is maintained and the particle velocity direction is reversed. Figure 3 also demonstrates the principle of superposition, i.e. the net stress and the net particle velocity at any point where waves meet or reflect, are the algebraic sums of the instantaneous stresses and the instantaneous particle velocities.

3.2 Numerical model

A one-dimensional finite difference method was implemented, using the age-dependent material properties of shotcrete, such as elastic modulus, compressive strength and adhesive strength, as variables.

The longitudinal wave velocities in the rock and shotcrete were assumed to follow elastic relations. The applied loads originated from acceleration histories in one direction, as obtained by the accelerometers in the tests described in section 2. The elastic numerical one-dimensional model tested, is based on an algorithm originally presented by Lundberg (1980), being an implementation of the elastic stress wave theory. The algorithm presented was reprogrammed, thoroughly tested and further developed by James (1998). The routine was tailor-made to

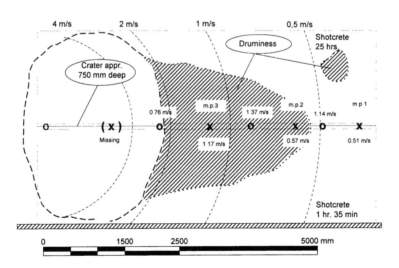

Figure 2. Rock and shotcrete damage at Test 3. Vibration velocities (ppv) on the rock surface are given as measured values next to accelerometer positions and theoretical values next to dashed lines (m.p. is a measurement point).

STRESS   PARTICLE VELOCITY

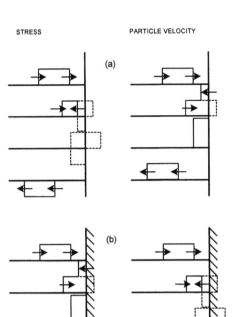

Figure 3. Reflection of stress and particle velocity of rectangular waves entering (a) a free surface and (b) a rigid boundary. At reflection, the magnitude of the wave is either doubled or temporarily zero due to superposition of waves of opposite senses.

solve the vibration of shotcrete problem described in this paper. The algorithm and program have been further developed and extensively tested (Ansell 1999), using realistic input data from the tests described in section 2.

The algorithm is presented in detail by Ansell (1999) and the following outlines its main features. The numerical process determines the stresses within a length-time plane, describing the stresses that propagate within a one-dimensional body, i.e. a column of shotcrete. The stresses from one iteration of the algorithm correspond to one shotcrete age. To determine the development of the stresses within the shotcrete, for identically applied loads, the process must be repeated for a number of shotcrete ages. As a first step, the wave propagation velocity for the rock and the shotcrete must be determined.

The second step defines the required grid of the length-time plane. The time uses the same scale as the acceleration record to be used as input load. The spatial coordinate refers to the position within lining thickness of the shotcrete, built up by identical elements. The reason for dividing the shotcrete into

elements is to facilitate the description of elastic wave fronts moving through the lining. In the third step, the impedances, i.e. density times wave propagation velocity and area, for the rock and all the shotcrete elements are determined. For the free space outside the shotcrete lining, the impedance is zero. The acceleration, given by an equally spaced vector, must be numerically integrated to give a vector of the incident particle velocities that are the actual load applied in the model. As a stress wave is assumed to be propagating in perfect rock, prior to reaching the shotcrete via the shotcrete-rock interface, accelerations registered on a rock surface have to be divided by 2 to account for the reflection at the free surface, as discussed in section 3.1. In the following step, the particle velocities are recalculated into incident stresses.

The actual calculation sequence begins with a seventh step, in which the wave fronts propagating in the positive and negative directions are updated through time-stepping. Within each shotcrete element, for each time-step, the wave fronts are then summarized to give the total stresses within each element, at each time-step taken. Since the model is one-dimensional, only the acceleration components perpendicular to the rock surface are used whereas the components parallel to the surface are omitted. The relation between the number of shotcrete elements in the model and time-steps limits the length of acceleration record that the model can operate on. Short approximate acceleration loads must therefore be used.

The approximate load used consisted of an extracted part of the measured acceleration record. Figure 4 shows a shaded area, chosen as the first approximate acceleration record. It is taken as the segment around the highest peak of the real acceleration record, followed by its mirror image to give equal compressive and tensile contributions. It contains almost all of the dominant frequencies of the spectrum of the real acceleration. Figure 4(b) shows this extracted approximate acceleration load.

The algorithm used can be summarized as follows. For all predetermined shotcrete ages t to be tested:

1. Determine the wave propagation velocities c_{rock} for the rock and c_{shcr} for the shotcrete:

$$c_{\text{rock}} = \sqrt{\frac{E_{\text{rock}}}{\rho_{\text{rock}}}} \qquad (4)$$

and

$$c_{\text{shcr}}(t) = \sqrt{\frac{E_{\text{shcr}}(t)}{\rho_{\text{shcr}}}} \qquad (5)$$

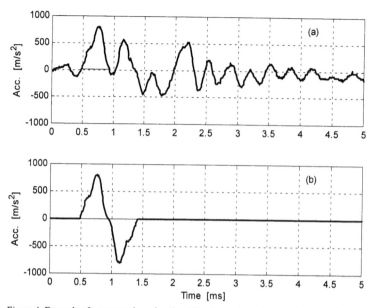

Figure 4. Example of a measured acceleration (a) and an extracted approximate acceleration (b) used in calculations.

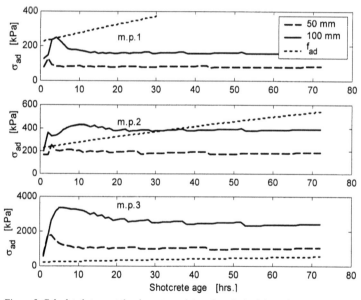

Figure 5. Calculated stress at the shotcrete-rock interface derived from three accelerometer records compared to assumed adhesive strength. The measurement points with accelerometers (m.p. 1-3) correspond to those in Figure 2.

2. Define the grid necessary in the length-time plane:

$$\Delta x = \Delta t c(t) \tag{6}$$

3. Compute the impedance for the rock and the shotcrete when $A=1$, according to:

$$Z_{rock} = A\rho_{rock}c_{rock} = \rho_{rock}c_{rock} \tag{7}$$

and

$$Z_{shcr}(t) = A\rho_{shcr}c_{shcr}(t) = \rho_{shcr}c_{shcr}(t) \tag{8}$$

4. Determine transmission and reflection coefficients.
5. Determine incident particle velocity v from integration of acceleration, numerically by the trapezoidal rule.
6. Determine incident stress from incident velocity:

$$\sigma = \rho_{rock} c_{rock} v \qquad (9)$$

7. Update for each time-step $n\Delta t$:
   7.1. Computation of waves travelling within each element at the free end, inside shotcrete lining and at shotcrete-rock interface.
   7.2. Assemble total stresses within each element.
8. Find maximum stress for each time step.

3.3 *Numerical results*

The in situ tests presented in Figure 2 have been modelled using the numerical stress wave model described above. The loads used were accelerometer records in the direction perpendicular to the length of the tunnel. Only accelerations from the rock surface were used. The calculated tensile stresses at the shotcrete-rock interface, for 50 and 100 mm thick shotcrete linings, are presented in Figure 5 together with curves representing the estimated adhesion strength, f_{ad}. The latter is taken as the average of tests reviewed by Ansell (1999) and approximated by the fourth order polynomial:

$$f_{ad} = -0.003t^4 + 0.28t^3 - 8.34t^2 + 130t + 222 \qquad (10)$$

where f_{ad} is the adhesive strength in kPa and t the shotcrete age in days. The modulus of elasticity E was assumed to follow the average relation:

$$E = 3.86 f_{cc}^{0.60} \qquad (11)$$

where f_{cc} is the age dependent compressive strength of shotcrete, which was measured in situ during the tests. The calculated results shown in Figure 5 refer to the first three surface mounted accelerometer points. The tensile stresses at the shotcrete-rock interface are compared to the adhesive strength as the concrete ages.

4 CONCLUSIONS

The most important conclusion from the numerical modelling is that a young thin shotcrete lining is less sensitive to vibrations than a thicker lining. Also, the shotcrete is less sensitive to vibrations when very young, or fresh, compared to when it is most sensitive at around 2 to 12 hours of age. When very young, a shotcrete lining is flexible and able to follow motions in the rock. As the material properties

such as E, f_{cc}, f_{ct} and f_{ad} develop, the lining becomes more capable of resisting the motion, also becoming able to restrain moving rock masses. During this curing process, the shotcrete goes through a stage where it is more vulnerable to vibrations than when either very young or fully mature.

A high early adhesion strength is vital to the vibration resistance of a shotcrete lining. A young lining that has been exposed to vibrations may appear to be undamaged, e.g. without cracks, but may yet have lost its supporting capacity due to adhesive failure. Therefore, hammer testing to detect drumminess is highly recommended. Under conditions similar to those in the Kiruna mine, a shotcrete lining is able to resist vibrations of around 500 mm/s if it is older than 24 hours. When estimating the vibration levels from a planned round of blasting in a tunnel, it is important to remember that the outer rock layer, that has been cracked during the excavation of the tunnel, may act as a damping filter, reducing the amplitude of the vibrations.

5 FURTHER RESEARCH

For further studies of young shotcrete, the adhesion properties must be addressed. In particular, the development of the adhesive bond between shotcrete and different types of rock, as well as the effect of preparation of rock surfaces prior to shotcreting must be examined. Investigations of the development of the adhesive strength is well suited to execution under ideal conditions in laboratory environments, but must be complemented by tests or observations in situ to account for the effect of imperfections in natural rock.

A further development of the elastic stress wave model, accounting for the effect of partially crushed rock, is possible. This should be combined with an extension of the model geometry to at least two dimensions, thereby making it possible to model the stress waves that occur at a tunnel front during blasting operations in tunnel excavations.

6 REFERENCES

Wood, D.F & Tannant, D.D. 1994. Blast damage to steel fibre-reinforced shotcrete. In Fiber reinforced Concrete – Modern Developments: 241–250. UBC Press.

Tannant, D.D & McDowell, G.M. 1993. Dynamic testing of shotcreted drifts. GRC Internal Report 93-12-IR, GRC, Laurentian University, Sudbury.

McCreath, D.R., Tannant, D.D. & Langille, C.C. 1994. Survivability of shotcrete near blasts. In Nelson & Laubach (eds.), Rock mechanics: 277–284. Rotterdam: Balkema.

Ansell, A. 1999. Dynamically loaded rock reinforcement, Doctoral thesis, Bulletin 52, Dept. of Structural Engineering, Royal Institute of Technology, Stockholm.

Meyers, M.A. 1994. Dynamic behavior of materials, New York: John Wiley & Sons.

Lundberg, B. 1980. Microcomputer simulation of longitudinal impact between nonuniform elastic rods. International Journal of Mechanical Engineering Education, 9: 301–315.

James, G. 1998. Modelling of young shotcrete on rock subjected to shock waves. Master thesis 106, Dept. of Structural Engineering, Royal Institute of Technology, Stockholm.

Shotcrete: Engineering Developments, Bernard (ed.) © 2001 Swets & Zeitlinger, Lisse, ISBN 90 5809 176 7

A Double Anchored (DD) steel fiber for shotcrete

N. Banthia and H. Armelin
University of British Columbia, Canada

ABSTRACT: Given the high deformability requirements in shotcrete in repair, rehabilitation, slope stabilization, ground support and other applications, the use of fibers in shotcrete is growing. Growing also are challenges surrounding the use of fibers in shotcrete not the least of which are a high fiber rebound in the dry process, lack of standardized test techniques and a poor understanding of reinforcement mechanisms. In the context of rebound, unfortunately, the requirements for a low rebound often clash with the requirements for a high material toughness. This paper describes a new fiber developed at the University of British Columbia where a rational balance was sought between the conflicting requirements of rebound reduction and a high material toughness. The new fiber is based on a double anchoring principle and is optimized for both reduced rebound as well as a high toughness in hardened shotcrete.

1 INTRODUCTION

Shotcrete, when used as ground covering or support, is subjected to both quasi-static ground movements as well as dynamic ground deformations often of large magnitudes. Not surprisingly, the requirements of material deformability, toughness, and energy absorption are often greater in fiber-reinforced shotcrete than in conventional fiber reinforced concrete.

One primary concern with the dry-process shotcrete is the high rebound; nearly 20-40% of material and up to 75% of fiber may be lost through rebound (Armelin 1997). During rebound, high proportions of the fibers fail to become embedded in the resultant concrete and thus are wasted. A loss of fiber through rebound translates into a major loss of fracture toughness, deformability, and the post-crack load carrying capacity in shotcrete.

While the issue of high rebound in dry-process shotcrete is well recognized, our understanding of the factors related to mix design (cement and silica fume contents, aggregate/cement ratio, etc) and/or placement variables (air pressure and volume, type of nozzle and distance and orientation of nozzle, etc.) that control rebound is far from adequate. Attempts have been made in the past to understand the kinematics of fast moving aggregate particles using high speed photography and to model the process of rebound (Armelin et al. 1997; Armelin et al. 1998a; Armelin et al. 1999). In the case of fibers, although fiber rebound has always been suspected to

be related closely to fiber geometry, the exact influence of fiber geometry on fiber rebound is not well understood. In a previous study (Banthia et al. 1992), fiber rebound was shown to be proportional to its specific projected area defined as the fiber projected area for a unit mass. In a later, more comprehensive study (Armelin et al. 1998b), a specific fiber parameter called the modified aspect ratio $(l/\sqrt{d})$ was shown to be linearly related to fiber rebound. The results of this latter study (Armelin et al, 1998b) became the basis for designing the novel fiber described here.

On the toughness side, pull-out of fibers across a matrix crack is recognized as the main mechanism that allows steel fiber reinforced shotcrete (SFRS) to be more ductile than unreinforced shotcrete. Thus, all commercial reinforcing fibers presently available in the market are deformed at the ends or along their length, to enhance the anchorage of the fiber with the cementitious matrix and generate a greater pullout resistance and energy. Unfortunately, the fiber geometrical requirements for higher toughness (i.e. a high aspect ratio) are in direct conflict with those required for a reduced rebound (i.e. a low aspect ratio).

2 FIBER REBOUND AND ITS REDUCTION

It is clear that the amount of fiber rebound seriously affects the toughness of the resulting in-situ fiber reinforced shotcrete and rebound reduction is key to obtaining a highly toughened shotcrete. For a rational design of a shotcrete fiber from the objective of reduced rebound, the results obtained by Armelin & Banthia (1998b) were considered. They conducted a series of experiments using circular cross section steel fibers having diameters of 0.5, 0.61, 0.65, 0.76 and 1 mm and lengths of 3, 12.5, 19, 24.5 and 40 mm. Fibers of each diameter were made to each length. Shotcrete was produced using the dry mix technique and the fiber rebound was evaluated for the above combinations of length and diameter. Their results are plotted in Figure 1, where it can be seen that there is a substantially linear relationship between fiber rebound R_f and a modified aspect ratio given by fiber length divided by the square root of fiber diameter, i.e.,

$$R_f = f \, l_f / \phi^{1/2} \tag{1}$$

where R_f = the fiber rebound; l_f = fiber length; ϕ = fiber diameter

It is apparent that a reduction in rebound R_f significantly increases the amount of fiber retained in the in-place shotcrete. For example, if the fiber rebound is reduced from the 75% figure that characterizes fibers presently in the market to 50%, the in situ fiber content is doubled for the final shotcrete produced. Further, to reduce the fiber rebound to below about 70%, which is less than that of conventional fibers, the ratio of fiber length over the square root of fiber diameter ought to be below about 30 $mm^{1/2}$.

3 FIBER ANCHORAGE MECHANISMS

The state-of-the-art in fiber design may be divided into two large groups with respect to their anchorage

mechanisms, namely fibers that rely on a "dead anchor" and those that rely on a "drag anchor". Dead anchors generally are produced by deforming the fiber with a hook or cone adjacent to each of its ends. Under stress, in an aligned fiber (i.e. under axial tension) the anchor is generally designed to fail (e.g. pullout) at a maximum resistance below the strength of the steel. However, these dead anchors, after failure, have a significantly reduced capacity to resist pullout displacement.

Drag anchors, on the other hand, generally are formed by enlarging the fiber adjacent to its end in such a way that during pullout, the enlargement generates friction with the matrix as the fiber is dragged out of the concrete. This type of fiber generally develops a lower maximum pullout resistance as compared to the dead anchor but its effect tends to last for a greater pullout displacement and therefore greater pullout energy is consumed by the end of the pullout process.

4 THE DOUBLE ANCHORAGE DD FIBER

In the new fiber developed at the University of British Columbia, the requirements for a lower rebound as discussed above were combined with the requirement for an optimal anchorage. The novel fiber called the Double Anchorage (or DD fiber) is different from other commercial fibers in that a low "$l_f/\phi^{1/2}$" ratio is maintained and the two anchoring mechanisms (dead and drag) are rationally combined in the same fiber. For a reduced rebound, the ratio of fiber length to the square root of fiber diameter is kept less than 30 $mm^{1/2}$ (Figure 1). The fiber thus may possess a length between 20 and 35 mm and a diameter of between 0.6 and 1 mm.

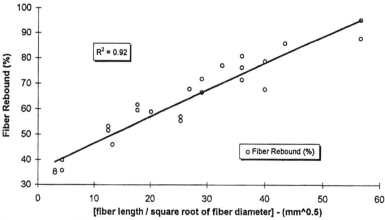

Figure 1. A plot of fiber rebound as percent by mass re-bounded vs. fiber length over the square root of the fiber diameter in millimeters.

One half of the Double Anchorage fiber (DD) is shown in Figures 2 and 3. The other half is essentially the same as each fiber is symmetrical on opposite sides of its mid length. The fiber comprises both a dead anchor, 18, and a drag anchor, 12, placed at the end of the fiber and separated by a zone of stress concentration weak link, 16 (Figure 2). The weak link, 16, is expected to fail under load. Failure of the weak link releases the dead anchor and activates the drag anchor. By properly proportioning the fiber, the weak link is constructed to fail at a load lower than the maximum load carrying capability of the fiber.

The drag anchor is formed by a pair of laterally projecting side flanges in the same plane but on opposite sides of the longitudinal axis of the fiber. The laterally extending side flanges are formed by reducing the fiber thickness from 'd' to 't_d' without producing areas of significant stress concentrations that would otherwise reduce the axial tensile strength of the fiber. The dead anchor is formed by a second pair of laterally projecting side flanges produced once again by reducing the thickness from 'd' to 't' such that $t < t_d$ and the dead anchor is wider than the drag anchor, i.e. $w > w_d$.

The drag anchor functions in essentially the same way as a conventional drag anchor in conventional reinforcing fibers. However, the maximum drag force or axial force applied to the fiber in order to permit the drag anchor to be dragged through the concrete is less than the maximum force necessary to break the dead anchor. The incremental added forces that are carried by the dead anchor under peak conditions cause the stress in the weak link to exceed the allowable such that either the weak link breaks off or the dead anchor gets deformed or folded resulting in its release. Thus the dead anchor functions to reinforce the concrete until its failure occurs either by breakage at the weak link or by its own folding or deformation. In either case, as will be seen later, the energy that can be absorbed by the fiber is substantially greater than can be absorbed by conventional anchor structures. The combined anchor system permits the application of a higher total pull out load without the risk of fiber breakage as the dead anchor releases before the stress in the remainder of the fiber, including the drag anchor exceeds its capacity. The drag anchor is designed to carry at least 80% of the peak load and preferably 90% or higher so that the incremental load carried by the dead anchor is small and the carrying capacity of the fiber is not reduced dramatically when the dead anchor is released.

Figure 4 shows pull-out test data and demonstrates the effectiveness of combining the two anchors as done in the DD fiber, in terms of an improved energy absorption capacity. The commercial fiber having only a drag anchor (Curve 1 in Figure 4 for the "Flattened-End", FE, fiber)

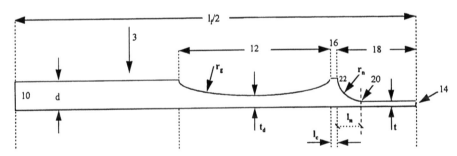

Figure 2. A side view of the DD Fiber

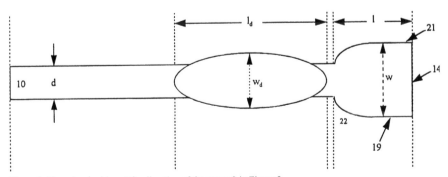

Figure 3. Plan view looking at the direction of the arrow 3 in Figure 2.

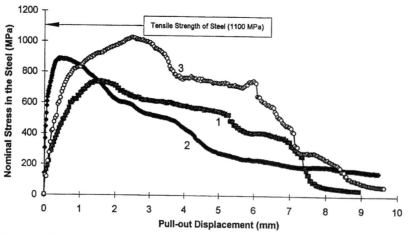

Figure 4. Pullout displacement versus nominal stress plots when only the 'dead' anchor is present (2), when only the 'drag' anchor is present (1), and finally when both anchors are combined (3) in a DD fiber (concrete compressive strength = 40 MPa).

provides a relatively gradual increase in stress as the displacement (pullout) is increased to about 1.5 mm. For a fiber with a dead anchor on the other hand, (Curve 2 in Figure 4 for the "Hooked-End" fiber), the peak or maximum stress that can be applied is significantly higher, approximately 900 MPa (tensile strength of the steel used in all cases is 1100 MPa), but the displacement that can be tolerated is less than approximately 0.5 mm. In both cases, the nominal fiber stress quickly diminishes (more so for the dead anchor than the drag anchor) as displacement is increased beyond the point of peak stress. For the fiber having a combination of the dead and drag anchors (Curve 3, Figure 4), notice a very significant increase in stress that can be tolerated i.e. the nominal stress for the fiber reaches above 1000 MPa while accommodating a displacement of about 2½ mm. After this, the fiber stress drops off but does not reduce to that of the commercial drag anchor per se until a very substantial amount of pullout has taken place, i.e. in the order of about 7 mm. At the peak values of applied load, either the weak link fractures or the dead anchor is deformed and released, and this is expected to occur prior to the fiber reaching its tensile capacity or a general failure. It is apparent that significant improvements in amount of pull out energy that can be absorbed is obtainable using the combination anchors as done in the DD fiber.

5 FURTHER OPTIMIZATION

As noted before, the requirements for a reduced rebound conflict with those for a high toughness. To find the optimal balance, an experimental route was

taken. Fibers were made from a fixed diameter wire with a 0.89 mm diameter formed with lengths of 12.5, 19, 25.4, and 40 mm, and all were tested at the rate of 60 kg/m^3 in dry-process shotcrete to determine their accumulated fracture energy under flexural loading of a standard ASTM C1018 test on 100×100×350 mm beam specimens (area under the flexural load versus displacement curve to a displacement of 2 mm). The results obtained are plotted in Figure 5 where it is apparent that a fiber length of somewhere between 20 to 40 mm, preferably about 25 mm, was found to be optimum. Next, after selecting an optimum length of 25.4 mm, fibers with diameters of 0.61, 0.76 and 0.89 were tested. The results of these tests are shown in Figure 6, where it is clearly indicated that a fiber diameter of about 0.75 mm (0.74 to 0.8 mm) was optimum. Based on these results, a length l_f = 25.4 mm and a diameter d = 0.76 mm was adopted, and the other fiber dimensions shown in Figures 2 and 3 were calculated. In this arrangement, the diameter 'r_g' of the indentation forming the drag section (12) was 10.7 mm, the thickness 't_d' was about 0.46 times diameter 'd', the width 'w_d' was 1.45 times the diameter 'd'. Based on the dimensions 'r_g' and 't_d' the length 'l_d' was derived. The length l of the dead hook section was set at 1.4 times the diameter 'd' of the fiber, and the thickness 't' was 0.23 times the diameter 'd', which produced a width 'w' of 2.36 times the diameter. The dimension 'l_c' was 0.2 mm and 'l_n' and radius 'r_n' for this example were equal and less than 0.5 mm, respectively. Other details of the fiber may be found elsewhere (Banthia et al, 1999).

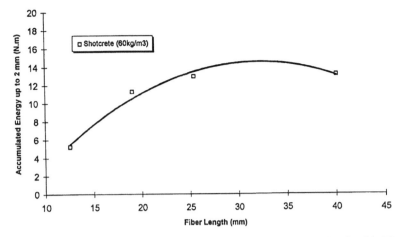

Figure 5. Plot of fiber length versus shotcrete fracture energy for four different lengths of the DD fiber.

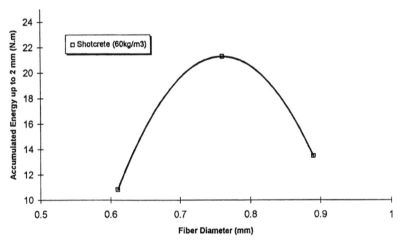

Figure 6. Plot of fiber diameter versus shotcrete fracture energy for three different diameter DD fibers.

6 DD FIBER IN DRY-MIX SHOTCRETE

Fibers as described in the above example were produced in sufficient quantity and compared with two other commercial fibers. For comparison, the ASTM C1018 tests were performed on at least five beam specimens (100×100×350 mm) for each fiber sawed from shotcrete. The results of these tests (averages of four to seven replicates) are plotted in Figure 7 wherein curve A is an average plot of the results obtained using the DD fiber, curve B is the average plot obtained using hooked-ended fibers, and curve C is the average plot using the pinched end (flattened-end or the FE) fiber. It is apparent that the DD fiber is able to accommodate more load carrying capacity and therefore consume more

fracture energy (the area contained by the curves in Figure 7) than either of the other two fibers. Some performance parameters are given in Table 1.

Table 1. DD fibre in dry process shotcrete.

Criterion	Hooked-End Fiber (DR30/50)	Flattened-End Fiber (N0730)	DD Fiber
ASTM I_{20}	3.28	4.09	8.37
ASTM I_{30}	5.43	7.28	12.83
ASTM I_{50}	9.01	13.29	20.27
JSCE, T (Nm)	9.82	14.25	18.15
JSCE, F (MPa)	1.47	2.14	2.72

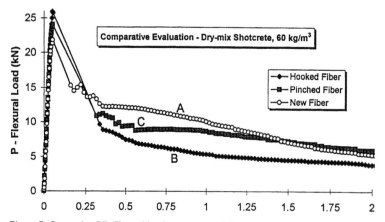

Figure 7. Comparing DD fiber with other commercial fibers in flexural toughness testing using ASTM C1018 (average rebound values for Hooked-end and Pinched-end (Flattened-end) fibers between 50 and 60%; for DD fiber: 40-50%.)

7 DD FIBER IN WET-MIX SHOTCRETE

The DD fiber was also recently investigated in wet-process shotcrete and the results are given in Table 2. Note that, as in the dry-process, the DD fiber demonstrates a superior performance over the leading commercial fibers based on both C1018 and EFNARC panel tests (LawGibb Group 2000).

Table 2. DD fibre in wet process shotcrete.

Fiber Type	40 kg/m³			Improvements in DD Fiber	
	DD Fiber	Hooked -End Fiber (DR30/ 50)	Flattened-End Fiber (N0730)	Over Hooked-End Fiber (DR30/50)	Over Flattened-End Fiber (N0730)
JCI (MPa)	2.84	2.46	1.76	16%	61%
EFNARC (Joules)	932	808	745	15%	25%

8 CONCLUSION

An improved reinforcing fiber for shotcrete is described. The fiber is based on a double anchoring principle where both dead and drag anchors are combined in the same fiber. The dead anchor breaks off as a result of ultimate stresses developed at a weak link in the fiber. This is then followed by the drag anchor frictionally resisting the pull-out without fiber breakage. The dead anchor improves the first crack strength and the drag anchor improves the overall energy absorption capacity. The fiber is also proportioned for a reduced rebound in the dry-process shotcrete. When the low rebound aspect is combined with the novel double anchoring concept, a superior shotcrete fiber is realized.

REFERENCES

Armelin, H.A. 1997. Rebound and toughening mechanisms in steel fiber reinforced dry-mix shotcrete, Ph.D. Thesis, University of British Columbia.

Armelin, H.S., Banthia, N., Morgan, D.R. and Steeves, C. 1997. Rebound in Dry-Mix Shotcrete, ACI Concrete International, 19(9): 54-60.

Armelin, H.S. and Banthia, N. 1998a. Mechanics of Aggregate Rebound in Shotcrete (Part 1), RILEM Materials and Structures (Paris), 31, March: 91-98; and Armelin, H.S. and Banthia, N. 1998. Development of a General Model of Aggregate Rebound in Dry-Mix Shotcrete (Part 2), RILEM Materials and Structures (Paris), 31, April: 195-202.

Armelin, H.S. and Banthia, N. 1998b. Steel Fiber Rebound in Dry Mix Shotcrete: Influence of Fiber Geometry, ACI Concrete International, 20(9): 74-79

Armelin, H.S., Banthia, N. and Mindess, S. 1999. Kinematics of Dry-Mix Shotcrete, ACI Materials J. 96(3): 1-8.

Banthia, N., Trottier, J.-F., Wood, D. and Beaupré, D. 1992. Steel Fiber Dry-Mix Shotcrete: Influence of Fiber Geometry, Concrete Int.: Design and Construction, American Concrete Inst., May (14)5: 24-28.

Banthia, N. and Armelin H.S. 1999. Concrete Reinforcing Fiber, US Patent #5965277, Issued October 12.

LawGibb Group. 2000. Technical Report, Report of EFNARC, C1018 and Compressive Strength Testing of DD Steel Fibers. Report No. 50157-0-3477-04, Atlanta, GA, September.

Shotcrete: Engineering Developments, Bernard (ed.) © 2001 Swets & Zeitlinger, Lisse, ISBN 90 5809 176 7

Recent developments in the field of shotcrete: the Quebec experience

D.Beaupré, M.Jolin & M.Pigeon
Department of civil engineering, Laval University, Quebec, Canada

P.Lacombe
Service d'expertise en matériaux (S.E.M.) inc., Quebec, Canada

ABSTRACT: This article presents an overview of the results of many research and demonstration projects in Quebec where innovative techniques and new developments were introduced into current shotcrete practice for repair and underground works. These new developments cover new types of shotcrete and equipment. Innovations such as air-entrained dry-process shotcrete, a high initial air content concept applicable to wet-process shotcrete, a new finishing technique using a motorized tool, and the development of exposed aggregate wet-mix shotcrete, are presented and described. Finally, the problem of reinforcement encapsulation using shotcrete is discussed, and potential test solutions are presented. Some experimental results are also presented on this matter.

1 INTRODUCTION

The Industrial Chair on Shotcrete and Concrete Repairs at Laval University has been involved in many new developments in shotcrete technology during the last seven years. The main goal of the shotcrete research program is to improve the quality of shotcrete repairs. This paper, divided in five parts, presents some of the developments related to shotcrete technology that have taken place in the province of Quebec.

The improvements presented concern: (1) the use of air-entraining admixtures, liquid or powdered, to enhance the frost durability of dry-mix shotcrete, (2) the use of a high initial air content concept in wet-mix shotcrete to facilitate pumping and shooting, (3) the use of a mechanical shotcrete finishing machine, (4) the production of exposed aggregate wet-mix shotcrete, and (5) advances in a research project aimed at understanding the effect of reinforcement encasement quality on shotcrete performance.

2 NEW TECHNOLOGIES

2.1 *Use of powdered air-entraining admixture in dry process shotcrete*

About ten years ago, the technique used to entrain air in dry-mix shotcrete was to add a liquid air-entraining admixture to the shooting water (Beaupré et al. 1994, Lamontagne et al. 1995). This technique is extremely efficient in improving the freeze-thaw

durability of dry-mix shotcrete and is still in use in many branches of the industry. However, this method has disadvantages: the amount of air-entraining admixture added to the shooting water is sometimes difficult to control on site and may vary as the nozzleman changes the shooting consistency.

A recent study conducted by the Industrial Chair on Shotcrete and Concrete Repairs of Laval University showed the effectiveness of *powdered* air-entraining admixtures at increasing the frost durability of dry-mix shotcrete. The use of the powdered form of admixture offers the possibility of being pre-bagged with the other oven dry materials (cement, sand, coarse aggregates and fibres). The dosage is therefore performed in a controlled environment. At a proper dosage rate, the powdered air-entraining admixture produces an excellent air void system, which generates a dry-mix shotcrete mixture resistant to freeze-thaw cycles and deicer salt scaling (Beaupré et al. 1996).

Table 1 and Figure 1 present results that show the effect of both liquid and powdered admixtures on the salt scaling resistance of dry-mix shotcrete. All of these tests have been made in accordance with the relevant ASTM specification. As shown in Table 1, it is possible to obtain very good spacing factors at low total air content by using either liquid or powdered admixtures in dry-mix shotcrete. The direct result is an improved salt scaling resistance, with no compressive strength reduction.

Table 1. Results on air-entrained dry-mix shotcrete (Beaupré et al. 1996)

Mixture	f_c at 28 d (MPa)	Air (%)	Spacing-factor (µm)	Salt scaling 50 cycles (kg/m²)
Plain [1]	28	4.7	415	8.8
Liquid AEA [2]	33	6.3	185	0.3
Powder AEA [3]	38	6.2	101	0.1

[1] Cement: 450 kg/m³, sand: 1500 kg/m³, coarse aggregate (10-2.5 mm): 265 kg/m³.
[2] As Plain, with 20 ml of *liquid* air entraining admixtures per liter of shooting water.
[3] As Plain, with 1.2 % of *powdered* air entraining admixtures by mass of cement.

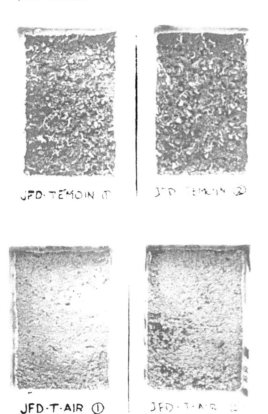

JFD·TÉMOIN ① JFD TÉMOIN ②

JFD·T·AIR ① JFD·T·A·R ②

Figure 1. Salt scaling resistance specimen (ASTM C 672) after 50 freeze-thaw cycles. Upper specimens: NON – Air entrained shotcrete mixture; specimens after 50 cycles of freezing and thawing with salt water on surface. Lower specimens: Air entrained shotcrete mixture; specimens after 50 cycles of freezing-thawing with salt water on surface.

2.2 Temporary high air content

Before 1994, the durability of shotcrete produced by the wet-mix process had been considered inadequate by the Quebec D.O.T. to allow its use in the repair of structures exposed to freeze-thaw cycles in the presence of deicer salts (Beaupré et al. 1994). Moreover, the use of wet-mix shotcrete is relatively more com-

plex since a compromise must be reached between pumping and shooting requirements (Beaupré 1994). Generally, the easier it is to pump the concrete, the more difficult it will be to obtain an adequate build-up thickness. Conversely, a concrete with a small slump value, although permitting a larger build-up thickness, is difficult to pump.

The concept of temporary high initial air content, introduced by Beaupré (1994), has been used on many projects to produce durable shotcrete repairs. A high initial air content (as high as 20%) is obtained by increasing the amount of air entraining admixture. The result is an increased fluidity of the fresh concrete, hence improving its pumpability. During pumping and particularly during shooting, a large volume of air is lost due to compaction. A stiffer in place material is then instantly obtained by this compaction effect, hence increasing the shootability of the shotcrete. Note that this volume of air can be incorporated with conventional equipment.

Table 2 shows the results obtained on several low water/cement ratio wet-mix shotcretes produced using the concept described above in a demonstration repair project of a parking structure. Using this concept, high slump shotcretes (150 mm to 220 mm) with initial air content ranging from 13% to 19% were pumped and placed. The in-place air content ranged from 5% to 7%, showing significant compaction. Also, the spacing factors were between 170 µm and 245 µm, which is in the range of acceptable values for good freezing and thawing durability as shown by the positive scaling results in Table 2. Typical wet-mix shotcrete shot with an initial air content between 5% and 8%, usually have spacing factors of 350 µm and more, led to higher scaling losses (as high as 10 kg/m²). These tests have been made in accordance to the relevant ASTM specifications. Since this demonstration repair job, the high initial air content concept has been used in more than one hundred shotcreting works, including underground projects.

Table 2 - Results on high initial air content wet-mix shotcrete (Beaupré et al. 1999)

Mixture's W/C	Slump (mm)	Initial air content (%)	In-place air content (%)	Spacing factor (µm)	f_c 7d 35d (MPa)		Scaling (kg/m²)
0.30	180	16	6.5	170	48	60	0.9
0.30	150	13	5.9	245	50	70	2.4
0.30	200	15	5.0	224	37	70	0.3
0.35	220	17	5.3	210	38	52	1.8
0.35	170	19	5.0	196	37	71	0.5
0.35	220	18	6.8	225	36	58	1.2

Composition: cement: 430 to 460 kg/m³, sand: 1105 to 1155 kg/m³, coarse aggregate (10-2.5 mm): 132 to 162 kg/m³.

2.3 Development of a shotcrete finishing machine

The intense use of shotcrete for repair has created an increased need for efficient finishing techniques and tools. Four years ago, an inventive shotcrete finisher from Quebec developed new equipment to accelerate finishing operations: the "shotcrete finishing machine".

The principle of this machine is a simple rotating wheel covered with a special rubber. It is powered by a small petroleum engine. The equipment, shown in Figure 2, does not only facilitate and speed up the finishing process, but it also produces a better looking surface, with reduced physical effort from the finisher. It reduces human efforts because of its petroleum-powered engine and its low weight (7 kilograms). The repetitive scratching/towelling movement made by the finisher is thus eliminated. With the shotcrete finishing machine, a qualified operator can finish approximately three times the surface he would achieve with conventional finishing tools.

2.4 Exposed aggregate shotcrete

The city of Quebec possesses an historic area in which several concrete structures require repair. The Quebec City urban planning department therefore needed an improved aesthetic appearance of its shot-

Figure 2. Shotcrete finishing machine.

crete repairs. The development of a technique to improve the visual aspect of shotcrete repairs was thus investigated. After a few trials with both processes, a wet-mix exposed aggregate surface was produced on an experimental surface of about 100 m² in Quebec City.

One of the challenges in this project was to incorporate a large proportion of coarse aggregate in order to obtain a uniform and dense aggregate distribution at the surface. The mixture composition used in the project is presented in Table 3. To facilitate the addition of a large volume of coarse aggregate, and also to ensure the frost durability of the material, the high initial air content concept was used (see Section 2.2). The high initial air volume helps the pumping by acting as a lubricant (equivalent to an artificial increase in the paste content).

To ensure an adequate exposed aggregate surface finish, a shotcrete thickness in excess of the desired final profile (approximately 25 mm extra) was applied. The excess concrete was cut off before finishing the surface. Because of the high coarse aggregate content, hand trowelling was difficult but feasible. For this reason, the surface was finished using the shotcrete finishing machine described above. This led to a more regular surface with minimum labour effort.

Shortly after the shotcrete finishing operations were completed, the surface was sprayed with a set-retarding admixture. After an 18-hour waiting period, the concrete surface was washed out with a high-pressure water jet in order to expose the top 2 to 3 mm of the aggregates. The appearance of the wall, presented in Figure 3, fully meets the city's requirements. Furthermore, the irregular rough surface finish tends to diffuse and hide the possible cracks which sometimes appear on thin repairs.

2.5 Reinforcement encapsulation using shotcrete

Shotcrete technology requires workers that are skilled and well trained in order to properly encapsulate reinforcement. To ensure that workers possess the required skills, nozzlemen qualification/ certification programs are often used. These programs generally implement some sort of visual grading of

Table 3. Shotcrete mixture composition use for exposed aggregate wet-mix shotcrete.

Components	Mixture (kg/m³)
Cement T10FS	450
W/B ratio	0.35
Coarse Aggregates 2-10 mm	858
Fine Aggregates	735
Water reducer	1.1
Superplasticizer	2.5
Air-Entraining Admixture	1.0
Air content before pumping	20 %
Slump	210 mm

Figure 3: Exposed aggregate wall.

Mechanical performances of correctly encased reinforcement (cast and shot) were compared to those of poorly encased reinforcing bars using pull-out tests. Preliminary results on a limited number of samples have shown that the average adhesion between steel and shotcrete can be reduced by about 75% when encapsulation quality goes from Grade 1 to approximately a Grade 2. Also, it seems that a good quality shotcrete application is comparable to similar conventional cast concrete with regard to reinforcement bond strength. Table 5 shows these results. For each type of concrete, 4 samples were tested.

Table 5. Results of reinforcement pull-out tests.

Type of concrete	Core Grade (ACI 506)	Bond stress (MPa)	Coefficient of variation
Cast concrete[1]	N.A.	24.2	0.8 %
Shotcrete[2]	1	24.7	2.1 %
Shotcrete[3]	2	6.6	17 %

Concrete UCS: 1 f_c = 52 MPa, 2 f_c = 38 MPa, 3 f_c = 43 MPa

cores (or sawed sections) to evaluate the workmanship quality achieved. However, there are some draw-backs associated with the use of visual criterion, the main one being subjectivity. Moreover, there are no agreed criteria stating what quality of reinforcement encasement is acceptable or not.

To address these issues, a research program has been started by the Industrial Chair on Shotcrete and Concrete Repairs in order to: (1) design a non-subjective test to evaluate the quality of reinforcement encapsulation, (2) evaluate the effect of reinforcement encapsulation quality on structural capacities and durability, and (3) propose acceptance criteria for nozzleman qualification/certification program in terms of reinforcement encapsulation quality.

So far, a new test that measures the amount of void around a reinforcing bar has been developed. It consists, after simple preparation, of measuring the amount of void around the reinforcement by water displacement. After putting a prepared core into a special container, water is allowed to fill the voids around the bar. The amount of water displaced during the test is recorded as the test result. Table 4 presents the results obtained for approximately 50 core samples. The encasement quality was evaluated according to ACI 506 Core Grade System.

Table 4. Results of reinforcement encasement test and Core Grading

Volume of displaced water (mL)	Assumed quality	Core Grade (ACI 506)
0 to 6	Excellent	1
6 + to 12	Good	2
12 + to 18	Fair	3
18 +	Bad	4 or 5

There are many aspects that require additional work in order to allow for a clear recommendation concerning the acceptable quality level of reinforcement encasement.

3 CONCLUSION

In recent years, several research projects have been conducted to enhance concrete resistance to freeze-thaw cycles (with and without deicer salts) and the aesthetic appearance of concrete repairs. Results are extremely positive. It has been shown that powdered air entraining admixtures are efficient at generating a good spacing factor without altering the compressive strength.

The concept of high initial air volume for wet-mix shotcrete yielded excellent field results. Supplementary to this concept, it is now possible to produce exposed aggregate shotcrete which improves the appearance of the finished surface.

Initial results obtained with the reinforcement encapsulation project are very promising. A new method to evaluate the quality of encapsulation has been identified and its validation is in progress through the establishment of acceptance criteria.

4 ACKNOWLEDGEMENT

The authors wish to thank the Natural Sciences and Engineering Research Council of Canada for its financial support to the Industrial Chair on Shotcrete and Concrete Repairs. The authors also thank the industrial partners: Quebec Department of Transport, St-Laurence Cement, Lafarge Canada, Béton Mobile du Québec, King Packaged Material, Rhodia, W.R.

Grace, Master Builders Technologies, City of Quebec and City of Montreal for their collaboration.

REFERENCES

ACI 506R-90, *Guide to Shotcrete*, American Concrete Institute, Detroit.

Beaupré, D. (1994), *Rheology of high performance shotcrete*, Ph.D. Thesis, University of British Columbia, 250 p.

Beaupré, D., Lacombe, P., Dumais, N., Mercier, S. Jolin, M. (1999) Innovation in the field of shotcrete repair, International Congress: Creating with concrete, Dundee, Scotland, 6-10 September.

Beaupré, D., Talbot, C., Gendreau, M., Pigeon, M., Morgan, D.R. (1994) Deicer Salt Scaling Resistance of Dry- and Wet-Process Shotcrete, ACI Materials Journal, 91 (5): 487-494.

Beaupré, D., Lamontagne, A., Pigeon, M., Dufour, J.F. (1996) Powder air-entraining admixture in dry mix shotcrete, ACI/SCA International Conference on Sprayed Concrete, Edinburg, 10-11 September.

Shotcrete: Engineering Developments, Bernard (ed.) © 2001 Swets & Zeitlinger, Lisse, ISBN 90 5809 176 7

Performance of multi-layered polymer and fibre reinforced concrete panels

E.S.Bernard

University of Western Sydney, Nepean, Australia

ABSTRACT: Spray-on polymer materials have recently become available as a means of improving the water-tightness of tunnel linings made with Fibre Reinforced Shotcrete (FRS). This type of material can be sprayed onto the surface of hardened shotcrete, and, after curing, can then be covered by a further layer of shotcrete to create a composite polymer/FRS tunnel lining. The highly ductile properties of the polymer allow it to bridge cracks that may occur in the FRS, thereby helping to improve resistance to the through flow of water. This investigation has sought to determine the structural capacity of composite polymer/FRS linings by testing the comparative toughness of Round Determinate panels made with neat cast Fibre Reinforced Concrete (FRC), and multi-layered polymer/FRC panels.

1 INTRODUCTION

1.1 Background

Tunnel linings made with Fibre Reinforced Shotcrete are often required to be water tight to prevent water flowing or dripping onto vehicles or equipment contained within the tunnel cavity. A common example of such an application is in electrified rail tunnels where the ingress of water can lead to corrosion of catenary or transmission equipment (Melbye 1994). Reliable water tightness is very difficult to achieve without the use of sheet polymer membranes and drainage outside of the FRS lining. Unfortunately, such measures are expensive and time-consuming to install.

An alternative method of improving the water-tightness of tunnel linings has recently become available in the form of a spray-on polymer membrane. This material consists of a highly deformable elastomer that can be sprayed onto the surface of an existing FRS lining. After curing, it can then be covered with a second FRS layer to produce a composite lining. The structural capacity of the complete lining can be considered equal to either the summed capacity of the individual layers of FRS, or that of a composite comprising interacting layers of polymer and FRS. However, the degree of composite behaviour that can be expected in this type of lining system is unknown. This investigation has been undertaken to provide insights into the influence of spray-on polymers on the overall structural capacity of composite polymer/FRS tunnel linings.

Several companies have recently developed acrylic-based spray-on water stop membranes that act to improve the water tightness of FRS linings. An example is MBT Masterseal 800A (MBT 1999), available in the form of a thick paste that can be sprayed onto the surface of hardened FRS using conventional equipment. Following curing, which takes about one day depending on thickness, temperature, and drying conditions, a second layer of FRS can be sprayed over the polymer resulting in a sandwiched lining of FRS and integral water stop membrane. The highly resilient and deformable properties of the polymer allow it to maintain continuity across cracks that may occur through the total lining, thereby helping to enhance water-tightness. However, the complete lining cannot be treated as a monocoque shell but instead must be considered to be either a series of individual linings separated by polymeric membranes or a composite system involving some degree of interaction between the component layers. The structural properties of the composite lining require investigation to determine the influence of the polymer membrane on stiffness, peak load capacity, and post-crack performance.

1.2 Objectives

To determine the influence of spray-on polymer water-stop membranes on the structural capacity of tunnel linings consisting of alternating layers of FRS and polymer. The behaviour of the composite lining system will be investigated experimentally through tests on Round Determinate panels made of neat

FRC and FRC cast in combination with intermediate layers of water-stop polymer.

2 EXPERIMENTAL PROGRAM

The Round Determinate panel test (Bernard and Pircher 2000) has been selected as the basis for examining the structural capacity of polymer and FRC composites. This test involves the application of a central point load on a round panel measuring 75×800 mm supported on three symmetrically arranged pivoting points (Figure 1). In recent trials involving comparative testing of beams and panels made with numerous shotcrete mixes, the Round Determinate panel test produced the most reliable estimates of post-crack performance (Bernard 1999). This test also presents the unique advantage that the pivoted supports mimic the action of rock bolt anchors in restraining movement of a continuous FRS lining.

Figure 1. Failure of a 37 mm thick monolithic FRC Round Determinate panel.

2.1 *Specimen Preparation*

Although the application for tests conducted in this investigation was sprayed concrete linings, the panels were cast using a Fibre Reinforced Shotcrete mix rather than sprayed. The polymer was similarly trowelled into place rather than sprayed. The reasons for this were twofold.

Firstly, consideration of the behaviour of a composite layered sandwich of polymer and FRS suggests that significant shear deformation is likely to occur within each layer of polymer in response to out-of-plane loading (Figure 2). Upon cracking of the FRS, a composite panel capable of supporting high shear strains at each polymer/FRS interface is likely to exhibit enhanced toughness compared to a

monolithic panel. This is because thin panels of FRS exhibit greater residual load capacity expressed as a ratio of peak load capacity than thick panels composed of the same material (Bernard 1998). Furthermore, the polymer is likely to exhibit some residual load carrying capacity itself. However, roughness at each polymer/FRS boundary may reduce the degree of shear strain possible and thereby diminish post-crack load carrying capacity. The present specimens were therefore made with cast FRC to produce smooth polymer/FRC boundaries so that the potential for toughness enhancement was maximised. If the enhancement in toughness for cast polymer/FRC composites is negligible compared to monolithic panels, then it is likely to be even poorer for sprayed composites.

The second reason for using cast FRC is that the size of panels used in the present investigation (75×800 mm diameter) and required accuracy in layer thickness (being as little as 25 mm thick) precluded the use of sprayed concrete and polymers for which thickness is difficult to control. The specimens therefore consisted of cast FRC and trowelled polymer layers.

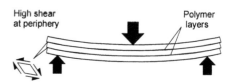

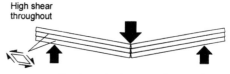

Figure 2. Sections through a sandwiched polymer/FRC composite Round Determinate panel showing expected shear straining within each polymer layer in response to a central load.

A total of 25 plywood moulds measuring 800 mm in diameter were prepared for the investigation by nailing sheet steel strip around the periphery of round disks to produce moulds. Fifteen moulds had a depth of 75 mm to permit full-depth panels to be produced. Of these, five were used to cast FRC specimens that did not include a polymer layer. These have been denoted *monolithic* panels. Five were used to produce double-layer panels incorporating a polymer membrane at mid-depth, and five were used to produce triple-layered panels incorporating two layers of polymer at one-third depths. These have been denoted as *composite* panels, all of which had a total thickness of 75 mm. Five moulds were produced with a total depth of 37 mm to permit half-thickness monolithic FRC panels to be cast, and five were 25 mm deep to permit one-third thickness monolithic FRC panels to be cast.

Because the FRC and polymer both required curing before a second layer of material could be placed over the surface, the specimens were produced over a period of several days. On the first day of casting, sufficient FRC was mixed to produce five monolithic panels in each of 25, 37, and 75 mm total thickness. The first layer of the multi-layered specimens were also produced. Following overnight hardening, a layer of polymer was trowelled onto the surface of the composite panels and left to cure for a second day. A second layer of FRC was then cast. This process was repeated again for the five specimens consisting of three layers of FRC and two polymer membranes. When all specimens were completed, they were stripped from the moulds and immersed in curing tanks for 2 months. Companion compression cylinders measuring ∅50×100 mm were cast with several batches of concrete and cured together with the panels.

2.2 Material Properties

The materials used in this investigation consisted of a dry-batched shotcrete mix produced by MBT (Australia) P/L and sold under the name *Shotpatch 20*, and a spray-on polymer called *Masterseal 800 A*.

Shotpatch 20 is intended to be used as a dry process shotcrete material for repair applications. The grading profile for this material consisted of a series of sands and pea-grit that resulted in a very uniform combined grading profile (see limits in Fig. 3). Using the recommended water addition rate, it had a UCS at 28 days of approximately 70 MPa. In the present investigation, a higher slump was required so additional water was added. This resulted in a mean compressive strength of 45 MPa. A total of 50 kg/m³ of Dramix RC65/35 fibres were used to produce the concrete used in this investigation.

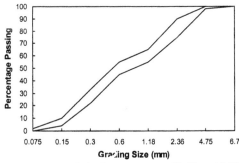

Figure 3. Grading limits for aggregates used in Shotpatch 20.

Masterseal 800A is a water-based acrylic compound normally used as a ventilation seal and water stop membrane. It is supplied in a paste form that can be painted, rolled, trowelled, or sprayed into place. Following a curing period of 1 to 2 days depending on thickness of application, it is claimed to develop a bond strength of about 1.6 MPa to almost any substrate. It will also develop a strong bond to concrete cast or sprayed over it once the concrete has hardened.

In this investigation, the Masterseal 800A was trowelled into place to produce layers with a thickness of approximately 2 mm.

2.3 Test Apparatus

All tests were undertaken using a purpose-built test rig placed inside an Instron 8506 Universal Test Machine. The test fixture consisted of three pivoted bearings located on a 750 mm diameter pitch circle diameter. Each pivot was secured to a 60 mm thick radial steel plate welded together at the centre to ensure very high stiffness in response to loading (Figure 1). The stiffness of the test apparatus was measured to be 250 kN/mm.

A test was typically carried out in the following manner. After being placed upon the three pivots, the specimen was carefully centred with respect to the loading piston. The test proceeded by lowering the actuator to within a small distance of the surface and commencing load application. Tests were conducted in accordance with T373 (RTA 1999), a procedure developed by the Roads and Traffic Authority of New South Wales to assess the post-crack energy absorbing capacity of FRS. Load was applied at a controlled rate of displacement equal to 2 mm/min up to 10 mm total deflection, after which the rate was increased to 10 mm/min up to a total displacement of 100 mm. Performance was measured as the load to cause first crack of the concrete matrix, peak load capacity, residual load capacity at 40 and 75 mm central deflection, and cumulative energy absorption at 40 mm central deflection. Energy absorption was determined by integrating the area under the load-deflection curve up to 40 mm deflection for each specimen.

3 RESULTS

3.1 Panel Data

The result of each panel test consisted of a load-deflection curve revealing performance between the onset of loading and 100 mm total central displacement. These curves have been plotted in Figures 4 and 5. Performance has also been summarised in terms of the peak and residual load capacity at 40 and 75 mm central deflection. This data is presented in Table 1.

The curves in Figure 4 show the performance of composite panels with one and two layers of polymer sandwiched between layers of FRC in comparison with the performance of 75 mm thick monolithic panels consisting of neat FRC. These curves reveal that the introduction of one layer of polymer results in a reduction in peak load capacity relative to monolithic FRC panels, and the introduction of two polymer layers causes a further reduction in capacity. However, residual load capacity at high levels of deformation expressed as a ratio of peak load capacity was much greater in the composite panels than in the monolithic FRC panels.

Observation of each specimen during testing revealed that shear straining occurred within the plane of each polymer layer in many of the specimens, but not all. An example of the resulting failure mode at a radial crack is shown in Figure 6. This shows a specimen with a mid-plane layer of polymer that suffered significant shear straining on only one side of a radial crack. Comparison of load-deflection curves and observations during testing indicated that high residual load capacity and significant shear straining was well correlated. This suggests that the mechanism of ductility enhancement described in Section 2.1 is valid, at least for specimens with smooth polymer/FRC boundaries.

It must be noted that inter-laminar shear straining appeared to be more prevalent in the double layered panels than the triple-layered panels, and that residual load capacity appeared to be higher as a consequence of this. The reason for this may be that shear stresses are highest at the mid-plane of a plate during bending, which would rationally lead to greater deformation at this depth through the section. However, it also appeared to be that one of the polymer layers in the triple-layered polymer/FRC composite panels was invariably thinner than the other and therefore offered greater resistance to slip. The resulting failure mechanisms in the triple-layer panels therefore consisted effectively of two FRC layers, one much thicker than the other, despite the presence of two polymer layers.

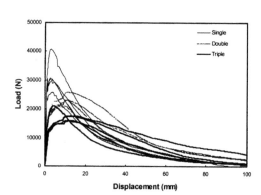

Figure 4. Load-displacement curves for all 75 mm thick specimens consisting of either a single layer of FRC, two 37 mm thick layers of FRC separated by polymer, or three 25 mm thick layers of FRC separated by two layers of polymer.

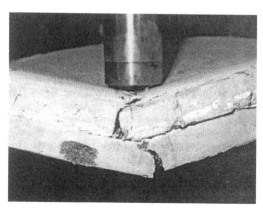

Figure 6. Side view of double-layered panel showing (at right) inter-laminar slip at polymer layer.

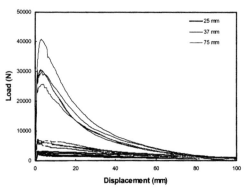

Figure 5. Load-displacement curves for all monolithic FRC specimens either 25, 37, or 75 mm thick.

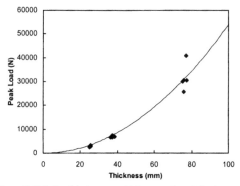

Figure 7. Relationship between thickness and peak load capacity for monolithic FRC specimens fitted with a quadratic expression.

The curves in Figure 5 show the influence of thickness on peak load capacity in the monolithic panels. This shows that peak load capacity increased rapidly with thickness, but that post-crack toughness was much lower for the thicker panels when measured relative to peak load capacity. The strong influence of thickness on peak load capacity is further revealed by the data in Figure 7. A least-squares adjusted curve-fit to this data found that the quadratic expression $P = 5.39 \times t^2$, where P is load capacity and t is panel thickness, was a good representation of the relationship.

4 DISCUSSION

Water tightness is generally only required in selected civil underground structures such as rail tunnels. This type of structure frequently has a low tolerance for deformation, so cracks are likely to be very narrow if they occur at all. If spray-on water stop polymer layers were to be used between layers of FRS to produce a composite lining, the degree of interlaminar shear that would occur in such structures would therefore be very small. The enhancement in residual load capacity at large deformations exhibited by composite polymer/FRC panels in this investigation would therefore be of little value.

In contrast to post-crack capacity at large deformations, peak load capacity is of considerable importance in civil applications, so the influence of polymer layers on the peak load capacity of a composite lining is of great interest. Referring to Table 1, it can be seen that a single 25 mm thick FRC panel exhibited a mean peak load capacity of 2930 N. In a simple approximation, three independent layers of 25 mm thick FRC could be summed to give a combined peak load capacity of 8790 N.

However, the composite panels consisting of two polymer layers between three 25 mm thick FRC layers exhibited an actual mean peak load capacity of 17662 N. This represents a 100 per cent increase in capacity over a simple summation of capacities of the component FRC layers.

The double-layered composite FRC and polymer panels had a peak load capacity 65 per cent greater than the summed capacity of two 37 mm thick FRC layers. Considering these results indicates that interaction occurred between the FRC components of the

polymer/FRC composite panels. The peak load capacity of the composite panels was intermediate between monolithic panels and a simple summation in capacities of the component layers.

When assessed in terms of energy absorption at 40 mm central displacement, the composite panels also performed much better than the sum of the component FRC layers, but not as well as a monolithic FRC panel of equal thickness. For example, a single 25 mm thick FRC panel absorbed 79 Joules at 40 mm central deflection, so three layers acting independently could be assumed to absorb 237 Joules. However, this would be a very conservative estimate considering the actual energy absorbed by the triple-layered polymer/FRC composite panels was 537 Joules. For both the double and triple-layered specimens, the composite panels absorbed about 100 per cent more energy than the sum of the component individual FRC panels. This suggests that the polymer layers either absorbed a significant amount of energy in their own right or caused the FRC acting in tension to 'work' more effectively during deformation. However, larger scale tests would need to be carried out to confirm whether this occurs in sprayed composite linings in which roughness at the polymer/FRS boundaries would probably play a major role in determining behaviour.

Considering residual load capacity at 40 and 75 mm central deflection as a ratio of peak load capacity, it is clear that performance increases with decreasing panel thickness (Table 1). It is also evident that at 40 mm central deflection, the polymer/FRC composite panels were as 'tough' as their component parts when expressed in this manner. However, by 75 mm deflection the polymer/FRC composite panels had begun to de-laminate and did not match the individual 25 and 37 mm thick panels for relative ductility.

Consideration of the residual load capacity of the polymer/FRC panels at large deflections showed that the presence of the polymer layers enhanced load carrying capacity relative to all the monolithic panel configurations. For example, at 40 mm central deflection, the triple layer composite panels had a mean residual load capacity of 8704 N, compared to 7142 N achieved by the 75 mm monolithic panels and $3 \times 1504 = 4512$ N summed capacity of the component 25 mm FRC panels. At 75 mm central deflection, the capacities were 3096 N, 2130 N, and

Table 1. Summary of specimen performance, all results are the mean of five specimens.

Specimen Set	Peak Load (N)	Load at 40 mm (N)	Load at 40 mm/ Peak Load (%)	Load at 75 mm (N)	Load at 75 mm/ Peak Load (%)	Energy at 40 mm (J)
Single layer 75 mm thick	31525	7142	22.9	2130	6.9	669
Double layer composite	22456	9496	41.6	3019	13.1	662
Triple layer composite	17662	8704	50.2	3096	17.8	537
37 mm thick monolithic	6797	2815	41.3	1384	20.3	180
25 mm thick monolithic	2930	1504	51.0	885	30.1	79

3×885 = 2655 N, respectively. This demonstrates the ductility enhancement possible at large deflections through the use of internal polymer layers.

5 CONCLUSION

Incorporation of a water stop polymer layer between layers of hardened Fibre Reinforced Concrete changes the structural characteristics of the resulting multi-layered composite relative to monolithic panels of similar thickness. While the peak load and energy absorbing capacity of the multi-layered panels is lower than that of monolithic panels, residual load carrying capacity is enhanced. The basis for the improvement in residual load capacity appears to be an increase in the effective ductility of the composite at cracks. However, this can only occur so long as sliding of the FRC layers relative to each other is possible. The potential for improved residual load carrying capacity in multi-layered shotcrete linings, which typically exhibit very rough surfaces, is therefore limited. However, the possibility that ductility can be enhanced through the use of polymers between layers of screeded FRS may merit further investigation.

ACKNOWLEDGEMENTS

The author gratefully acknowledges the support of MBT (Australia) P/L, through their representative John Gelson, for their generosity in supplying Masterseal 800A and Shotpatch 20 pre-batched shotcrete mix for this investigation.

REFERENCES

Bernard, E.S., 1998, "The Behaviour of Round Steel Fibre Reinforced Concrete Panels under Point Loads", *Engineering Report No. CE8*, Department of Civil and Environmental Engineering, University of Western Sydney, Nepean.

Bernard, E.S., 1999, "Correlations in the Performance of Fibre Reinforced Shotcrete Beams and Panels", *Engineering Report No. CE9*, Department of Civil and Environmental Engineering, University of Western Sydney, Nepean.

Bernard, E.S. and Pircher, M., 2000, "Influence of Geometry on Performance of Round Determinate Panels made with Fibre Reinforced Concrete", *Civil Engineering Report CE10*, School of Civic Engineering and Environment, UWS Nepean.

Melbye, T. A., 1994, *Sprayed Concrete for Rock Support*, MBT, Zurich.

MBT *Product Catalogue*, 1999, MBT (Australia) P/L, Sydney.

RTA Specification T373, 1999, "Round Determinate Panel Test", from B-82 *Shotcrete Work*.

Shotcrete: Engineering Developments, Bernard (ed.) © 2001 Swets & Zeitlinger, Lisse, ISBN 90 5809 176 7

The influence of curing on the performance of fibre reinforced shotcrete panels

E.S.Bernard
University of Western Sydney, Nepean, Australia

M.J.K.Clements
Jetcrete Australia P/L, Sydney, Australia

ABSTRACT: Panels have recently emerged as a more reliable and economical alternative to beams for Quality Assurance testing of Fibre Reinforced Shotcrete (FRS). The large size of panels relative to beams results in a more representative assessment of post-crack performance, but the size of the specimens also presents a challenge to contractors and testing agencies in developing adequate handling procedures. At present, flexural beams are generally continuously moist cured, but this is seldom the case for panel specimens. Instead, panel specimens are cured under wet hessian or sprinklers during the early stages of hydration, but are then transported to the laboratory in a dry state. This investigation will examine the affect of this on first-crack and post-crack performance.

1 INTRODUCTION

1.1 Background

The importance of curing to the performance of conventional concrete and shotcrete is well documented (Neville 1996). Curing is essential to the proper hydration of cement. Failure to adequately cure concrete can result in low strength and abrasion resistance, high permeability, and poor durability. Nevertheless, curing practices are seldom adequate on site and the properties of *in situ* concrete are therefore unlikely to equal those of Quality Assurance (QA) specimens. Such specimens, which normally take the form of compression cylinders, are typically moist cured up to the age of testing. This promotes optimal hydration and thereby ensures the highest possible performance for a given mix. However, the use of panels to determine the post-crack performance characteristics of Fibre Reinforced Shotcrete (FRS) makes it difficult to achieve the same high standard of curing that is routinely expected for cylinders. This is because panels are much larger and heavier than other types of QA specimen. Handling and provision of room in curing tanks and fog rooms for such bulky specimens is both challenging and expensive.

On site practice in many projects has been to cure panels under wet hessian during the early stages of hydration, and then transport them to the testing laboratory in a dry state. Due to limitations on space in curing tanks, panels are often left dry up to the age of testing. The rationale behind this approach is that adequate curing is believed to occur within the first 7-14 days, and that further wet curing will provide minimal benefit. The present investigation was instigated to examine the validity of this assumption.

1.2 Objectives

The objective of this investigation was to examine the value of curing on the performance of Fibre Reinforced Shotcrete panels used for Quality Assurance testing. The investigation has specifically addressed the influence of continuous wet curing on post-crack performance compared to dry curing and partial wet curing. The effect of curing was assessed through a comparison of the load to cause first crack of the concrete matrix, and energy absorption during the deformation process, in FRS tested using the Round Determinate panel test (Bernard and Pircher 2000).

2 EXPERIMENTAL PROGRAM

2.1 Selection of Procedures

A number of different panel geometries are available for the purpose of QA testing for toughness assessment in FRS. The most commonly used is the EFNARC panel (EFNARC 1996), consisting of a square simply supported specimen measuring 100×600×600 mm loaded at the centre. Although the size and shape of this specimen are relatively easy to produce and handle, this test suffers the disadvantage that performance is dependent on the degree of flatness achieved in the base upon which the specimen rests (Bernard 1998). Under field conditions it is very difficult to produce specimens that are free of distortions, so the consistency of results using EFNARC panels can be compromised. It is also inconvenient to cut off the inclined sides that are included to avoid rebound being incorporated into the sides of the specimen. Other specimens tested on a simply supported base (eg. Marti et al. 1999) suffer the same problems as the EFNARC panel so this type of test has been excluded from this investigation.

The South African water bed test (Morgan et al. 1999) is an interesting alternative to the EFNARC panel. This test involves subjecting a 1600×1600 mm shotcrete panel to a distributed load via a pressurised water bed. Although the loading mechanism used for this test simulates *in situ* loading within some tunnels very well, the bulkiness of the specimen precludes its use as a routine QA tool, and it has therefore not been used in the present investigation.

The Round Determinate panel test was recently developed as attempt to over come the problems inherent in the alternatives described above (Bernard and Pircher 2000). This test involves the application of a central point load to a round panel measuring 75×800 mm supported on three symmetrically arranged pivoting points. In large scale trials involving comparative testing of beams and panels made with numerous shotcrete mixes, the Round Determinate panel test produced the most reliable estimates of post-crack performance (Bernard 1999). Correction factors that account for the variations in thickness and diameter that are inevitable in production are also available (Bernard and Pircher 2000). The many advantages of this test have lead to its selection as the basis of experimental work in this investigation.

2.2 Test Apparatus

Panel tests were undertaken in a purpose-built apparatus consisting of a servo-hydraulic 250 kN MTS 244 actuator mounted within a rigid cylindrical housing surrounded by three vertical supports upon which the specimens were mounted (Fig. 1). After being placed upon the three pivots, each specimen was tested by lowering the actuator to within a small distance of the surface and commencing load application. Tests were conducted in accordance with T373 (RTA 1999), a QA procedure for FRS incorporating Round Determinate panels, developed by the Roads and Traffic Authority of New South Wales. Load was applied at a controlled rate of displacement equal to 10 mm/min up to a total displacement of 100 mm. Performance was measured as the load to cause first crack of the concrete matrix, and cumulative energy absorption at 40 mm central deflection. Energy absorption was determined by integrating the area under the load-deflection curve for each specimen.

2.3 Specimen Preparation

The concrete used to make the specimens was produced in a commercial ready-mixed plant by dry-batching 4 m³ of materials into a 6 m³ agitator truck. Fibres were manually added when the truck arrived on site, after which the concrete was mixed for 15 minutes before use. This exceptionally long period of mixing was deemed necessary to ensure thorough and uniform distribution of fibres throughout the mix. The mix design for the shotcrete is listed in Table 1. Three types of steel fibre were used in combination to achieve a high performance mix with exceptional crack containing qualities at low levels of deformation. A hydration accelerator was not used.

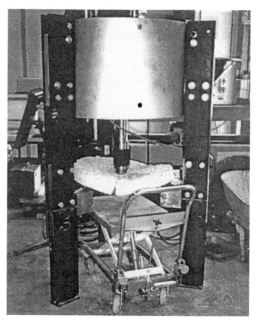

Figure 1. Testing of a Round Determinate Panel.

Formwork consisted of round form ply moulds around which sheet steel had been nailed to produce a rigid dish into which the concrete was sprayed. All specimens were produced by spraying FRS into the forms as they were inclined at 45° against a supporting frame. Following spraying, each specimen was laid flat on the ground and screeded to achieve a level surface and uniform thickness. They were then covered with plastic and left to harden overnight. Although performance correction factors are available that make it possible to adjust apparent performance to account for variations in thickness and diameter (Bernard & Pircher 2000), maintenance of dimensions close to those specified will help to achieve more reliable and repeatable results.

Table 1. Mix design for shotcrete.

Ingredient	Quantity (kg/m³)
Coarse aggregate (10/7 mm)	620
Coarse sand (5 mm)	615
Fine sand (2 mm)	410
Cement (ASTM Type 1)	380
Fly ash	40
Silica fume	20
Water reducer	1.9 L
Dramix RC65/35 fibres	40
BHP 256EE fibres	18
Novotex 0730 fibres	14
Slump	90 mm

A total of 30 specimens were produced in this investigation. The influence of curing was examined by varying the conditions under which subsets of 10 specimens were cured. Following initial hardening and stripping of moulds, ten specimens were simply stacked together and cured in air under normal atmospheric conditions for the entire 28 day curing period. These specimens have been referred to as the 'dry cured' panels. Wooden spacers were placed between each specimen to promote ventilation. The remaining twenty panels were immersed in lime water within two adjacent curing tanks. Ten of these were removed after 14 days curing and left to dry in air next to the dry cured specimens. These have been referred to as the 'partially wet cured' panels. The other ten were continuously 'wet cured' for the remaining 14 days, and were tested in a surface-saturated condition. The dry cured and partially wet-cured specimens were tested in a surface-dry condition. No curing compounds were applied to the surface of any of the specimens.

3 RESULTS

The result of testing each specimen consisted of a load-displacement curve of the type shown in Fig. 2. The results for the continuously wet cured panels are shown together in this figure. The origin of loading has been translated for some of these specimens so that the ascending part of each curve maintained registration with the rest. The variation evident within this set of curves is quite typical of results for panel tests, and represents a Coefficient of Variation

equal to 7% in the load to cause first crack of the concrete matrix, and 11% in energy absorption up to 40 mm total central deflection (see Table 2). Notice that each of the wet-cured panels suffered a slight drop in capacity following first crack of the concrete matrix.

Comparison of performance between the three sets of panels has been facilitated by finding the mean and variation in performance within each set throughout the loading envelope. This has been done using a computer program that samples each set of 10 curves and produces an estimate of the mean and standard deviation at each of a number of increments throughout the displacement profile. The result for the wet cured specimens is presented in Fig. 3, which shows the averaged load-deflection curve as a dark line with error bars at one standard deviation above and below this line. These curves reveal that the panels behaved in a very similar manner up to the point of first crack of the concrete matrix, after which performance diverged, especially around the peak in load capacity, but then converged again as the level of deformation became severe.

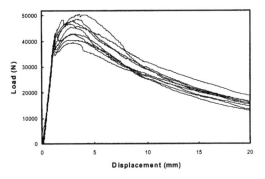

Figure 2. Load-displacement curves for wet cured Round Determinate panels.

A mean load-deflection curve was similarly determined for each of the dry cured and partially wet cured specimen sets. The three results are presented for comparison in Fig. 4. This graph reveals all the most important results of this investigation. Firstly, it is apparent that the wet cured specimens exhibited superior performance to both the dry and partially-wet cured panels. The difference in performance was

Table 1. Summarized test results.

Specimen Set	Load to cause first crack		Peak Load		Energy Absorption (40 mm)	
	Mean (N)	COV (%)	Mean (N)	COV (%)	Mean (J)	COV (%)
Wet Cured Specimens	37706	7.3	41047	9.2	729	10.9
Partially Wet Cured Specimens	31762	6.9	36060	7.4	661	7.7
Dry Cured Specimens	31807	8.0	34800	6.3	678	7.3
Dry/Wet	0.844		0.848		0.930	
Partially Wet/Wet	0.842		0.879		0.907	
Dry/Partially Wet	1.001		0.965		1.025	

manifested as an increase in the load to cause first crack of the concrete matrix, and higher residual load capacity up to 10 mm central deflection. By 20 mm central deflection, the difference in load capacity between the three sets was negligible. A noteworthy feature of the results is the similarity between the dry and partially wet cured specimens. Not only did these two sets of panels exhibit almost the same load to cause first crack and peak load, but load capacity was very similar throughout the load-deflection record.

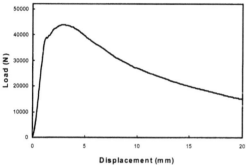

Figure 3. Mean load-displacement curve and error bars for wet cured specimens.

The relative performance of each set of specimens is revealed by the ratios listed in Table 2. This shows that the wet cured panels exhibited a load to cause first crack approximately 16% higher than either the dry or partially wet cured specimens, and energy absorption at 40 mm central deflection approximately 8% higher. Using a statistical difference test based on a t-distribution for small sample sizes, there is a 99.5% chance that the mean load capacity of the wet cured specimens differed from that of the dry and partially wet cured specimens. There is also a 96% chance that the difference in energy absorption data between these sets of specimens was significant. In contrast, there is a less than 50% chance that the difference in performance between the dry and partially wet cured specimens was significant.

4 DISCUSSION

The mean load-deflection record for the wet-cured specimens differed most noticeably from the other two sets around the point of first cracking. The wet cured specimens exhibited a significantly greater load to cause first crack of the concrete matrix, and the majority revealed a small drop in load capacity at the point of first cracking before further increasing to a rounded peak. The load-displacement record before first crack of the concrete matrix was strongly linear for almost all the wet cured specimens, suggesting an absence of cracking and softening. In

contrast, the dry and partially-wet cured specimens all exhibited a more rounded curve with progressive softening as the peak in capacity was reached. The rounded character of the load-displacement curves, and absence of an abrupt drop in load that is normally associated with cracking, suggests that these specimens failed by a more gradual formation of cracks.

The most rational explanation for this is that cracks existed in the dry and partially-wet cured specimens prior to the application of load, probably as a result of drying shrinkage at the tensile surface. Since the behaviour of the dry cured panels was almost identical to the panels that had been cured under water for 14 days prior to air curing, it appears that the presence of shrinkage cracks at the tensile surface may play a more important role in the behaviour of panels than the extent of wet curing. Surface imperfection limited performance is a phenomenon known to occur in other types of plate structure, such as architectural glass (Griffith 1924). Surface cracks that arise from drying shrinkage act as stress raisers that decrease the effective tensile strength of the concrete and reduce the load carrying capacity of the panel. Although not visible to the naked eye, the presence of drying shrinkage cracks is highly likely given the high cementitious content of the FRS mix used. Given that the very high fibre content of this mix did not eliminate the difference between the wet and dry cured specimens, the results of this investigation suggest that lightly reinforced FRS panels will show an even greater sensitivity to curing conditions. However, use of accelerators and curing membranes may reduce the significance of dry curing.

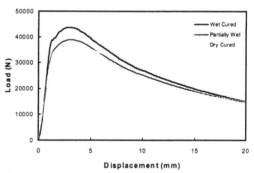

Figure 4. Mean load-displacement curves for wet, dry, and partially wet cured specimens.

The difference in performance between the wet and dry cured specimens tested in this investigation is not surprising. The benefits of curing have been known for many years. However, the similarly in behaviour between the dry and partially-wet cured specimens was not expected. Partial wet curing prior to transport of specimens to a laboratory is almost universally practiced on account of the practical difficulty involved in moving large test specimens to a place where they can be tested whilst also keeping them wet. However, this investigation has revealed that there is a price to pay for this that can be measured in terms of lost potential performance. If mix designs are to be carefully optimised on the basis of post-crack performance in panels, then this factor may have to be taken into account.

The difference in performance between the wet and partially wet cured specimens is of significance to research into FRS behaviour. If panels are to be used to assess the performance of the concrete matrix as well as post-crack fibre performance, then the load to cause first crack is the primary parameter of interest. There is therefore a clear need to prepare and cure specimens in a consistent manner to ensure apparent differences arise as a result of changes in the mix and not handling and curing of the specimen. Beam specimens have been required to be continuously wet cured and tested in a surface-saturated condition for many years. This investigation has revealed the same may be required for panels.

5 CONCLUSION

An investigation was undertaken into the influence of curing conditions on the performance of Round Determinate FRS panels used as Quality Assurance tools. Sets of panels were produced under identical conditions, but were subjected to different curing regimes consisting of either continuous dry curing, continuous wet curing, or wet curing followed by dry curing. The results indicate that continuous wet curing together with testing in a surface-saturated condition results in superior performance both with respect to the load to cause first crack of the concrete matrix, and post-crack energy absorption. Furthermore, specimens subjected to wet curing followed by air curing produced results that strongly resemble those of continuously dry cured specimens. Inspection of the load-deflection curves suggests that shrinkage cracks on the tensile surface of dry specimens may have contributed to the disparity in these results.

6 ACKNOWLEDGEMENTS

The authors gratefully acknowledge the support of CSR Readymix Concrete, through their representative Dr. Dak Baweja, for their generosity in providing concrete for this investigation. Material support was also provided by Jetcrete Australia P/L through their representative Matthew Hicks. Fibres were kindly donated by the respective suppliers.

REFERENCES

Bernard, E.S., 1998,"Measurement of Post-cracking Performance in Fibre Reinforced Shotcrete", *Australian Shotcrete Conference*, Sydney, October 8-9.

Bernard, E.S., 1999, "Correlations in the Performance of Fibre Reinforced Shotcrete Beams and Panels", *Civil Engineering Report CE9*, School of Civic Engineering and Environment, UWS Nepean, July.

Bernard, E.S. and Pircher, M., 2000, "The Use of Round Determinate Panels for the Assessment of Flexural Performance of Fiber Reinforced Concrete", *Cement, Concrete, and Aggregates,* ASTM (submitted February 2000).

European Specification for Sprayed Concrete, 1996, European Federation of National Associations of Specialist Contractors and Material Suppliers for the Construction Industry (EFNARC).

Griffith, A.A. 1924, "The Phenomenon of Rupture and Flow in Solids", *Philo. Trans. Royal Soc.*,Vol. 221, Oct., p163-179

Marti, P., Pfyl, T., Sigrist, V. and Ulaga, T., 1999, "Harmonized Test Procedures for Steel Fiber-Reinforced Concrete", ACI Materials Journal, Vol. 96, No. 6, Nov-Dec., pp 676-685.

Morgan, D.R., Heere, R., McAskill, N. and Chan, C., 1999, "Comparative Evaluation of System Ductility of Mesh and Fiber Reinforced Shotcretes", *Shotcrete for Underground Support VIII Conference*, Campos do Jordão, Brazil, April 11-15.

Neville, A., 1996, *Properties of Concrete*, Longman, London,.

RTA Specification T373, 1999, "Round Determinate Panel Test", from B-82 *Shotcrete Work*.

Shotcrete: Engineering Developments, Bernard (ed.) © 2001 Swets & Zeitlinger, Lisse, ISBN 90 5809 176 7

Punching a rock bolt assembly through a fibre reinforced concrete panel

E.S. Bernard & S. Harkin
University of Western Sydney, Nepean, Australia

ABSTRACT: Rock bolts are commonly used in conjunction with Fibre Reinforced Shotcrete (FRS) linings to stabilise ground around excavations. A steel cover plate is often placed over the protruding end of a rock bolt to dissipate load from the bolt into the FRS lining. The process of drilling and installing rock bolts takes place after the lining is sprayed, so a hole must be made through the lining to insert the bolt. To examine the effect of this on lining behaviour, an investigation has been carried out to determine the influence of a central hole on the punching capacity of a FRC panel. Grout or concrete is commonly placed between the cover plate and lining to distribute the load more evenly, so the effect of this aspect of rock bolting practice on punching behaviour has also been examined.

1 INTRODUCTION

Rock bolts and Fibre Reinforced Shotcrete (FRS) have become established and accepted means of ground stabilisation in underground excavation technology (Kaiser & McCreath 1992). The rock bolts are used to secure ground in a zone of influence around the excavation, whilst the FRS serves to stabilise the ground at the surface of the excavation.

Because newly excavated ground can become unstable very quickly, the FRS is often applied soon after the face has been mucked out during the excavation cycle. Holes for the rock bolts are later drilled through the hardened FRS soon after spraying, and additional material may be sprayed over this to protect the bolt from corrosion (Melbye 1994). A steel cover plate is often placed over the end of the bolt to dissipate load from the bolt into the FRS lining, should such loads arise. Causes of loads on a bolt can include distributed earth pressure on the lining from nearby ground, or straining of the bolt due to deeper movement within the ground.

A large number of studies have examined the shear capacity of concrete reinforced with both fibres and conventional steel bars (Regan 1986, Kuang & Morley 1992, Menétrey et al. 1997). The majority have focussed on punching capacity through continuous panels or slabs that do not include a hole under the point of loading. Since the hole is an integral feature of FRS linings around rock bolts, the first part of the current investigation has examined the influence of a hole on punching behaviour in FRS. The purpose of this is to permit a comparison between theoretical predictions of behaviour in a concrete plate and empirical behaviour in a lining with a hole located at the point of load application. Panels made with Fibre Reinforced Concrete (FRC) have been used to simulate behaviour in FRS linings.

Common construction practice is to place a dab of wet concrete between a cover plate assembly and the surface of the hardened FRS lining immediately prior to installation. This is based on the premise that the bedding will improve the transfer of load between the rock bolt and lining and thereby reduce the likelihood of lining failure immediately under the cover plate should high loads arise. The second part of this investigation has sought to examine the influence of this practice by comparing the punching behaviour of grouted and non-grouted cover plate assemblies.

2 EXPERIMENTAL PROGRAM

2.1 *Test Apparatus*

Punching tests were conducted in a purpose-built apparatus consisting of a thick steel annular base onto which a FRC specimen panel was placed, and a servo-hydraulic loading piston. Load was imposed on the upper surface of each panel so as to cause punching failure through the centre. The specimen was rigidly restrained around the periphery during each test by a number of bolts that passed through a fixed top plate and abutted the surface of the specimen (Fig. 1). All the specimen panels were cast inside thick steel rings that prevented radial dilation and edge rotation during punching (Fig. 2). The use of the steel rings and restraining bolts has been shown to effectively prevent the specimen edges rotating or translating during a punching test (Bernard & Curnow 1999).

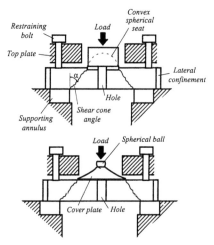

Figure 1. Sections through loading configuration for investigation of effects of (a) central hole, and (b) grout between cover plate and hardened FRC panel surface.

The annulus under the panel had a diameter that exceeded that of the loading device by twice the thickness of the specimen. This ensured that the conical punching surface generated during each test had sides inclined at 45°. Although the resulting load capacities and energy absorption may not be representative of natural failure surfaces, the minimum size of specimen required to produce results consistent with those of a continuous plate is thereby reduced (Bernard & Curnow 1999). Punching was achieved by imposing a load at the centre of the specimen using an Instron 8506 Universal Test Machine. The load was imposed at a controlled rate of displacement equal to 2.0 mm/min, up to a total central displacement of 20 mm.

2.2 Test Series 1: Effect of a Central Hole

Two series of specimens were produced as part of this investigation. The method of loading the specimen to induce punching failure differed between the two sets. For the first set of 14 panels produced to examine the influence of a central hole, load was imposed through a 100 mm diameter spherical seat and the diameter of the annulus under each specimen was 200 mm. The spherical part of the seat was convex, and matched the design used by Bernard & Nyström (2000). This type of seat allowed rotation of the punching cone during failure, thereby simulating the rotational freedom believed possible in cover plates that are punched through a FRS lining. A 25 mm diameter central hole was cast into approximately half the specimens by placing a wooden peg in the centre of the mould prior to casting. This was removed after hardening to leave a hole. A 5 mm thick ground steel plate was bedded into the surface of the specimens to achieve intimate transfer of load (Bernard & Nyström 2000) between the spherical seat and the specimen (see Fig. 2).

Figure 2. Casting of specimen with no hole inside steel ring.

2.3 Test Series 2: Effect of Grouting

For the second series of tests, 20 specimens were cast with a central hole included but no ground steel bedding plate on the surface. Instead, the surface was trowelled flat and left to harden. Half of the specimens were then tested with a steel cover plate loaded in compression directly against the hardened FRC surface. The other half had 'grout' in the form of wet concrete placed as a bedding layer between the cover plate and panel surface. This was left to harden prior to testing. Load was imposed through a proprietary steel ball and round cover plate assembly of type *VZN*, produced by Ørsta Stålindustri AS, Norway. The conical cover plates had a nominal diameter of 150 mm, a height of 20 mm, and a thickness of 6 mm. For this series of tests, the annulus under each specimen had a diameter of 250 mm.

2.4 Specimen Preparation

Although the intended application for the data produced in this investigation was sprayed concrete tunnel linings, production of specimens by spraying is relatively expensive and logistically complicated. The present specimens were therefore cast using

Table 1. Mix design for FRC used in Series 2.

Ingredient	Quantity (kg/m³)
Coarse aggregate (10/7 mm)	620
Coarse sand (5 mm)	615
Fine sand (2 mm)	410
Cement (ASTM Type I)	380
Fly ash	40
Silica fume	20
Water reducer	1.9 L
Deformed fibres	50
Slump	70 mm

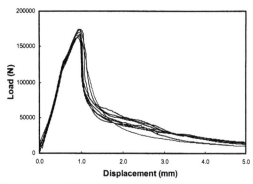

Figure 3. Load-displacement curves for specimens with no central hole, loaded with spherical seat.

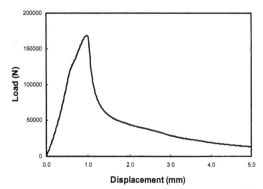

Figure 4. Mean load-displacement curve and error bars for specimens with no central hole.

FRC that would normally have been sprayed. Consolidation was assisted by vibration, and the upper surface of all specimens was screeded and trowelled flat. All specimens were produced with a nominal thickness of 50 mm. The panels in Test Series 1 had a diameter of 400 mm, and those used in Test Series 2 had a diameter of 500 mm.

All the specimens incorporated a 10 mm thick steel ring around the periphery to provide restraint against in-plane dilation during failure. This ring also proved a convenient form into which the concrete could be cast during production (Bernard & Curnow 1999). A carefully ground steel plate was used as the base onto which each specimen was cast. This was necessary to achieve a very flat lower surface that has been found useful in generating consistent and reliable results (Bernard & Nyström 2000). Most specimens were cured under plastic for the first two days, after which they were transferred to a lime bath and immersed continuously for several weeks. The age at testing was about 70 days. It must be noted that FRS linings are typically only a few days old when a bolt is installed, so the punching characteristics of the present specimens is likely to be more brittle than in situ should a failure load be introduced immediately.

The mix design for the fibre reinforced concrete differed between the first and second sets of specimens. For the first set, MBT pre-batched Shotpatch 25 was used with 50 kg/m³ of hooked-end fibres. For the second set, a normal shotcrete mix with 10 mm maximum sized aggregate was used. The mix design for this concrete is listed in Table 1. Companion compression cylinders measuring Ø50×100 mm were cast together with both sets of panels. These resulted in an average compressive strength of 44.5 MPa for the first set, and 43.2 MPa for the second. Brazil splitting tests were also performed. These produced an average tensile splitting strength of 4.97 MPa for the first set, and 5.55 MPa for the second.

3 RESULTS

3.1 Test Series 1

The result of each punching test consisted of a load-displacement curve, with load measured in Newtons and displacement in millimetres. The load capacity was linearly scaled to account for small differences in thickness between specimens; the standardised thickness was taken to be 50 mm. The energy under each curve was integrated between the onset of loading and 20 mm total deflection to determine energy absorption during the failure process. The mean peak load capacity and energy absorption are reported for each set of nominally identical specimens in Table 2.

Load-displacement curves generated for the first set of specimens with no central hole are shown in Fig. 3. The mean curve, with error bars corresponding to one standard deviation, are shown in Fig. 4. These results were remarkably consistent compared to data obtained from similar tests in the past (Bernard & Curnow 1999, Bernard & Nyström 2000). The results for the specimens with a 25 mm diameter central cast hole were less consistent. The mean curve for these specimens is compared to the specimens with no central hole in Fig. 5. Several of the specimens in the second set failed at a much lower load than average, and displayed multiple peaks in load capacity. The standard deviation in peak load

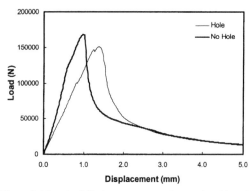

Figure 5. Mean load-displacement curves for panels with and without a central hole.

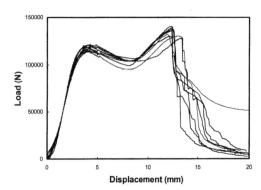

Figure 6. Load-displacement curves for panels without grout between cover plate and FRC surface.

capacity and energy absorption was therefore considerably higher than for the specimens with no central hole (Table 2).

3.2 Test Series 2

Load-displacement curves for the specimens tested with no grout between the cover plate and FRC surface in the second series of specimens are shown in Fig. 6. These results exhibited a very consistent pattern of behaviour characterised by two peaks in load capacity and extensive ductility. The first peak in capacity coincided with the onset of yielding in the steel cover plate, whilst the second peak coincided with contact between the centre of

Figure 7. Punched out cone with non-grouted cover plate showing radial cracking and inverted plate.

the deforming cover plate and the surface of the FRC panel. The reduction in capacity following the second peak occurred as a result of combined flexural and shear failure of the FRC. Radial and circumferential cracks occurred across the base of the punched out cone, and a large number of fine radial cracks occurred around the punching zone (Fig. 7). This suggests that failure of the FRC originated near to the central hole, possibly as the deformed cover plate came into contact with the panel suffering predominantly radial cracks. The cover plate became 'inverted' by the end of the test, indicating the severe degree of deformation experienced.

The mean result for those specimens in which grout was placed between the cover plate assembly and the surface of the FRC panel is compared to the mean result for the non-grouted specimens in Fig. 8. This figure revealed a much higher peak load capacity for the grouted specimens than the non-grouted specimens, but only a single peak in capacity and lower overall ductility. Indeed, failure was abrupt and violent in these specimens. The punched out cones remained largely free of cracks on the opposite face, and concrete surrounding the failure zone also remained free of visible cracks.

Mean results for the second series of specimens are listed in Table 2. The performance figures differ from the first set of tests because the size of the specimens was considerable greater.

Table 2. Summary of specimen performance for Test Series 1 and 2.

Specimen Set	Number of Specimens	Peak Load		Energy Absorption	
		Mean (N)	COV (%)	Mean (J)	COV (%)
No Central Hole	8	169082	2.8	283	8.7
Central Hole	6	159554	15.5	292	16.2
Grouted Cover Plate	10	215659	5.6	701	15.5
Non-grouted Cover Plate	10	133411	3.4	1536	5.9

4 DISCUSSION

The present investigation has indicated that a central hole has a relatively minor influence on punching behaviour in FRC panels compared to specimens with no central hole. The introduction of a hole resulted in a drop in the stiffness of the specimen in response to initial loading, a 5 per cent decrease in peak load capacity, and a minor rise in energy absorption. However, the differences in load capacity and energy absorption were relatively insignificant given the high variability evident in the results for those specimens with central holes. Based on a *t*-distribution analysis incorporating the variances determined for each sample (see Table 2), there was an 85% chance that the mean population load capacity for these two sets of data differed, but a lower than 50% chance that the means in energy absorption differed. Observations of failures during testing indicated that the introduction of a central hole increased the incidence of radial cracking, suggesting that the failure became more flexural in nature. This was corroborated by numerous large cracks evident in the punched out cones from these specimens.

The second series of tests revealed a much greater difference in behaviour brought about by the introduction of grout between the cover plate and FRC panel surface. The grouted specimens exhibited a 61 per cent increase in peak load capacity compared to the non-grouted specimens, but energy absorption during failure was halved. The non-grouted specimens sustained much greater deformation up to the peak in load capacity prior to punching (average 12 mm, compared to 2.5 mm), and suffered a more ductile drop in capacity. It is interesting to note that the grouted specimens absorbed an average of 218 J following the single peak in load capacity, whilst the non-grouted specimens absorbed an average of 292 J following the second peak. This suggests that the cracked FRC was capable of absorbing more energy when flexural deformations occurred in combination with shear. However, the post-crack energy absorption of the FRC was relatively low compared to the absorption of the whole system during the failure process, reinforcing the fact that FRC is not a particularly ductile material in shear.

It appeared that the hooked-end fibres used in the present investigation did not suffer straightening during pull-out, and therefore did not absorb energy as they effectively do during flexural failures. Instead, the fibres acted as tiny dowels that were bent sideways. Since the specimens were only 50 mm thick, very few fibres were oriented in the direction of principal tensile stress across the conical failure surface. Few were therefore available to pull-out and deform in the manner they have been designed for.

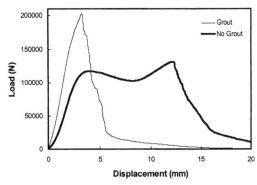

Figure 8. Mean load-displacement curves for panels with grouted and non-grouted cover-plates.

The grouted specimens absorbed an average of 483 J prior to cracking of the FRC matrix, whilst the non-grouted specimens absorbed 1245 J. Yielding of the steel cover plate therefore contributed approximately 762 J to energy absorption during the failure process, which represents an increase of 160 per cent over that absorbed by the grouted specimens prior to failure of the concrete. This feature of a non-grouted assembly illustrates the value of cover plates as energy absorbers that may deform whilst helping to prevent failure of a FRS lining. Although the peak load capacity is reduced by not grouting, ductility is greatly improved. However, the degree of pre-cracked ductility available will depend on the design of the cover plate.

5 CONCLUSION

An investigation into two aspects of punching failure in Fibre Reinforced Concrete panels loaded through cover plate assemblies was undertaken. A series of laboratory tests were used to examine the influence of a central hole on punching resistance, and the influence of grouting between a steel cover plate assembly and the surface of FRC.

The presence of a central hole was found to increase the frequency of radial cracks in and around the punching cone, but induced only a minor change in performance as measured by peak load capacity and energy absorption. Specimens with central holes were also found to behave in a less consistent manner. In contrast, the presence of grout between the cover plate assembly and the surface of the FRC panel effected a major change in behaviour. Grout dramatically increased the load required to initiate punching failure, but decreased ductility and overall energy absorption during the failure process.

6 ACKNOWLEDGEMENTS

The authors gratefully acknowledge the support of the following organisations and individuals for this investigation: CSR Readymix Concrete, through their representative Dr. Dak Baweja, for providing concrete; Rock Technology P/L for their support in constructing the test apparatus; Jetcrete Australia P/L, through their representative Matthew Clements, for support and advice; Ørsta Stålindustri AS of Norway for cover plates; Bekaert for a donation of fibres; BHP Steel for providing steel pipe to manufacture the restraining rings, and MBT (Australia) P/L for donating Shotpatch 25.

REFERENCES

Bernard, E.S. & Curnow, M.C.P., 1999, "Experimental Factors Affecting Punching Resistance in Fibre Reinforced Concrete Panels", *Third International Symposium on Shotcrete for Ground Support*, Gol, Norway, September 26-28.

Bernard, E.S. & Nyström, E.M.C., 2000, "The Influence of Loading Conditions on Punching Resistance in Fibre Reinforced Concrete Panels", *International Workshop on Punching Shear Capacity of Reinforced Concrete Slabs*, KTH, Stockholm, June 8-9.

Kaiser, P.K. & McCreath, D.R., 1992, *Rock Support in Mining and Underground Construction*, Pub. A.A. Balkema, Rotterdam.

Kuang, J.S. and Morley, C.T., 1992, "Punching Shear Behaviour of Restrained Reinforced Concrete Slabs", *ACI Structural Journal*, Vol. 89, No. 1, January-February, pp 13-19.

Melbye, T. A., 1994, *Underground shotcreting*, MBT, Zurich.

Menétrey, P., Walther, R., Zimmermann, T., Willam, K.J., and Regan, P.E., 1997, "Simulation of Punching Failure in Reinforced Concrete Structures", *J. Structural Engineering*, ASCE, Vol. 123, No. 5, May, pp 652-659.

Regan, P.E., 1986, "Symmetric punching of reinforced concrete slabs", *Magazine of Concrete Research*, Vol. 38, No. 136, September, pp115-128.

Shotcrete: Engineering Developments, Bernard (ed.) © 2001 Swets & Zeitlinger, Lisse, ISBN 90 5809 176 7

Numerical analysis of ground - shotcrete interaction in tunnelling

D.Boldini
University of Rome "La Sapienza", Italy

A.Lembo-Fazio
University of "Roma III", Italy

ABSTRACT: The increasingly frequent use of shotcrete as primary support in tunnelling is not always accompanied, at the design stage, by analyses capable of describing its mechanical behaviour. In the first part of this paper the experimentally observed behaviour of shotcrete and its main properties as a supporting material are outlined, together with a critical overview of the general approaches adopted in modelling ground-support interaction in tunnelling. The results of three-dimensional parametric analyses of tunnel excavation and support are then presented. The analyses have been performed in axi-symmetric conditions, assuming an elastic constitutive law characterised by a time-dependent stiffness for the shotcrete. The numerical results are then compared in order to determine the influence of the following factors: constitutive law, strength parameters, tunnel advancement rate, relative ground-support stiffness. In the final part of the paper, simplified techniques are proposed to introduce the main features that emerged from the axi-symmetrical approach in the standard two-dimensional analysis.

1 INTRODUCTION

Shotcrete is a fundamental support element for tunnels built in accordance with the principles of the New Austrian Tunnelling Method (Rabcewicz, 1964). Installing the shotcrete close to the tunnel face, with rockbolts and steel arches, if any, reduces the release of ground stresses hence improving excavation stability. Shotcrete is widely utilized in the excavation of underground openings in widely differing ground conditions. Many investigations have been undertaken to improve the mechanical characteristic of the material and for optimizing construction technology.

Modelling ground-lining interaction in tunnelling is a very complex matter, especially when the primary support is realised with shotcrete, because the mechanical properties of this material change as it is loaded. Near the tunnel face the state of stress and deformation is typically three-dimensional, and hence only a three-dimensional numerical analysis, in which a suitable shotcrete constitutive model is considered, can successfully assess the ground-lining interaction.

In professional practice, the design of tunnel support is based on a number of approximate procedures. In almost all cases, simplified plain strain analyses are carried out because three-dimensional numerical analyses are time-consuming and require

a good deal of ability and experience (e.g. the Convergence-Confinement Method, Lombardi 1973). The support effect produced by the proximity of the face is taken into account by considering a fictitious internal pressure applied to the tunnel surface:

$$P_f(y) = (1 - \lambda(y)) \cdot P_\infty \qquad (1)$$

where P_f is the fictitious pressure; $P_\infty =$ is the initial isotropic pressure; λ is the so-called ground stress release; and y is the tunnel face distance.

The fictitious pressure is the pressure that results in the same value of tunnel convergence as in the three-dimensional analyses. It is generally calculated considering an unlined tunnel (Panet & Guenot 1982). The main approximation of the plain strain analyses is related to the evaluation of the fictitious pressure, or the ground stress release, from numerical analyses of unlined tunnels. The correct choice of the ground stress release before installing the lining is instrumental in producing a successful lining design. Several results of three-dimensional numerical analyses emphasize how the presence of the support structure may produce different ground stress release values depending on the relative stiffness of the ground and the lining (Kielbassa & Duddeck 1991; Bernaud & Rousset 1992). In recent years, many authors have tried to provide simplified techniques to improve the accuracy of plain strain analyses through the evaluation of three-dimensional numerical results (e.g. New Implicit Method, Bernaud & Rousset 1992). But these methods often involve a lining with constant stiffness and the shotcrete is modelled as an elastic material with an equivalent constant elastic modulus. More precise constitutive models for the shotcrete (Pottler 1990, Swoboda & Moussa 1992, Aydan et al. 1992, Mang 1995) are instead applied only in plain strain condition analyses.

This work can be considered a first step in the investigation of the ground-shotcrete interaction in tunnelling because a simplified constitutive model for the shotcrete is considered.

2 EXPERIMENTALLY OBSERVED BEHAVIOUR OF SHOTCRETE

The structural behaviour of shotcrete may be dependent on ground conditions. Where it is possible to consider the ground as a continuous medium, the action of the shotcrete consists in the application of a confining pressure over the whole of the excavation, forming a bearing ring. In discontinuous media, like hard rock, the shotcrete confinement support is applied on single blocks: the main function of the shotcrete is to prevent blocks from becoming detached from the tunnel surface. In this case, bending behaviour of the shotcrete and its adhesion to the rock are the most important characteristics to be considered.

In the next part of the paper, reference will be made to shotcrete behaviour in ground conditions which can be represented as a continuous medium and so the emphasis will be placed on the mechanical properties resulting from compression and creep tests on shotcrete specimens.

Uni-axial compressive tests carried out on shotcrete specimens at different ages are presented in Figure 1 (Sezaki 1989, in Chang 1994). During its

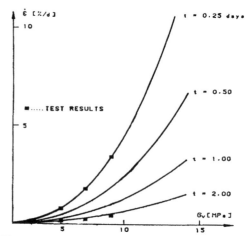

Figure 2. Strain rate as a function of stress and time (Pottler, 1990).

first day the shotcrete exhibits ductile behaviour, with significant residual strength associated with the development of large strains. The high deformability of fresh shotcrete is an important characteristic for excavation stability, because it allows the development of controlled ground deformations without excessively loading the lining structure. At later ages, the material becomes more brittle, and stiffer, and the strength increases.

Many empirical expressions have been proposed to model age-dependent behaviour. A widely used expression for representing the increasing stiffness of shotcrete as a function of time is:

$$E_c = a_1 \cdot E_{28} \cdot e^{c_1/t^{b_1}} \qquad (2)$$

where t is time in days; E_c is the elastic modulus at time t, in GPa; E_{28} is the elastic modulus at 28 days, in GPa; and a_1, b_1, and c_1 are empirical coefficients. Chang (1994) suggested the following values for these coefficients: a_1=1.062, b_1=0.6, c_1= -0.446.

A similar expression can be written for the uniaxial strength:

$$\sigma_c = a_2 \cdot \sigma_{c28} \cdot e^{c_2/t^{b_2}} \qquad (3)$$

where σ_c is the uni-axial strength at time t, in MPa; σ_{c28} is the uniaxial strength at 28 days, in MPa; and a_2, b_2, c_2 are empirical coefficients. On the basis of many empirical results, Chang (1994) proposed the following values for these coefficients: a_2=1.105, b_2=0.7, c_2= -0.743. The uni-axial strength at 28 days may vary, generally from 30 to 55 MPa, depending on the shotcrete mix design and on the quantity and quality of the concrete additives.

Another aspect to be taken into account in order to gain further insight into the main features of the mechanical properties of shotcrete is its viscous be-

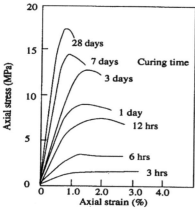

Figure 1. Uniaxial compressive behaviour of shotcrete (Sezaki 1989 in Chang 1994).

haviour. Experimental results from traditional compressive creep tests indicate the importance of viscous strains during the initial curing of the shotcrete. The viscous strain rate is influenced by the deviatoric stresses and by time, as indicated in Figure 2 (Pottler 1990). An increase in deviatoric stresses entails a more than linear growth for the viscous strain rate, while time has the effect of reducing the viscous behaviour. At this stage of the study the creep behaviour of shotcrete is not considered and thus this aspect of behaviour will be ignored.

3 THREE-DIMENSIONAL NUMERICAL ANALYSES

To better understand the shotcrete support behaviour in tunnelling, three-dimensional parametric analyses have been performed with the help of F.L.A.C., a Finite Difference Method programme (Itasca 1996). The present investigation has concentrated on difficult ground conditions, where the shotcrete support is fundamental for excavation stability. An axisymmetric state of stress and geometry has been adopted. This is a common hypothesis with analytical plain strain methods (e.g. Convergence-Confinement Method and New Implicit Method). In the analyses, the initial stress in the ground, which is considered isotropic given its hypothesised symmetry, has been set at 1500 kPa. The tunnel radius is 5 metres and the lining thickness 200 millimetres. A detail of the mesh close to the tunnel face is illustrated in Figure 3.

Several numerical analyses have been carried out in order to determine the influence of the following factors:

1. ground constitutive law;
2. ground strength parameters;
3. tunnel advancement rate;
4. relative ground-support stiffness.

The ground has been modeled both as an elastic and as an elasto-plastic material, with a Mohr-Coulomb strength criterion. It is possible to define a stability coefficient, N_s, as:

$$N_s = \frac{2 \cdot P_\infty}{R_c} \tag{4}$$

where R_c is the uni-axial compressive strength, and P_∞ the initial isotropic stress. The uni-axial compressive strength is equal to:

$$R_c = \frac{2 \cdot c \cdot \cos\varphi}{1 - \sin\varphi} \tag{5}$$

where c is the cohesion and φ the friction angle. The stability coefficient has been set to either 2 or 4 in the present analyses. By setting the friction angle at a value of 20°, the cohesion is 525 and 263 kPa, respectively.

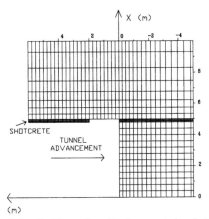

Figure 3. Detail of the mesh used in the numerical analysis (the final position of the tunnel face is set at y = 0 m).

The excavation and support installation has been modelled using a step-wise procedure. Three different rates of tunnel advance v have been considered, namely 2, 3, and 4 m/day, with an unsupported tunnel length l_u of 1.0, 1.5 and 2.0 m respectively. These low advance rates are typical of excavation in difficult ground conditions. An elastic constitutive law characterised by a time-dependent stiffness has been used for the shotcrete. The development of the elastic modulus has been expressed by the empirical relationship in Eqn. (2). Poisson's ratio has been taken to be constant and equal to 0.3. The curing time for each lining ring has been calculated in accordance with the tunnel advance rate.

Finally, the ground elastic moduli have been made to vary across a large range of possible values to investigate the effect on ground-shotcrete interaction. In particular, elastic moduli equal to 90, 900, 9000 and 90000 MPa have been considered, leading to a relative ground-support stiffness parameter:

$$\beta = \frac{E \cdot R}{E_{28} \cdot e} \tag{6}$$

equal to 0.094, 0.94, 9.4, and 94, respectively. In this expression, E is the elastic modulus of the ground; E_{28} is the elastic modulus of the shotcrete at 28 days, taken to be 24 GPa; R is the tunnel radius, and e is the lining thickness.

An example of calculated convergence, as a function of distance from the tunnel face, is shown in Figure 4 for different values of relative ground support stiffness. To compare the results of the various analyses, each curve has been normalized to the final radial displacement of an unsupported tunnel having the same ground stiffness.

The effect of the lining on convergence increases as the ground support stiffness parameter β decreases. A comparison with the unsupported tunnel

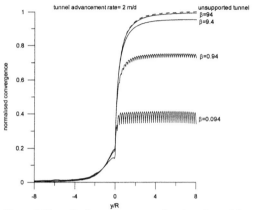

Figure 4. Normalized convergence as a function of tunnel face distance y over the tunnel radius R.

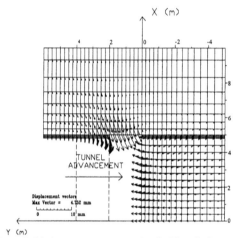

Figure 5. Displacement vectors associated with a single excavation step (for a tunnel advancement rate of 4 m/d).

Changing from elastic to elasto-plastic ground behaviour, convergence increases due to the development of plastic strains in the ground. For equal values of β, the effect of the lining on convergence is greater compared to the case of ground exhibiting elastic behaviour. The displacement vectors have a more differentiated character within a lining segment installed in a single step.

The hoop stresses in the ground and in the lining are closely related to the convergence values (Fig. 6). For a single excavation step, the maximum hoop stress in the shotcrete lining is recorded in the area closest to the tunnel face. Comparing the results of the elastic and elasto-plastic analyses, the lining is found to be subjected to almost the same hoop stress intensity, ground stiffness and tunnel advancement rate being equal. Moreover, the largest ground displacements develop before the lining is installed.

Figure 7 represents the radial stress distribution close to the tunnel face. Note the formation of a longitudinal arch which bypasses the unsupported tunnel up to the first shotcrete element. In the Convergence-Confinement Method (Lombardi 1973) the three-dimensional ground-lining interaction is simplified as a plain strain problem. The fictitious internal pressure applied to the tunnel surface is dependent only on distance from the face. When the internal pressure decreases, convergence increases in a monotonic way as one moves away from the face

analysis (which produces the same normalized result for all ground moduli) underlines the influence of the in-place shotcrete on the area closer to the tunnel face. The saw-toothed shape of each curve reflects the effects of different distances from the excavation zone of the various parts of a lining element of the same age. In each step, a new lining section composed of many elements at different distances from the tunnel face is installed, each of which undergoes different deformations and stresses. The largest displacement increments, especially the radial ones, are concentrated in the unlined tunnel sections (as shown in Figure 5, where displacement vectors associated with a single excavation step are represented). The presence of the installed shotcrete in the previously excavated sections reduces the radial displacement increments nearer the tunnel face compared to recorded values in an unlined tunnel.

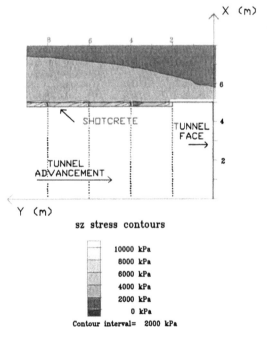

Figure 6. Hoop stress distribution in the ground and in the shotcrete close to the tunnel face.

according to a relationship that depends on the mechanical properties of the ground. However, in 3D numerical analyses, the situation is quite different. In the unlined sections close to the tunnel face, the deformation of the ground surface is not restrained by any internal pressure. Major displacements may occur if the mechanical parameters of the ground are poor, and the equivalent ground stress release based on these convergence values could be overestimated. The pressure on the lining is not solely a function of the deformation level of the ground, but is also determined by the stress concentration arising from the bearing arch that is formed close to the face. These ground elements, closer to the face and in contact with the shotcrete, experience a simultaneous increase in convergence and in radial tension which is not in agreement with the Convergence-Confinement framework.

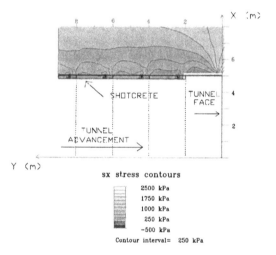

Figure 7. Radial stress distribution in the ground and in the shotcrete close to the tunnel face.

Longitudinal displacements and stresses are considered to be relatively unimportant for the study of ground-lining interaction, but the numerical analyses give interesting results that should not to be overlooked. The ground is always compressed in the direction of the tunnel axis, while the shotcrete lining presents a tensile stress throughout its length. In fact, the stress in the lining is primarily influenced by the direction of the displacement vectors relative to a single step, and not the total displacements experienced by the ground. The displacement vectors associated with a single step have a longitudinal component directed towards the tunnel face (Fig. 5). No tensile cracks are generally recorded in situ.

4 3D/2D COMPARISON

Through a three-dimensional numerical analysis, major improvements can be introduced over simplified methods operating under plain strain conditions. In this case it was decided to make reference to the Convergent-Confinement Method framework. The tunnel convergence curve can be defined as the relationship between the pressure acting on the boundary of the excavation and convergence, while the lining confinement curve is given by the relationship between pressure applied at the lining extrados and the lining convergence itself (Fig. 8). At the moment of lining installation, the tunnel convergence is equal to U_0. The internal pressure acting at the excavation boundary is the sum of the fictitious pressure resulting from the support provided by its proximity to the face plus the pressure afforded by the lining. In terms of convergence and pressure, the ground-lining system can be said to reach an equilibrium condition when the two curves intersect (P_{eq} and U_{eq}). We have spoken about the need to better evaluate the ground stress releases before installing the lining. Moreover, in the calculations, the Convergence-Confinement Method also requires a lining with a constant stiffness. When shotcrete is used for the support, an equivalent constant stiffness needs to be defined for it.

A choice must to be made as to the modality of information transfer from the axi-symmetric to the plain strain conditions. Firstly, the ground stress release is calculated considering the radial tunnel displacement at the section where the lining is to be installed. Secondly, beginning from this ground stress release, the equivalent stiffness is obtained by checking which lining elastic modulus gives, by the Convergence-Confinement Method, almost the same equilibrium results as those provided by the numerical analyses. The convergence and stress values adopted are those for the central node or the zones of single lining elements installed in the same step. This way of structuring the comparison has been

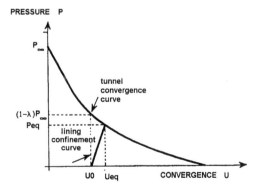

Figure 8. Convergence-Confinement Method (Lombardi 1973).

found to be the best solution in situations where the mechanical properties show variability along the tunnel axis.

As a result of the present analyses, the values of the ground stress release before installing the lining, λ, are reported in Figure 9 as a function of the relative ground-support stiffness, β, for three different unsupported tunnel lengths and for an elastic constitutive model of the ground. For the most extreme value of β = 94, the ground stress release is practically equal to the value defined through the convergence curve for an unsupported tunnel due to very limited ground-lining interaction. This is represented by (Panet & Guenot, 1982):

$$\lambda(y_0) = 0.27 + 0.73 \cdot \left[1 - \left(\frac{0.84 \cdot R}{y_0 + 0.84 \cdot R} \right)^2 \right] \qquad (7)$$

where y_0 is $l_u/2$ (the mid-point of the unsupported tunnel length).

As the ground support stiffness parameter β decreases, the magnitude of the ground stress release, λ, also decreases. For the same relative ground-support stiffness, the ground stress release is larger for larger unsupported tunnel lengths (it should be noted that the different unsupported tunnel lengths are also characterized by a different tunnel advancement rate). The curves obtained by connecting the λ values for identical unsupported tunnel lengths present the same shape, but are merely translated one above another.

The equivalent elastic modulus of shotcrete determined with the Convergence-Confinement Method is reported in Figure 10, in which the influence of the relative ground-support stiffness and of the unsupported tunnel length is illustrated. The equivalent elastic modulus to be assigned to the shotcrete lining increases as β increases, and it varies

in an interval from 5 to 15.5 GPa. Passing, for example, from a β value equal to 0.094 to 0.94, the reduction of the final load on the lining is accompanied by a large convergence decrease in the presence of the shotcrete support. Indeed, in this latter situation, the ground stress release is larger and convergence before installing the lining is closer to the equilibrium value. These factors must also be considered valid for other cases. Moreover, for the same relative ground-support stiffness, the equivalent elastic modulus for shotcrete is lower for larger unsupported tunnel lengths. When the unsupported tunnel length increases, the shotcrete becomes less stiff, face distance being equal.

When an elasto-plastic constitutive model is considered for the ground, the radial displacements increase and as a consequence the λ values become greater. Considering the same unsupported tunnel length, the ground stress releases are more differentiated when the relative ground-support stiffness varies, but the general behaviour is the same as described for the elastic case.

Problems are encountered in determining the equivalent elastic modulus for the shotcrete if the ground has high deformability (for example β = 0.094). The unsupported tunnel surfaces exhibit a very large deformation, and this does not agree with the equilibrium values of stress and convergence at great distances from the face. The ground stress release thus calculated does not make it possible to correctly reproduce the ground-lining interaction. In fact, the equivalent elastic modulus required in the Convergence-Confinement Method to obtain the numerical solution, is, for a stability coefficient N_s equal to 2, greater than that of the shotcrete at 28 days. It cannot be obtained at all for N_s equal to 4 because the ground undergoes a deformation that is larger than the deformation at equilibrium. These re-

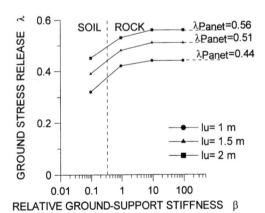

Figure 9. Values of the ground stress release as a function of the relative ground-support stiffness for selected values of unsupported tunnel length l_u.

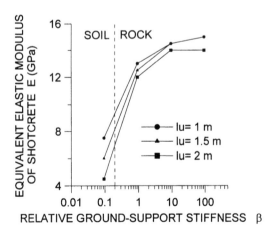

Figure 10. Equivalent elastic modulus of shotcrete determined using the Convergence-Confinement Method.

sults underline the problems associated with adopting the Convergence-Confinement model for difficult ground conditions and the difficulty of achieving agreement between three-dimensional and plain strain analyses.

5 CONCLUSION

Ground-lining interaction in tunnelling is a typical three-dimensional problem, but plain strain analyses are almost always used for support design. The Convergence-Confinement Method gives a powerful framework for representing the ground and lining behaviour up to equilibrium, but the quality of the result depends on the choice of the ground stress release value when the support is installed. If shotcrete is used as the primary support close to the tunnel face, it is necessary to evaluate equivalent support properties that are not time-dependent.

Numerical analyses in axisymmetric conditions have been performed assuming elastic constitutive behaviour and a time-dependent stiffness for the shotcrete. It has been possible to evaluate the influence of several factors, such as the choice of the constitutive law of the ground and strength parameters, relative ground-support stiffness, and tunnel advance rate, on ground stress release and the equivalent elastic modulus for the shotcrete. This offers the possibility of improving the prediction capacity of simplified methods.

The results of this investigation indicate that the Convergence-Confinement Method is not always a useful framework for reproducing many important three-dimensional topics, especially when difficult ground conditions are considered.

ACKNOWLEDGEMENTS

Research supported by MURST project "Tunnelling in difficult ground conditions".

REFERENCES

Aydan, O., Sezaki, M. & Kawamoto, T. 1992. Mechanical and numerical modelling of shotcrete. In G.N. Pande & S. Pietruszczak (eds), *Numerical Models in Geomechanics*: 757-765. Swansea: Balkema.

Bernaud, D. & Rousset, G. 1992. La nouvelle méthod implicite pour l'étude du dimensionnement des tunnels. *Rev. Fran.. Géotechnics* 60: 5-26.

Chang, Y. 1994. *Tunnel support with shotcrete in weak rock – A rock mechanics study*. Ph.D. Thesis, Royal Institute of Technology. Stockholm.

Itasca 1996. *Fast Lagrangian Analysis of Continua. User's Manual*.

Kielbassa, S. & Duddeck, H. 1991. Stress-strain field at the tunnelling face – three-dimensional analysis for two-dimensional technical approach. *Rock Mech. & Rock Eng.* 24: 115-132.

Lombardi, G. 1973. Dimensioning of tunnel linings with regard to constructional procedure. *Tunnel & Tunnelling* 7: 340-351.

Mang, H. 1995. Computational mechanics of the excavation of tunnels, *New Austrian Tunnelling Method, Summercourse*, Vienna, July 2-8.

Panet, M. & Guenot, A. 1982. Analysis of convergence behind the face of a tunnel. *Tunnelling* 82: 197-204.

Pottler, R. 1990. Time-dependent rock-shotcrete interaction. A numerical shortcut. *Computers and Geotechnics* 9: 149-169.

Rabcewicz, L.V. 1964. The New Austrian Tunnelling Method. *Water Power*, November.

Swoboda, G. & Moussa, A. 1992. Numerical modelling of shotcrete in tunnelling. In G.N. Pande & S. Pietruszczak (eds), *Numerical Models in Geomechanics*: 717-727. Swansea: Balkema.

Shotcrete: Engineering Developments, Bernard (ed.) © 2001 Swets & Zeitlinger, Lisse, ISBN 90 5809 176 7

Mode of action of alkali-free sprayed shotcrete accelerators

T.A.Bürge
Product Technology Director, Sika AG, Switzerland

ABSTRACT: Accelerating admixtures for shotcreting in general commercial use today are mixtures of sodium carbonate and sodium aluminate. While they produce acceptable early strengths relative to plain concrete, they typically reduce the ultimate strength by more than half, requiring the use of thicker layers than would be necessary in the absence of the accelerator to achieve the same structural strength. Additionally, these admixtures are highly caustic and great care must be taken in handling them. The new generation of accelerators that are free of alkali metal ions and chlorides contain at least one water-soluble sulfate and further constituents. The sulfate is mostly aluminium sulfate. The sulfate ions react with calcium hydroxide and with the surface of the tricalcium aluminate particles, producing a compound called ettringite. The mode of action of aluminium sulfate on the hydration of different cements from the time of mixing up to 7 days was investigated using X-ray diffraction.

1 INTRODUCTION

It is well known that the setting and hardening of a hydraulic binder or of a mixture which contains a hydraulic binder can be accelerated through the addition of a setting accelerator or a hardening accelerator. A typical example of a binder is portland cement.

The most commonly used setting accelerators are substances with strongly alkaline properties, like alkali metal hydroxides, alkali metal carbonates, alkali metal silicates and alkali metal aluminates (Figure 1). Substances with strongly alkaline properties are undesired due to the health risks they can cause when persons are handling such substances, burning of the skin of the worker, harmful effects on the respiratory system, especially the lungs, and irritation of the eyes and etching of the cornea can occur.

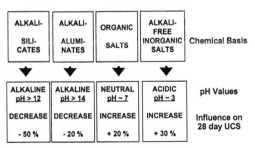

Figure 1. General overview of shotcrete accelerators

Furthermore the introduction of alkali metal ions into mortar or concrete has an adverse effect onto the properties of the construction material. The final compressive strength of the hardened construction material is lowered, and the shrinkage is increased, which may result in the development of cracks in the construction material. Thus the stability and durability of the construction is reduced.

2 ALKALI-FREE ACCELERATORS

The rapid hardening of shotcrete seems to result mainly in early formation of ettringite, a reaction product of lime and aluminium sulfate.

Most alkali-free shotcrete accelerators contain aluminium sulfate as a main component. Therefore its chemical reactions in laboratory tests and its practical performance in shotcrete tests were investigated.

3 THEORETICAL BACKGROUND

The mechanism and kinetics of the hydration of cement have been investigated (Havlica & Sahu 1992; Gabrisova et al. 1991) and some progress has been made towards an understanding of the main factors governing the hydration process. The conditions under which ettringite forms, and its stability, are of importance with regard to the setting characteristics

Accelerated with 3% Al$_2$(SO$_4$)$_3$

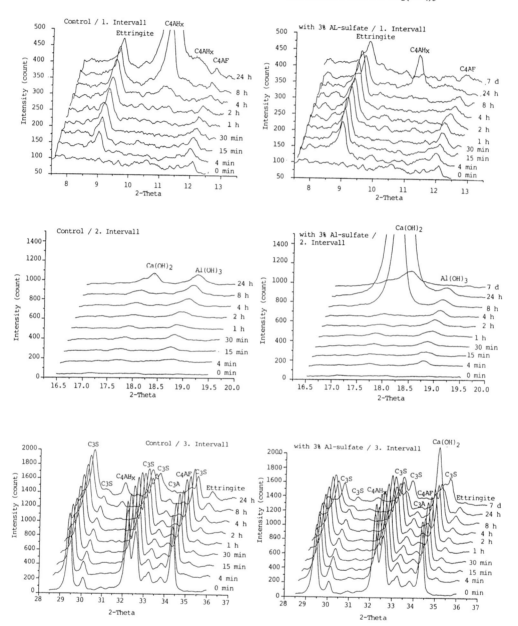

Figure 2. X-ray diagrams of cement Type III with and without accelerators.

CEMENT Type V
Control

Accelerated with 3% Al₂(SO₄)₃

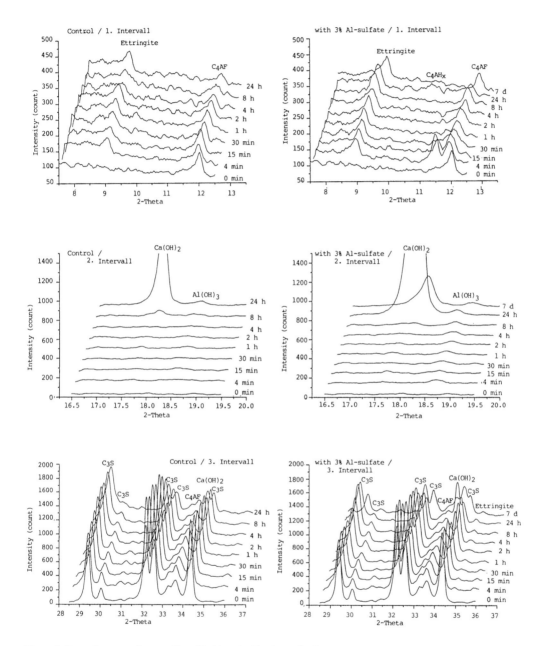

Figure 3. X-ray diagrams of cement Type V with and without accelerators.

and the durability of concrete. Numerous studies indicate that the factors influencing the conditions of ettringite formation in a cement system include the pH of the liquid phase, the specific surface of cement, presence of admixtures, etc. There are two main theories for the hydration mechanism: the topo-chemical and the through-solution mechanism. According to the former, ettringite is formed on the surface of C_4A_3S and grows as slender needles (Ogawa & Roy 1982). According to the latter, ettringite is formed from a liquid super-saturated with respect to Ca^{2+}, Al^{3+} and SO_4^2. The process of Al^{3+} transfer into solution and the solubility of the C_4A_3S phase depend on the pH of the surrounding liquid. Al^{3+} is stable only in an environment of suitable pH. Upon an increase of the pH of the surrounding liquid, a hydrolytic reaction leading to $Al(OH)_4$ formation occurs, with precipitation in the form $Al(OH)_3$ as the pH decreases.

Earlier investigations have studied the effects of set-modifying admixtures (generally inorganic or organic electrolytes) on the hydration of PC (Sahu et al. 1992, Deng et al. 1993). Ettringite is always precipitated when Ca^{2+}, SO_4^{2-} and Al^{3+} are present in the stoichiometric ratio:

$$6Ca^{2+}+2Al(OH)_4^-+3SO_4^{2-}+4OH^-+26H_2O \Rightarrow$$
$$3CaO \cdot Al_2O_3 \cdot 3CaSO_4 \cdot 32H_2O \qquad (1)$$

In this study, the influence of aluminium sulfate on the setting and hardening of Portland cement was investigated.

4 SAMPLE PREPARATION AND ANALYSIS

Ettringite was prepared according to the following method:

1. Aluminium sulfate and calcium hydroxide in molar ratios suitable to form ettringite were mixed in water under a stream of nitrogen.

2. Samples were extracted at 5, 30, 60, 120 minutes and at one day and 3 days. The samples were filtered under suction, washed with methanol, and dried under vacuum. Solutions were analysed for their calcium and aluminium contents by atomic absorption spectroscopy and sulfate content by conventional methods.

The composition of the aqueous phase during the formation of ettringite is shown in Table 1. The aqueous Ca^{2+} ion concentration is constant at about 14 mmolL^{-1} during the first 60 minutes, and then decreases rapidly to 0.20 mmolL^{-1} at 120 minutes where it remains constant up to 3 days. The decrease in Ca^{2+} concentration presumably relates to ettringite formation.

Aluminium concentrations increase up to 30 minutes then decrease to below detectable limits (<0.01

mmolL^{-1}) at 3 days. Aqueous sulfate solutions show a similar trend to the Ca^{2+} ion concentration where the values are reasonably constant up to 30 minutes at 0.020 molL^{-1}, then decrease rapidly giving 0.002 molL^{-1} at 120 minutes with a further slight decrease to less than 0.001 mmolL^{-1} at 3 days. Ettringite can be readily prepared from reaction of aluminium sulfate and calcium hydroxide in solution.

Table 1: Ca^{2+}, Al^{3+} and SO_4^{2-} content

Hydration Time (min)	Ca^{2+} (mmolL^{-1})	Al^{3+} (mmolL^{-1})	SO_4^{-2} (molL^{-1})
5	14.2	0.11	0.022
30	13.5	1.14	0.020
60	13.9	0.73	0.012
120	0.1	0.12	0.002
1440 (1d)	0.2	nd	0.001
4320 (3d)	0.2	nd	nd

nd: Not detected

5 ETTRINGITE FORMATION

Aluminium sulfate is known as a chemical with high solubility. It acts in a similar way to calcium sulfate by accelerating the formation of ettringite. (Lukas 1999).

Ettringite formation during hydration was investigated by X-ray Diffraction, using ASTM cement Types III and V. Diffraction occurs as x-ray waves interact with the structure of minerals. The angle of diffraction is the theta angle. For practical reasons, twice the theta angle is measured, so it is termed the angle 2-theta. This is significantly different for different minerals and can be used for the assessment of ettringite, portlandite, etc. All tests were executed at the University of Innsbruck, Austria. X-ray diffractograms were recorded with and without the addition of 3% aluminium sulfate immediately after mixing and up to 7 days, and at 3 different Degree 2-Theta intervals. The line diagrams are shown in Figures 2 and 3.

The fundamental reactions are as follows:

$$3CaO \cdot Al_2O_3 + 3CaSO_4 + 32H_2O \Rightarrow$$
$$3CaO \cdot Al_2O_3 \cdot 3CaSO_4 \cdot 32H_2O \qquad (2)$$

(tricalcium aluminate) (gypsum) (water) $\Rightarrow$
(trisulfate; ettringite; AFt-phase)

$$Al_2(SO_4)_3 + 6Ca(OH)_2 + 32H_2O \Rightarrow$$
$$3CaO \cdot Al_2O_3 \cdot 3CaSO_4 \cdot 32H_2O + Al(OH)_3 \quad (3)$$

(aluminium sulfate) (calcium hydroxide) (water) $\Rightarrow$
(ettringite; aluminium hydroxide)

Table 2. Formation of mineral compositions according to the X-ray diffraction analysis (quantities measured in mm^2)

TYPE III CEMENT

Control	Portlandite	Ettringite	C4AHx	Al(OH)$_3$	3% Al Portlandite	Ettringite	C4AHx	Al(OH)$_3$
4 min	6.1	17.6	-	9.5	7.3	34.0	-	13.4
15 min	7.5	28.5	-	14.0	5.6	31.9	-	13.5
30 min	8.5	30.0	-	17.7	7.5	36.8	-	15.5
60 min	8.8	34.8	-	14.5	11.5	34.4	-	24.2
120 min	10.5	36.0	6.9	26.6	10.8	46.9	-	25.8
4 h	12.5	35.7	20.2	33.6	40.0	43.5	-	20.3
8 h	19.6	28.8	80.2	42.2	370	51.7	-	24.5
24 h	41.1	26.0	297	42.7	480	20.4	34	30.5
7 d	-	-	-	-	530	37.7	67	38.2

TYPE V CEMENT

Control	Portlandite	Ettringite	C4AHx	Al(OH)$_3$	3% Al Portlandite	Ettringite	C4AHx	Al(OH)$_3$
4 min	-	15.7	-	-	4.4	19.3	-	13.6
15 min	-	15.0	-	-	6.5	18.2	-	11.4
30 min	-	19.3	-	-	-	18.3	-	17.6
60 min	5.4	17.2	-	9.2	9.3	19.7	-	10.3
120 min	-	27.2	-	9.1	3.9	21.1	-	12.2
4 h	-	14.1	-	5.8	-	21.9	-	12.4
8 h	20.1	15.0	-	10.1	17.1	23.4	-	14.0
24 h	231	22.2	-	14.1	1264	25.9	16	16.2
7 d	253	22.2	-	16.8	118	25.1	52	19.2

The line diagrams have a different scale, therefore it is difficult to compare the mineral formation. For a better understanding the peak surfaces (in mm^2) were determined and recorded. The results are given in Table 2.

The peak surfaces represent the relative concentration of the minerals formed. The X-ray diffractograms show amazing results. It can clearly be seen, that the ettringite, portlandite and aluminium hydroxide formation is accelerated by an addition of 3% aluminium sulfate. In Type III cement the ettringite formation is nearly doubled whereas Type V cement shows an increase of about 25%. Also the amount of ettringite formed is higher for the Type III cement than for Type V cement.

The portlandite content shows a very strong increase in both accelerated cements representing a fast hydration of the cements and rapid strength gain. The calcium aluminate C$_4$AH$_x$ content decreases, maybe in favour of the increased ettringite formation. Aluminium hydroxide formation is increased at early ages and is much higher for Type III cement.

6 PRACTICAL RESULTS

6.1 Setting time

Setting times were determined using a laboratory penetrometer produced by RMU Bergamo, Italy. 3% and 4% of alkali-free accelerator by weight of cement added to a mixture of 1 part cement Type III or Type V and 5 parts of fine sand (0-2 mm). The mortar was stabilized with 1% of an acrylic based superplasticizer, w/c 0.55. A Type V cement can be accelerated similar to a Type III cement, as shown in Figure 4.

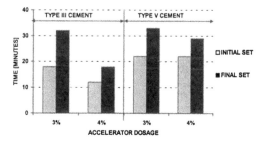

Figure 4. Setting time test with alkali-free accelerator.

6.2 Early compressive strength

The following graph shows the influence of an alkali-free accelerator in comparison to an aluminate-based accelerator in a wet process shotcrete. Cement Type III was used at a content of 450 kg/m^3, sand 0-8 mm, w/c 0.50. 500 Proctor Units are equal to 1 MPa compressive strength, which is reached after only 1 hour with the alkali-free accelerator at a dosage of 5% by weight of cement. The aluminate-based accelerator with 5% addition showed 170 Proctor Units, equal to 0.34 MPa after 2 hours (Figure 5).

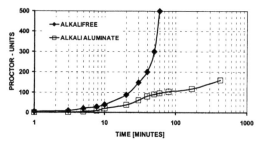

Figure 5. Early strength determined by Proctor Needle Test (Type III Cement).

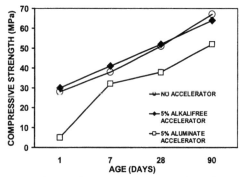

Figure 6. Compressive strength gain (Type III Cement).

6.3 *Compressive Strength Gain*

Compressive strength gain from 1 to 90 days was determined using drilled cores of 50 mm diameter and a length of 100 mm. The use of an alkali-free accelerator resulted in equal strength compared to a non-accelerated shotcrete whereas the aluminate-based accelerator showed a strength decrease of 22% (Figure 6).

6.4 *Leaching Test*

Figure 7 shows a leaching test cell for the determination of soluble ions in shotcrete. From a drilled core, 10 discs of 100 mm diameter and a thickness of 10 mm were cut, resulting in a total surface area of 1450 cm^2.

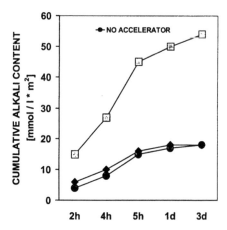

Figure 8. Soluble alkali content from leaching tests.

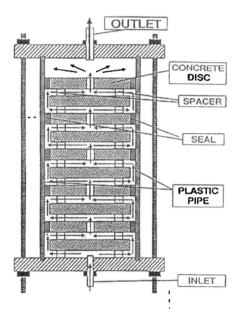

Figure 7. Leaching test cell (Breitenbücher et al. 1999).

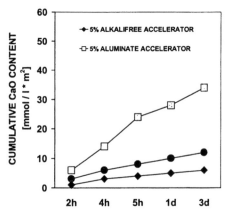

Figure 9. Soluble calcium oxide contents from leaching tests.

De-ionized water was forced through the cell at a flow rate of 300 mL/hour. The resulting solution was analyzed and the content of alkali and calcium ions were determined by flame photometry. It could be clearly shown, that an alkali-free accelerator does not increase the leaching of alkalis compared to shotcrete without any accelerator, and that the soluble calcium oxide content is even reduced. An aluminate-based accelerator shows an increase of soluble alkalis by a factor of 3 and of soluble calcium oxide by a factor of more than 6 (Figure 8).

The total soluble alkali content of shotcrete is responsible for potential Alkali Aggregate Reactivity (AAR). Since alkali-free accelerators do not increase the alkali content (Figure 8), they do not increase the potential for AAR (Figure 10).

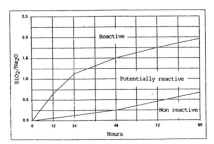

Figure 10. Alkali reactivity of aggregate as a function of the SiO₂/Na₂O ratio, according to the French Standard AFNOR P 18-589.

7 DISCUSSION

It was shown that aluminium sulfate reacts together with lime and leads to the formation of ettringite. Ettringite is responsible for the rapid setting of accelerated concrete. As a by-product of the reaction, aluminium hydroxide is formed which is also known to be an accelerator. The increased formation of calcium hydroxide is a sign of rapid hydration of cement and strength gain.

Practical shotcrete results show the acceleration effect of aluminium sulfate as an alkali-free accelerator for normal and sulfate resistant cement. The advantage of an alkali-free accelerator is demonstrated by measurement of early and final compressive strength and by leaching tests.

8 CONCLUSIONS

Alkali-free accelerators react with lime and lead to the rapid formation of ettringite, aluminium hydroxide, and calcium hydroxide This is the basis of the accelerated hydration of cement. Rapid setting can be obtained by this mechanism with both normal and sulfate resistant cements. Early compressive strength develops faster, and final compressive strength is not reduced.

Leaching of alkali ions is not increased in concrete accelerated with alkali-free accelerators, and leaching of calcium ions is lower compared to a control concrete. Field experience indicates that alkali-free accelerators significantly reduce the clogging of drainage pipes and the discharge of highly alkaline waste water into drainage systems. Moreover, alkali-free accelerators do not increase Alkali-Aggregate-Reaction (AAR).

REFERENCES

Breitenbücher, R. et al, Auslaugbarkeit von Beton, *Beton- und Stahlbetonbau 89* (1999) Heft 9, pp. 237

Deng Min & Tang, M. 1994. Formation and expansion of ettringite crystals. *Cement and Concrete Research*, 24 (1): 119.

Gabrisova, A. Havlica, J. & Sahu, S. 1991. Stability of calcium sulphoaluminate hydrates in water solutions with various pH values. *Cement and Concrete Research*, 21 (6): 1023.

Havlica, J. & Sahu, S. 1992. Mechanism of ettringite and monosulphate formation. *Cement and Concrete Research*, 22 (4): 672.

Lukas, W., 1999. R+D Cooperation with Sika AG, CH-8048 Zürich and University of Innsbruck, Austria.

Ogawa, K. & Roy, D.M. 1982. C₄A₃S hydration, ettringite formation, and its expansion mechanism. *Cement and Concrete Research*, 12 (2): 247-256.

Sahu S., Majling, J. & Havlica J. 1992. Influence of anhydrite particle size on the hydration of sulphoaluminate belite cement, 9th ICCC, IV, New Delhi, pp443-448.

Shotcrete: Engineering Developments, Bernard (ed.) © 2001 Swets & Zeitlinger, Lisse, ISBN 90 5809 176 7

Use of recycled aggregates in fibre reinforced shotcrete: mechanical properties, permeability and frost durability

T.Farstad
Norwegian Building Research Institute, Oslo, Norway

C.Hauck
Veidekke ASA, Heavy Construction Division, Oslo, Norway

ABSTRACT: As part of a construction project in Oslo, Norway, wet-mixed steel fibre reinforced shotcrete was used substituting up to 20% of the aggregate with recycled aggregate. The shotcrete was used outdoors for the main purpose of protecting a lightweight bridge foundation. This paper presents some of the results from laboratory testing of shotcrete specimens made with recycled aggregates. The investigation was performed as a part of RESIBA, an ongoing research project in Norway demonstrating the use of recycled aggregates in various applications. The results indicate a reduction of approximately 15-20% relative to the control mix for compressive and flexural strength of shotcrete with up to 20% recycled aggregate. The use of recycled aggregate however does not appear to have any effect on the residual flexural strength or the flexural toughness. Furthermore, results from durability testing (freeze/thaw-test, water intrusion, and capillary absorption) do not show any clear reduction in durability for shotcrete produced with up to 20 % recycled aggregate.

1 INTRODUCTION

1.1 Background

A demonstration project using recycled aggregates in shotcrete was started in the summer of 1999. The project is part of the Norwegian research project RESIBA. The objective of RESIBA is to find suitable applications for the use of Recycled Concrete Aggregates (RCA) in the construction industry in Norway, (Mehus et al. 2000). The wet-mix shotcrete made with recycled aggregates and used in this study, was produced the same way as an ordinary shotcrete at a ready-mix concrete plant.

2 MATERIALS AND SPECIMENS

2.1 Mix design

Four different shotcrete-mixtures with 0% (control or "reference" mix), 7.5%, 15%, and 20% RCA (by weight of the total aggregate) were used. The RCA was a blend of concrete, tile, brick, etc. produced at a recycling plant from building and construction debris from the city of Oslo. Typical RCA properties from Oslo are given by Jacobsen (1998). The shotcrete was mixed using a water-reducing admixture and 40 kg/m³ of steel fibre reinforcement. The shotcrete had a theoretical compressive strength of 45

MPa. During spraying of the shotcrete, a sodium silicate based accelerator was used. Test specimens were made on-site from the shotcrete mixes with 0%, 15%, and 20% recycled aggregate. Mix design and laboratory testing prior to full scale mixing, together with on site documentation, is given by Hauck & Farstad (2001). Table 1 shows the mix design.

Table 1 Shotcrete mixes (kg/m³)

Material	Shotcrete with RCA			
	0%	7.5%	15%	20%
Portland cement	485	485	485	485
Silica fume	30	30	30	30
Aggreg. A (0- 8 mm)	415	359	303	247
Aggreg. B (0- 8 mm)	1071	1003	935	866
Recycled aggreg. (0- 4 mm)	0	99	198	298
Water	239	239	238	237
Steelfibre	40	40	40	40
Water-reducing admixture	3.50	3.50	3.50	3.50
Superplasticizer	2.67	3.50	4.00	4.50
Accelerator	23	20	20	20
w/b	0.46	0.46	0.46	0.45

2.2 Preparation of test specimens

Three test panels were sprayed on the construction site with the control mix (0%) and the mixes with 15% and 20% recycled aggregates. The dimensions of each panel were approximately 1.2×0.8×0.2m (L×W×H). After spraying, the panels were covered with plastic and cured for 14 days before transportation to the laboratory for further preparation. Table 2 shows the different specimens cut from each panel and which test methods they were used in.

Table 2 Specimen from each panel.

Specimen	Number	Size in mm L×W×H (D)	Test method
Beam	3	550×125×75	Flexural toughness
Prism	4	150×150×150	Frost resistance
Cylinder	3	150×150	Water intrusion depth
Cyl. slice	3	100×20	Capilary absorption

Details regarding the preparation of specimens for each of the test methods are given in the next section.

3 EXPERIMENTS AND RESULTS

3.1 Mechanical properties

The mechanical properties of the shotcrete involved testing of: compressive strength, flexural strength, and residual flexural strength at 1 and 3 mm deflection, and determination of flexural toughness, according to ASTM C 1018.

3.1.1 Compressive strength

The compressive strengths of 0 and 20% RCA-mixes were tested according to the Norwegian Standard NS 3668. The test was performed on cubes cast on site, stored and cured outdoors for the first 14 days together with the sprayed panels and brought to the laboratory together with the panels. The cubes were stored in water from arrival to the day of testing at 28 days. The results are shown in Table 3.

Table 3. Compressive strength

Mix	Density (kg/m³)	Comp. strength (Mpa)
Reference	2266	61.8
20% Recycled aggregate	2155	50.7

3.1.2 Flexural strength

The flexural strength was tested according to the Norwegian Concrete Association Publication no. 7 (NCA Publ. no. 7). Three beams from each panel were tested. After sawing, the beams were stored in water for 3 days prior to testing. In Figure 1 the test-setup for the flexural strength and toughness test is shown. The results from the flexural strength tests are shown in Tables 4 and 5.

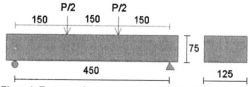

Figure 1. Test setup, beams according to NCA Publ. No. 7.

Table 4. Flexural tensile strength.

Mix	Flexural tensile strength (MPa)
Reference	5.8
15% Recycled agg.	4.9
20% Recycled agg.	4.9

Table 5. Residual flexural strength

Mix	Residual flexural strength (MPa) Deflection	
	1 mm	3 mm
Reference	2.2	0.9
15% Recycled agg.	1.9	1.3
20% Recycled agg.	1.9	1.0

In addition to the flexural strengths obtained in accordance with NCA Publ. no 7, toughness indices according to ASTM C 1018 were calculated. The results are shown in Table 6.

Table 6. Toughness indices according to ASTM C 1018

Mix	Toughness indices				
	I_5	I_{10}	I_{20}	I_{30}	I_{50}
Reference	3.8	6.2	9.9	12.5	16.0
15% Recycled agg.	3.8	5.9	9.6	12.9	18.7
20% Recycled agg.	3.7	5.8	9.6	12.8	17.3

3.2 Durability

Investigations were performed on the wet mixed sprayed shotcrete for durability. The tests are carried out according to the standards/procedures described in Table 7.

Table 7. Method of testing

Method	Test method used
Water intrusion depth	pr EN-ISO 7031
Capillary absorption	Nordtest NT Build 368
Freeze/thaw test	Swedish Standard SS 13 72 44 *

* Modified to include surface- and internal damage

3.2.1 Water intrusion depth

The water intrusion depth was measured on cores with a diameter/length of 150mm, drilled out from the panels. The specimens were tested on a sawn surface approximately 10mm from the sprayed surface of the panels. Prior to testing, the cores were stored for 3 days in minimum 95% RH. The water pressure used on the test surface was 300kPa, 500kPa and 700kPa with 24 hours at each pressure level. The specimens were then split in half, perpendicularly to the face on which the water pressure was applied. The water penetration front was then marked on the specimen and the maximum depth of penetration under the test area was recorded. The results from the water intrusion test are shown in Table 8.

Table 8. Water intrusion depth

Mix	Maximum water intrusion (mm)
Reference	45
15% Recycled agg.	38
20% Recycled agg.	30

3.2.2 Capillary water absorption

The capillary water absorption tests were performed on 20 mm thick slices sawn from cores with a diameter of 100 mm drilled from the sprayed panels. The water absorption was performed on sawn surfaces 10 mm, 35 mm and 60 mm from the top of the sprayed panel. The water absorption was measured after 10 and 30 minutes, and 1h, 2h, 3h, 4h and 6h, and 1d, 2d 3d and 4d. Based on the water uptake the following constants were calculated:

$$C = Q_{kap}/ t_{cap} \qquad (1)$$

where Q_{kap} is the amount of absorbed water in $kg/m^2s^{0.5}$;and t_{cap} is the time exposed for water absorption in seconds. This is an expression of the rate of water absorption (Capillary number). Also, the resistance number, R, was calculated as

$$R = t_{cap}/h^2 \qquad (2)$$

where h is the specimen height in mm. This is an expression of the relative time the water absorption front takes to reach a height h (Resistance number). The results of the capillary water absorption tests are shown in Table 9.

3.2.3 Freeze/thaw test

The freeze/thaw test SS 13 72 44 is a test method for durability of concrete surfaces against frost and frost/salt scaling damage. The test was performed on sawn surfaces from prisms measuring 150×150×50

Table 9. Results capillary water absorption

Mix	R s/m²•10⁷	C kg/m²•√s•10⁻²
Reference	5.68	2.38
15% Recycled agg.	7.50	2.29
20% Recycled agg.	7.40	2.47

Table 10. Freeze/thaw results, scaled material.

Mix	Accumulated scaled material in kg/m² (cycles)				
	7	14	28	42	56
Reference	0.003	0.006	0.010	0.014	0.016
15% Recycled agg.	0.006	0.010	0.017	0.021	0.024
20% Recycled agg.	0.04	0.009	0.015	0.020	0.023

Table 11. Freeze/thaw results, water uptake.

Mix	water uptake in g (cycles)				
	7	14	28	42	56
Reference	11.75	17.51	26.76	31.77	34.52
15% Recycled agg.	13.26	19.51	30.77	36.27	37.53
20% Recycled agg.	18.50	25.76	38.52	43.69	45.36

Table 12. Freeze/thaw results, ultrasonic pulse velocity.

Mix	% of UPV at 0 cycles (cycles)				
	7	14	28	42	56
Reference	103.6	104.8	105.4	106.0	106.6
15% Recycled agg.	103.7	104.5	105.2	106.0	106.5
20% Recycled agg.	103.6	105.0	105.6	107.2	107.2

mm cut from the sprayed panels. The tests performed in this investigation used demineralized water on the surface during freezing and thawing. This was chosen in order to simulate actual conditions on site. In addition to the standard SS 13 72 44, uptake of water during freezing and thawing was measured. Wet freeze/thaw cycling leads to a much higher water uptake than in normal capillary suction without freeze/thaw (Jacobsen et al. 1999). To monitor internal crack formation measurements of ultrasonic pulse velocity (UPV) were also carried out. This was measured normal to the test surface with 50 mm transducers operating at 54 kHz. The results from the freeze/thaw tests are shown in Tables 10, 11 and 12.

4 DISCUSSION OF TEST RESULTS

4.1 Mechanical properties

A reduction in the compressive strength of approximately 18% was observed for the specimens with 20% RCA. This was mainly due to the increased porosity and lower strength of the RCA. The flexural tensile strength was also reduced by approximately 15% for both mixes (15% and 20%) sprayed with RCA. Results from calculations of the residual flexural strength according to NCA Publ no. 7 show a reduction in flexural stress of approximately 15% at 1mm beam deflection. However, when the beam deflection reached 3 mm, no reduction in flexural strength was observed in the RCA concrete. The specimens with 15% recycled aggregate showed a 40% increase in flexural strength. The reason for this result was not apparent in the tests performed in this research program.

Toughness indices according to ASTM C 1018 show no significant effect related to the use of recycled aggregate. The values are, however, low because of the relatively high compressive strength and the low dosage of fibre used in the mixes (40 kg/m³).

4.2 Durability

The results from the water intrusion test show a reduction in water intrusion depth when the normal aggregate was replaced with up to 20% recycled aggregate. The reason for this is most likely the effect of a reduction in the w/b ratio caused by suction of water into the porous RCA in the fresh concrete. However, these are rather small differences and all exceeded the maximum criteria required in Norway (25mm).

The capillary water absorption tests showed very small differences between the reference concrete and the concrete with RCA. The somewhat higher resistance number R, was caused by the higher t_{cap} in the RCA specimens.

The freeze/thaw test ranks the three mixes equally and with very good frost durability according to SS 13 72 44. The somewhat higher scaling in the RCA concrete is probably due to the scaling of RCA particles at the cut surfaces.

The water uptake is higher for the specimens sprayed with recycled aggregate because of the increase in porosity of these aggregates. This increased water uptake has, however, not lead to a clear increase of deterioration in the form of scaling or internal cracking. Measurements of the ultrasonic pulse velocity show an increase of UPV for all the mixes. The increase of UPV is most probably caused by the water uptake for the specimens.

5 CONCLUSION

Use of recycled aggregates in fibre reinforced shotcrete results in a reduction in the compressive strength of 20%. When 15% and 20% of the normal aggregate was replaced by recycled aggregate, the flexural tensile strength was reduced by 15% for both mixes. The residual flexural beam tests showed a reduction in performance of approximately 15% at 1 mm beam deflection. However, when the beam deflection reached 3 mm, no reduction in flexural strength was observed in the RCA concrete. Toughness indices according to ASTM C 1018 show no effect on post-crack capacity when replacing the normal aggregate with 20% recycled aggregate.

Replacing the normal aggregate with up to 20% of recycled aggregate results in no reduction of durability in the tests performed. The tests show an increase in water absorption, but no signs of internal cracking measured as loss of ultrasonic pulse velocity UPV, and no clear increase of surface damage in the freeze/thaw test.

6 REFERENCES

Hauck, C. & Farstad, T., 2001. Use of recycled aggregates in fibre reinforced shotcrete. Mix design, properties of fresh concrete and on-site documentation. *International Conference on Engineering Developments in Shotcrete*. April 2-4, Hobart, Tasmania, Balkema.

Jacobsen, S. & Rommetvedt, O-E. & Gjengstø, KT., 1998. Properties and Frost Durability of Recycled Aggregate from Oslo Norway. Sustainable Constructions, Use of Recycled Concrete Aggregate, 189-196. Thomas Telford.

Jacobsen, S. & Bager, D. & Kukko, H. Luping, T. & Nordström, K., 1998, Measurement of internal cracking as dilation in the SS 13 72 44 frost test, NBI-project report 250, ISBN 82-536-0645-1.

Mehus, J. & Lahus, O. & Jacobsen, S. & Myhre, Ø., 2000, Use of RCA in the building and construction industry. NBI-project report 287, ISBN 82-536-0705-9, 64p. (In Norwegian).

pr EN-ISO 7031-1994. Testing concrete – Determination of the depth of penetration of water under pressure

NT BUILD 368 1991. Concrete, Repair Materials: Capillary Absorption.

Swedish Standard SS 13 72 44., Concrete testing – Hardened concrete – Scaling at freezing.

Shotcrete: Engineering Developments, Bernard (ed.) © 2001 Swets & Zeitlinger, Lisse, ISBN 90 5809 176 7

Determining the ultimate displacements of shotcrete tunnel linings

N.N.Fotieva & A.S.Sammal
Tula State University, Tula, Russia

ABSTRACT: A method of determining the ultimate displacements at which a shotcrete tunnel lining acting in combination with anchors ceases to possess sufficient load bearing capacity is proposed. The method is based on an analytical solution of the plane contact problem using elasticity theory for a double-layer ring of arbitrary shape. The internal layer simulates a shotcrete lining, the external one simulates a rock zone strengthened by anchors. The shotcrete lining is taken to support the opening within a linearly elastic medium simulating the rock mass subject to the action of gravitational forces.

1 INTRODUCTION

At present the New Austrian Tunnelling Method (NATM), based on the application of shotcrete in combination with anchors as a temporary tunnel lining, is widely used in the underground construction industry. The necessary parameters for the lining are usually determined on the basis of numerical modelling and are corrected taking the results of monitored lining displacements into account.

The ultimate admissible displacements of a shotcrete lining are, as a rule, determined from an elasto-plastic analysis of a one-dimensional problem without consideration of the real shape of the opening, or on the basis of practical experience. This is why an analytical method for the determination of maximum admissible displacements in shotcrete tunnel linings of arbitrary cross-sectional shape has been developed at Tula State University.

2 THE DESIGN METHOD

The method proposed is based on the analytical solution of the plane contact problem using elastic theory. The design scheme for this problem is given in Figure 1.

In this diagram, the linearly deformable medium S_0, with an elastic modulus described by E_0 and a Poisson's ratio equal to ν_0, simulates the rock mass. The structure is simulated by a double-layered ring of arbitrary shape, with one axis of symmetry.

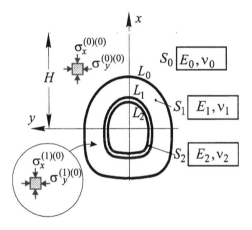

Figure 1. The design scheme.

The rock zone, discretely strengthened by anchors, is simulated by a homogeneous layer S_1. The effective deformation modulus of this material is given by $E_1 = \beta E_0$. The relationship between the coefficient β and n_s, the anchor installation compactness (number of anchors per 1 m² of opening surface) and the bearing capacity of the lining P_a, obtained on the basis of experimental investigations (Timofeev & Trushko 1984) is shown in Figure 2. Poisson's ratio ν_1 is assumed to be the same as in the intact rock mass.

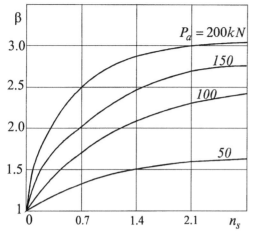

Figure 2. Relation between β and the parameters n_s and P_a.

The mechanical properties of the sprayed concrete lining represented by the layer S_2 are described by the elastic modulus E_2, and Poisson's ratio, v_2.

The action of gravitational forces on the rock mass is simulated by the presence of initial stresses in the S_0 medium and S_1 layer determined by the formulae

$$\sigma_x^{(j)(0)} = -\gamma H \alpha^*, \quad \sigma_y^{(j)(0)} = -\lambda \gamma H \alpha^* \qquad (1)$$

$$(j = 0, 1)$$

where γ = rock unit weight; H = tunnel depth; λ = lateral pressure coefficient in an intact rock mass; and α^* = correcting multiplier introduced as an approximation of the influence of the distance l between the location where the shotcrete is being sprayed and the opening face. This multiplier may be determined by the following empirical expression (Bulychev & Fotieva 1991):

$$\alpha^* = 0.64 e^{-1.75 \, l/R_1} \qquad (2)$$

where R_1 is the average radius of the opening.

The S_j ($j = 1, 2$) layers and the S_0 medium undergo deformation together, i.e. conditions of continuity of vectors of stresses and displacements are satisfied on the L_j ($j = 0, 1$) contact lines. The L_2 internal outline is free from loads.

The described problem of elasticity theory has been solved through the application of the complex variables analytic functions theory (Muskhelishvili 1966) using the method of conform mapping and complex series.

After representing complete stresses in the rock mass as the sums of initial stresses (1) and additional stresses $\sigma_x^{(j)}, \sigma_y^{(j)}, \tau_{xy}^{(j)}$, caused by the pres-

ence of the opening, and introducing the complex potentials $\varphi_j(z)$, $\psi_j(z)$ ($j = 0, 1, 2$) in the areas S_j ($j = 0, 1, 2$) related to stresses by the method of Muskhelishvili (1966), the contact problem for the determination of the additional stresses and displacements is transformed into a boundary problem using complex variable analytical functions theory for the following boundary conditions:

$$\varphi_{j+1}(t) + t\overline{\varphi'_{j+1}(t)} + \overline{\psi_{j+1}(t)} = \varphi_j(t) + t\overline{\varphi'_j(t)} + \\ + \overline{\psi_j(t)} - \lambda_{j,1} \gamma H \alpha^* \left[\frac{1+\lambda}{2} t - \frac{1-\lambda}{2} \bar{t} \right]$$

$$\text{on } L_j \; (j=0, 1) \qquad (3)$$

$$\text{æ}_{j+1} \varphi_{j+1}(t) - t\overline{\varphi'_{j+1}(t)} - \overline{\psi_{j+1}(t)} = \\ = \frac{\mu_{j+1}}{\mu_j} \left[\text{æ}_j \varphi_j(t) - t\overline{\varphi'_j(t)} - \overline{\psi_j(t)} \right]$$

$$\varphi_2(t) + t\overline{\varphi'_2(t)} + \overline{\psi_2(t)} = 0 \qquad \text{on } L_2$$

where $\lambda_{v,k} = \begin{cases} 1 & at \; v = k \\ 0 & at \; v \neq k \end{cases}$, $\text{æ}_j = 3 - 4v_j$,

$$\mu_j = \frac{E_j}{2(1+v_j)}, \quad (j=0, 1, 2), \quad t = \text{affix of a point of}$$

the corresponding outline.

To solve the problem, a conform mapping of the exterior of a circle having a radius $R_2 < 1$ in the plane of the ζ variable on the exterior of the outline L_2 in z, mapping so that the circumference of the $R_0 = 1$ radius turns into the outline L_0, is fulfilled by applying a rational function

$$z = \omega(\zeta) = \sum_{v=0}^{\infty} q_v \zeta^{1-v} \qquad (4)$$

where q_v ($v = 0, ..., \infty$) are coefficients determined using, for instance, the method by Melentiev (1962). Then the circle of radius R_1 turns into the corresponding outline L_1.

After some operations the boundary conditions (3) in the transformed area become the following:

$$\overline{\varphi}_{j+1}(\frac{R_j}{\sigma}) + \frac{R_j}{\sigma} \frac{\overline{\omega}(\frac{R_j}{\sigma})}{\omega'(R_j\sigma)} \varphi'_{j+1}(R_j\sigma) + \psi_{j+1}(R_j\sigma) = $$

$$= \overline{\varphi}_j(\frac{R_j}{\sigma}) + \frac{R_j}{\sigma} \frac{\overline{\omega}(\frac{R_j}{\sigma})}{\omega'(R_j\sigma)} \varphi'_j(R_j\sigma) + \psi_j(R_j\sigma) - \quad (5)$$

$$-\lambda_{j,1}\gamma Ha *\left\{\frac{1+\lambda}{2}\,\overline{\omega}(\frac{R_j}{\sigma})-\frac{1-\lambda}{2}\,\omega(R_j\sigma)\right\},$$

$$(j=0, 1)$$

$$\text{æ}_{j+1}\overline{\varphi}_{j+1}(\frac{R_j}{\sigma})-\frac{\overline{\omega}(\frac{R_j}{\sigma})}{\omega'(R_j\sigma)}\varphi'_{j+1}(R_j\sigma)-\psi_{j+1}(R_j\sigma)=$$

$$(6)$$

$$=\frac{\mu_{j+1}}{\mu_j}\left[\text{æ}_j\overline{\varphi}_j(\frac{R_j}{\sigma})-\frac{\overline{\omega}(\frac{R_j}{\sigma})}{\omega'(R_j\sigma)}\varphi'_j(R_j\sigma)-\psi_j(R_j\sigma)\right]$$

$$\overline{\varphi}_N(\frac{R_N}{\sigma})+\frac{\overline{\omega}(\frac{R_N}{\sigma})}{\omega'(R_N\sigma)}\varphi'_N(R_N\sigma)+\psi_N(R_N\sigma)=0,\,(7)$$

where $\sigma = e^{i\vartheta}$.

The complex potentials $\varphi_0(\zeta)$, $\psi_0(\zeta)$ are regular in the area S_0 and become zero at infinity. They are represented in the transformed area as complex series on negative degrees of the variable ζ.

The complex potentials $\varphi_j(\zeta)$, $\psi_j(\zeta)$ (j=1, 2), regular in corresponding ring layers, are represented in the form of Laurent series. Due to the presence of force symmetry, all the unknown coefficients of the series mentioned above are real values. Substituting those series into (5) and (6) conditions it is possible to obtain the recurrent relationships connecting the coefficients of series characterising the complex potentials in contacting areas S_{j+1} and S_j (j = 0, 1). Then introducing those relationships into the boundary condition (7) we obtain a system of linear equations with respect to the unknown coefficients of the expansion of $\varphi_0(\zeta)$ and $\psi_0(\zeta)$ complex potentials. This allows the stresses and displacements at points in the area S_0 simulating the rock mass to be obtained. Furthermore, on the basis of the recurrent formulae mentioned above, the complex potentials $\varphi_j(\zeta)$, $\psi_j(\zeta)$ (j=1, 2) can be determined. This allows the stresses and displacements at points within the area S_2, simulating the tunnel lining, to be calculated.

The solution describe above allows the determination of stresses and displacements at any point in a lining, which may be represented in the form

$$\sigma = \tilde{\sigma}\cdot\gamma Ha*, \quad v = \tilde{v}\cdot\gamma Ha*, \tag{3}$$

where the symbols σ and v signify all the components of the stress tensor and the displacements vector; and σ, v are the stresses and displacements

obtained from the contact problem solution at $\gamma Ha* = 1$.

By applying the lining strength conditions

$$\sigma^{(t)}_{\theta\max} \le R_t, \quad \left|\sigma^{(c)}_\theta\right|_{\max} \le R_c \tag{4}$$

where

$$\sigma^{(t)}_{\theta\max}, \text{ and } \left|\sigma^{(c)}_\theta\right|_{\max}$$

are the maximum tensile (positive) and compressive (negative) circumferential stresses occurring in the internal lining cross-section outline, and R_t, R_c are limiting stresses for shotcrete in tension and compression, one can obtain a value of $\gamma H_{\max}\alpha^*$, at which these conditions are still satisfied:

$$\gamma H_{\max}\alpha^* = \min\left(\frac{R_c}{\left|\tilde{\sigma}^{(c)}_\theta\right|_{\max}}, \frac{R_t}{\tilde{\sigma}^{(t)}_{\theta\max}}\right) \tag{5}$$

Substitution of this value into the expression

$$v_{\max} = \tilde{v}\cdot\gamma H_{\max}\alpha^* \tag{6}$$

allows the determination of the ultimate admissible values of normal $v^{in}_{\rho\max}$, vertical $v^{in}_{x\max}$ and horizontal $v^{in}_{y\max}$ displacements, or $\Delta v^{in}_{y\max}$ convergences of points of the lining cross-section internal outline.

It should be observed that the distribution of displacements along the outline depends strongly on the value of the lateral pressure coefficient λ in the intact rock, which may be determined on the basis of measurements according to the method by Fotieva & Sammal (1999) taking the presence of the rock layer strengthened by anchors into account.

After this, the method allows the determination of ultimate values of displacement or convergence at any points of the lining internal outline and the location of points where measurements must be carried out particularly carefully.

3 EXAMPLE OF DESIGN

As an example of design, a shotcrete lining of non-circular shape in combination with anchors installed in an opening of 7.9 m height and 10.2 m width is considered for the following input data: E_0 = 8000 MPa, v_0 = 0.25, λ = 0.33, the lining thickness is 0.1

m, the elastic modulus and Poisson's ratio of sprayed concrete are $E_2 = 30000$ MPa, $v_2 = 0.20$, each anchor has a length of $l_a = 3$ m, and their bearing capacity is $P_a = 100$ kN, the compactness of anchor installation is $n_s = 1$ piece/m² (therefore $E_2 = 25200$ MPa), allowable stresses $R_c = 26.5$ MPa, $R_t = 2.45$ MPa.

The distribution of circumferential stresses $\widetilde{\sigma}_\theta^{in}$ in the lining cross-section internal outline is shown in the left part of Figure 3. For comparison, the same stresses obtained for $\lambda = 1$ are shown on the right side of the same figure.

$\widetilde{\sigma}_\theta^{in}$

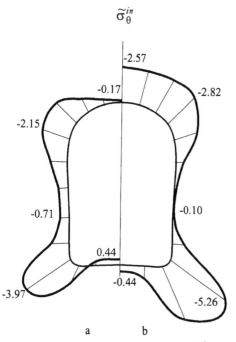

Figure 3. Distribution of circumferential stresses $\widetilde{\sigma}_\theta^{in}$: a) for $\lambda = 0.33$, b) for $\lambda = 1$.

$v_{\rho\,max}^{in}$, mm

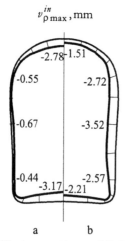

Figure 4. The ultimate admissible normal displacements: a) for $\lambda = 0.33$ and, b) for $\lambda = 1$.

$v_{x\,max}^{in}$ mm

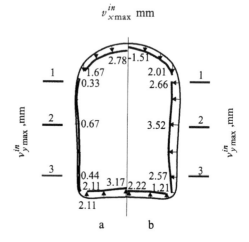

Figure 5. The ultimate vertical and horizontal displacements of points in the lining arch and walls: a) for $\lambda = 0.33$, b) for $\lambda = 1$.

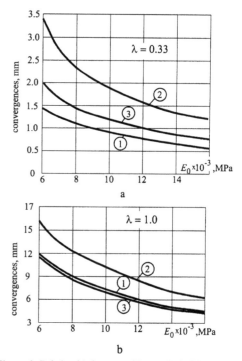

Figure 6. Relationship between ultimate admissible convergence and the elastic modulus E_0: a) at $\lambda = 0.33$, b) at $\lambda = 1$.

On the basis of the results obtained we have from formula (6), $\gamma_{max}\alpha^* = 5.568$ MPa for $\lambda = 0.33$ and $\gamma_{max}\alpha^* = 5.038$ MPa for $\lambda = 1$. Then the ultimate $v^{in}_{\rho\,max}$ normal displacements will be the same as those shown in Figure 4. The ultimate vertical $v^{in}_{x\,max}$ displacements in the arch of the lining and the horizontal $v^{in}_{y\,max}$ displacements in the lining walls are given in Figure 5.

The ultimate horizontal convergences $\Delta v^{in}_{y\,max}$ of points in the lining walls at levels 1, 2 and 3 (shown in Figure 5) are 0.66 mm, 1.34 mm, and 0.88 mm for $\lambda = 0.33$, and 5.32 mm, 7.04 mm, and 5.14 mm for $\lambda = 1$.

The relationship between these convergences and the elastic modulus E_0 of the rock for different lateral pressure coefficients λ in intact rock are given in Figure 6 (a and b) (the curves 1, 2, and 3 correspond to the convergences of the lining at levels 1, 2 and 3).

4 CONCLUSIONS

The method proposed allows the ultimate displacements of the internal surface of a shotcrete tunnel lining to be determined at which load bearing capacity exceeds admissible limits based on an elastic model of material behaviour. This method is applicable to linings of arbitrary cross-sectional shape, and may be applied in conjunction with monitoring systems for underground structures.

ACKNOWLEDGEMENTS

The paper describes the part of research works supported by Russian Ministry of Education.

REFERENCES

Bulychev, N.S. & Fotieva, N.N 1991. Basic Problems of Underground Structures Mechanics. *Podzemnoye i shakhtnoye stroitelstvo*, 1: 19-24.

Fotieva,N.N. & Sammal, A.S. 1999. Inverse problems for interpretation of full-scale measurements data in tunnels. *Proc. Of Int. Congress on Rock Mechanics*. Paris, France: 1363-1364. Rotterdam: Balkema.

Melentiev, P.V.1962. *Approximate calculations*. Moscow: Fizmatgiz.

Muskhelishvili, N.I. 1966. *Some basic problems of mathematical theory of elasticity*. Moscow: Nauka.

Timofeev, O.V. & Trushko, V.L. 1984. Effect of Monolithic Structure Strengthening by Anchors before and after Limiting Stage. *Interaction of support and rock mass in complicated conditions*. 101-106. Leningrad.

Shotcrete: Engineering Developments, Bernard (ed.) © 2001 Swets & Zeitlinger, Lisse, ISBN 90 5809 176 7

A specialised shotcrete system for refractory and mining applications

F.T.Gay & AJ.S.Spearing
UGC Americas, MB. Inc, Cleveland, Ohio, USA

R.First
ChemRex Inc, Cleveland, Ohio, USA

ABSTRACT: A unique chemical system has been developed for use in shotcreting in which high viscosity cementitious materials may be more easily pumped and sprayed. This system involves the incorporation of an organic foaming agent in the mixture to facilitate easier pumping and a viscosifying additive introduced at the nozzle that imparts thixotropy to the mix. The use of large volumes of compressed air at the nozzle and proper nozzle selection ensure that the foam incorporated within the mix is dissipated when the mix impacts and adheres to the target surface. Although originally developed for refractory applications, the system is applicable to mixes that have previously been considered unpumpable either due to dilatancy of the mixture or to the inherent qualities of low water content mixes. Testing to date with refractory concrete demonstrates that the system imparts no deleterious effects to the final properties of these materials.

1 INTRODUCTION

Over the past 10 to 15 years, many refractory suppliers have developed castable concrete formulations to compete with brick and pre-shaped products (The Refractories Institute 1987). Such cast-in-place formulations are economically advantageous because of reduced installation time, and labour costs; however, lower in-place densities of such castable formulations have long been a concern.

More recently, wet shotcrete techniques have been investigated as a viable alternative installation technique that allows greater installation cost reductions over castables. Low in-place densities have however been of even greater concern with the wet shotcrete method. Moreover, highly alkaline accelerating additives are needed at the nozzle to ensure that the refractory concrete will adhere to the target surface. Such alkaline chemistries function as fluxes at operating temperatures, leading to decreased service life.

Both castables and wet shotcrete mixes pose several problems with respect to the ease of pumping. Calcium aluminate binder systems are inherently dilatant (i.e. shear-thickening) and therefore may require higher pumping pressures for transportation. Moreover, it is always preferable to reduce the water content of a refractory castable because the water must eventually be burned out of the system when fired. Despite recent developments in powdered

superplasticizer chemistry that could permit very low water contents in castable refractories, minimum water contents must be maintained in the mixes to retain pumpable consistencies. Specialty refractory products have been specifically produced for shotcreting with optimised gradations and binder ratios; but, because these products are very well engineered, they are typically very expensive. As a result, many of the castable refractory products are not modified for use in refractory shotcrete applications but are used as produced. Consequently, many of these products may not be optimized for pumping and placement as shotcrete.

The system described in this article consists of an organic foaming agent that intentionally produces foam in the mass of material to be pumped and sprayed. At the nozzle, the foamed stream of material is broken up by a high volume and velocity of compressed air that destroys the foam and facilitates the spraying of the material. Rather than adding a traditional, highly alkaline accelerator at the nozzle, another organic additive imparts thixotropy to the mixture, and prevents sagging of the material when applied to vertical and overhead surfaces. By employing large volumes of compressed air at the nozzle to ensure a high velocity stream, the resulting material will have in-place properties comparable to those of their castable counterparts.

The primary application of this system was intended to be castable formulations for wet shotcreting where conventional refractory castings or brick are used such as linings of steel ladles, tundishes and refining furnaces. The system would offer application cost efficiencies as well. In some segments of industry, it is possible that more conventional Portland cement-based shotcrete could benefit by using this system. The system would facilitate placing a material either at reduced water contents, or permit improved pumpability of poorly graded mixtures by the use of the internal foaming agent and the consistency control mechanism. Neither of these chemicals has any significant effect on the hydration properties of the cementitious components.

2 RHEOMETRIC PROPERTIES

The rheometric properties of a shotcrete mix may increase production expenses caused by requirements for pumping and placing the material. A material that typically requires a high pumping pressure must be transported using equipment that may be more expensive because of the capacity required. In contrast, the use of an additive in the mix that results in reduced pumping pressure with the same water content may require less expensive equipment because less work is required to transport the material. Another option is the reduction in the water content of the mixes that may improve the in-place performance of the mixture by virtue of the reduced water content.

The rheometric properties of a shotcrete mixture can be described by the use of an instrumented mixer (Beaupré 1994). The mixer measures the torque on a mixer blade as the speed of the mixer blade changes according to programmed speeds. Three basic types of rheometric behavior are shown schematically in Figure 1. Curve 1 shows a linear (i.e. Newtonian) relationship between torque and speed. The slope of the curve represents the viscosity of the material. Curve 2 shows shear-thinning behavior; as the speed of the blade increases, the slope of the curve decreases implying a reduction in the rate of viscosity change with increasing shear. Curve 3 shows shear-thickening (i.e. dilatant) behavior; as the speed of the blade increases, the slope of the curve increases, implying an increasing rate of viscosity change with increasing shear.

Refractory cement pastes and refractory castable mixes are notoriously shear-thickening, while Portland cement-based shotcrete mixes are typically shear-thinning by nature. In wet process shotcreting, there may be pumping issues that cannot be easily or economically avoided with regard to the gradation and transportation of the mixtures through the pipes

and hoses of a shotcrete unit. The addition of other components may produce more viscous materials. If, for example, sufficient amounts of ground granulated blast furnace slag are added to a portland cement based mixture, the mass may become shear-thickening or dilatant. The use of the proposed chemical system would promote easier pumping for any of the situations described above.

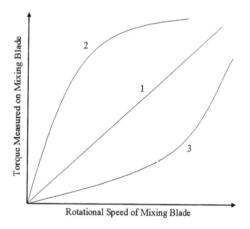

Figure 1. Three basic rheometric modes.

3 LABORATORY AND FIELD TESTING

The material used for these investigations was a 90% spinel castable refractory. Spinel is an aggregate mineral with appropriate physical properties at refractory temperatures. The binder was high alumina cement.

3.1 Laboratory Testing

Laboratory mixes were produced and tested to assess the pumping pressure of the same mixture both with and without the foaming agent. Each mix was composed of a 25 kg (55 lb) batch with a water content of 8.7% by mass. The results of a laboratory test using the instrumented mixer comparing the rheological behaviour of un-foamed versus foamed refractory mixes are shown in Figure 2. The specific gravity of the un-foamed mix was 2.96 and that of the foamed mass was 2.11.

The greater slope of the line of the un-foamed mix indicates a higher viscosity without the foaming agent. The reduced slope of the foamed mix indicates that relative viscosity is reduced. When pumping a dilatant material, the greater the shear and pumping effort, the more viscous the material becomes. For the present mixes, the shear rates and elevated water content of the mix resulted in a less obvious dilatant character. Other mixes of lower water content show more pronounced shear-thickening than was indicated by these mixes. The slope of the linear portion of the un-foamed torque-speed curve (i.e. speed greater than 0.2 revolutions/second) was 3.34 Nms versus the slope of the foamed mix, which was 1.34 Nms. While viscosity is typically reported in units of Centipoise, the instrumented mixer measured viscosity in Newton metre seconds (Nms). Because the ratio of the two slopes represents the ratio of the viscosities and is unitless, the units used to report viscosity are immaterial. As such, the viscosity of the foamed mix is approximately 40% of the viscosity of the un-foamed mix.

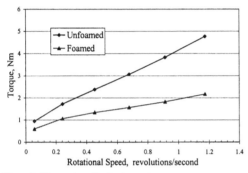

Figure 2. Rheometry of refractory mix.

From these data, it is expected that the pumping pressure of the foamed mix should be approximately 40% of the unfoamed mix.

3.2 Field Testing

To demonstrate that the effect is transferable to real applications, a similar test was performed in the field to verify the results of the laboratory study. Two mixes were made; each mix was composed of two 155 kilogram batches with a water content of 8.7% by mass. The mixes were mixed using an Allentown AP 10 shotcrete mixer and pump.

After the mix was wetted out, the mixture was pumped through 7.6 meters (25 feet) of pre-slicked 38 mm (1.5 in) diameter rubber hose while the pumping pressure was monitored. The pumping pressure of the un-foamed mix was initially 13.7 MPa (2000 psi) and the specific gravity of the material was determined to be 2.98. As pumping continued, the pressure stabilized to a lower pressure of approximately 11.7 MPa (1700 psi).

A separate mix of identical composition was used to evaluate the effect of the foaming agent. For this mix, the dry, powdered foaming agent was added

after the mix was wetted out. Mixing continued for approximately three minutes to produce the foamed mass. The material was then pumped and recirculated while the pumping pressure was monitored. The initial pumping pressure of the foamed mix was 6.2 to 6.6 MPa (900 to 950 psi) and the specific gravity of the mix was 2.35. With continued pumping, the pressure stabilised at 5.5 MPa (800 psi). In these field experiments, the pumping pressure was reduced to 47%. This is consistent with the laboratory rheometry experiments that indicated the viscosity was reduced to 40% with the addition of foam to the mass. The slight difference between the laboratory and field results can be explained by considering the densities of the two foamed mixes, which were 2110 kg/m^3 and 2340 kg/m^3, respectively. As a result, the amount of foam generated (i.e. air content) in the field experiment was slightly less than in the laboratory.

4 FOAMING ADDITIVE

The described system incorporates a dry powder foaming additive in the product that facilitates the production of a foam structure when the material is mixed with an appropriate amount of water. This "in situ" foaming feature is seen as less complicated and more convenient than separate pre-generation of foam. Subsequent addition of pre-generated foam adds a sequence that may lead to variability of the foaming and mix density that could complicate production, transport, and quality control of the final product.

The incorporated foam reduces the density of the mix and facilitates the break-up of the pumped material when it enters the high-energy air circulating in the nozzle. The dynamic volume change of the air voids within the mass of the transported material tends to reduce the impact of pump pulsation, and produce a more smoothly flowing material during transport through the hose. The dynamic response to changing pressure regimes also tends to prevent segregation within the transport line at vulnerable areas such as reducers. In such places, gap-graded materials tend to produce sand bridging in the more coarse components of the sand. Also, the binder and finer aggregate particles in the concrete can be pressed through a gap-graded aggregate to aggravate bridging and clogging. It is believed that the voids of the foam tend to fill in the gradation gaps of the aggregate and thus correct the potential problems of poor gradation. In poorly graded refractory materials, such blockages can result in downtime, as they usually require thorough cleaning of the pump and hoses before pumping and spraying can be resumed.

Use of the foaming agent then can also serve to alleviate such blockage problems.

5 PUMPING PRESSURE REDUCTION

Other studies of the effect of foaming agents on granular materials have shown that two mechanisms operate in the transport of foamed masses through pipes. The most obvious mechanism is density and viscosity reduction by the incorporation of fine voids within the mass of the material. The incorporation of voids reduces the density and cohesive properties of the pumped material. Another mechanism is the reduction in friction within a thin layer of foam at the material/pipe wall interface. This annular foam layer facilitates the transportation of the mass by lubrication of the material whether or not the mass itself is foamed. The reduced viscosity, caused by foaming of the mass and/or the development of the annular foam layer, contribute to the general reduction of the pumping pressure of the material through the transport system.

6 THE NOZZLE

In any shotcreting activity, the nozzle is a key element in the successful application of the material. For these tests, the nozzle used was a prototype of a proprietary nozzle that was specifically designed to handle larger than normal volumes of air for the placement of refractory materials. In the nozzle, a high volume of compressed air breaks up the foamed extrusion when it arrives at the nozzle and then promotes mixing of the material containing the consistency control nozzle additive. This promotes uniform distribution of the consistency control additive into the disintegrated, extruded material and is considered essential to the success of the system.

7 CONSISTENCY CONTROL ADDITIVE

Beyond the initial reduction of pumping pressure required to transport the material to the nozzle, there is a further need to promote adhesion of the sprayed material to the substrate. In conventional shotcrete practice, this may be accomplished with water-reducing admixtures that promote cohesion of a mix or by adding a traditional accelerator at the nozzle that promotes rapid hardening of the cementitious material and adhesion to the substrate.

The addition of a water-reducing admixture to a refractory mix usually aggravates dilatancy if the water content of the mix is substantially reduced. Alternatively, excessive use of accelerators may alter the final properties of the sprayed material, either

through reduced ultimate strengths in conventional Portland cement-based materials or compromised thermal properties in the case of refractory materials. In both cementitious systems, the addition of alkalis through traditional accelerating admixtures such as sodium silicate can result in detrimental consequences.

In the chemical system described in this investigation, an organic additive at the nozzle increases the viscosity and cohesion of the mix and prevents sagging upon impact with the substrate. This is accomplished by a reaction of the nozzle additive with the water within the mixture. The resulting change in viscosity re-establishes the cohesion of the mix and facilitates the adhesion of the sprayed material to the surface. This is not a cementitious effect but a real change in the consistency of the mix on contact with the surface.

The field evaluation compared the properties of a cast sample of refractory material to the properties of the refractory shotcrete with the aid of the chemical system described in this paper. The pumping pressure of the mix without the foaming agent was 27.6 MPa (4000 psi) whereas the same mixture with the foam additive was pumped at 19.3 MPa (2800 psi). The physical properties of the two refractory mixes are shown in Table 1. The term $\Delta L/L$ in this table represents the change in length of the material after firing and HMOR represents the "hot modulus of rupture."

Table 1. Physical Properties

State	Shot Properties		Cast Properties	
	Porosity (%)	Strength (MPa)	Porosity (%)	Strength (MPa)
110°C	N/A	8.41	N/A	7.62
1000°C	11.6	7.22	11.6	6.23
1500°C	15.3	34.1	15.1	33.7
	$\Delta L/L$ (%)	HMOR (MPa)	$\Delta L/L$ (%)	HMOR (MPa)
1000°C	-0.04	N/A	-0.09	N/A
1500°C	-0.03	20.8	-0.15	19.2*

*This value for a mix with 4.8% water, all other data is for 5.0% water

8 DISCUSSION

The chemical system described in this paper was developed in response to limitations presented by refractory materials and their current application methods. The consistency control system is routinely used in the application of steel ladle linings and tun-dishes for steel industry. Currently, in the U.S. Steel Gary Works in Gary, Indiana, the system is used with robotic shotcreting equipment for relining steel ladles. It is likely that the use of this system in the smelting and metals processing industry will surpass its potential use in the steel industry.

Other sectors of the construction industry may also find the system useful in more conventional situations. In materials where the aggregate gradation may not be optimized or where the binder may contain components such as ground granulated blast furnace slag that are not ideal for pumping, the system provides a means of more practical application.

9 CONCLUSIONS

A chemical system has been developed for shotcrete applications to facilitate the pumping and placement of highly viscous materials, while keeping the water content of the material to a minimum. Furthermore, the proposed system facilitates pumping of dilatant materials, especially materials containing poorly graded or non-optimally engineered aggregates.

The system also facilitates adhesion of sprayed materials to substrates without excessive use of traditional accelerating admixtures that can negatively affect the long-term physical properties of the in-place material.

ACKNOWLEDGEMENTS

Jeff Champa, Dontave Cowsette, Brad Hulvey, Mike Urbas, and Dwight Wilson participated in the data collection for this article. Bill Allen of Allentown Pump and Gun acquired the refractory material used in the tests for the article.

REFERENCES

Beaupré, D. 1994. Rheology of High Performance Shotcrete. PhD. Thesis, University of British Columbia
Refractories Institute, The, 1987, *Refractories*. Pittsburgh, Pennsylvania, USA.

Shotcrete: Engineering Developments, Bernard (ed.) © 2001 Swets & Zeitlinger, Lisse, ISBN 90 5809 176 7

An evaluation of repair mortars installed by worm-pump spraying

C.I.Goodier
Building Research Establishment, Watford, Herts., UK

S.A.Austin & P.J.Robins
Loughborough University, Loughborough, Leics., UK

ABSTRACT: This paper examines the fresh and hardened performance of wet-process sprayed mortars and the influence of rheology on the pumping and spraying of these mortars. Seven commercially available pre-blended repair mortars designed for hand application, together with a laboratory-designed fine mortar, were investigated using the Tattersall two-point rotational viscometer, the pressure bleed test, the slump test, and a vane shear strength test. The mortars were pumped and sprayed with a small diameter worm pump and the build thickness determined. Hardened properties measured include compressive strength, tensile bond strength, hardened density and drying shrinkage. Tests were conducted on cast and in-situ specimens and, where possible, on specimens produced by spraying directly into a cube or beam mould. Initial findings for predicting the pumpability and sprayability of the mortars are presented and this is linked together with the hardened performance. These results show that the majority of proprietary pre-blended repair mortars designed for hand application are suitable for wet-process application with a worm pump.

1 INTRODUCTION

Sprayed mortar can be defined as a mortar conveyed through a hose and pneumatically projected at high velocity from a nozzle into place. The rheological properties of the mix in the wet process are obviously critical. These properties have been examined by the authors in previous papers (Austin et al. 1999a and 1999b). The mortar's hardened properties are of equal importance, so that a durable and long lasting repair can be obtained, and these properties have also been investigated by the authors (Austin et al. 2000a). The findings of this work have also contributed to a Technical Report be published by the authors in conjunction with the Concrete Society (Austin et al. 2000b).

This paper describes some of the findings of a three year UK Government and industry funded research programme into wet-process sprayed concrete for repair, and more specifically the fresh and hardened properties of a range of mortar mixes, which are defined as mixtures of cement, aggregate with a maximum particle size of 3 mm, water and any admixtures. The mortars tested include seven commercially available pre-blended concrete repair mortars and a generic mix design consisting of crushed Portland stone, Portland cement (PC), silica fume and a styrene butadiene rubber liquid additive (SBR).

2 WET PROCESS SPRAYED MORTARS

Wet process sprayed application offers a number of advantages over cast and hand-applied repairs, including the reduction or elimination of formwork, the construction of free form profiles and faster and more efficient construction (Austin 1997). It can also provide enhanced hardened properties if properly placed.

Previous work published by the authors have discussed both the materials, installation and physical properties of sprayed concrete (Austin 1995a) and the associated application methods and quality considerations (Austin 1995b).

Browne & Bamforth (1977) showed that it is possible to change from a saturated to an unsaturated state by excessive loss of mix water due to pressure, thus increasing frictional stress, and even blockage. A concrete that de-waters quickly under pressure will be prone to blocking in a pipeline. Beaupré (1994) investigated the rheological properties of sprayed concrete and the relationship between pumpability and sprayability, including the development of predictive models based on yield and flow resistance determined from tests conducted with a rotational viscometer.

Tattersall (1976) developed a rheometer, termed the two-point test apparatus and found that when the torque (T) was plotted against the speed (N) for decreasing results only, the relationship was almost linear:

$$T = g + hN \qquad (1)$$

where g is the intercept on the torque axis and h the slope of the line. Beaupré (1994) referred to g as the flow resistance, and h as the torque viscosity. This equation is of the same form as the Bingham model ($\tau = \tau_0 + \mu.\gamma$) and thus it can be said that g is a measure of yield value, and h of plastic viscosity.

Hills (1982) conducted tests on both wet- and dry-process sprayed concrete, and compared results with those from cast concrete. He concluded that the performance of the sprayed concretes did not appear significantly different from those of properly compacted cast mixes of similar composition and he argued that it was the modified mix design needed for sprayed concretes that altered the hardened properties, not the method of placement. However, Banthia et al. (1994) have argued that cast and sprayed concrete are of a different nature, with the spraying process affecting the internal arrangement of constituents and hence the strength and durability.

Work conducted by Gordon (1991) on wet-sprayed pre-blended repair mortars concluded that the wet process achieves greater compaction than hand application and that the materials tested achieved compressive strengths approximately 30% higher when wet sprayed than when hand applied. Increases in fresh wet density, bond strength and build were also recorded.

3 EXPERIMENTAL PROGRAMME

A number of pre-blended proprietary mortars designed for hand application were investigated (Table 1). In order to obtain the aggregate/cementitious ratio the mortars were sieved and the particles collected on each sieve were examined under a x40 magnification microscope. The approximate proportion of aggregate, lightweight filler (approx. 75-300μm in diameter) or cementitious material on each sieve (to the nearest 10 %) was determined and the weight of each calculated accordingly.

The laboratory-designed mortar D1w was a 3:1 crushed Portland stone : Portland cement mix with 5% silica fume (by weight of cement) and an SBR in a 1:3 solution with water. All the mortars were mixed to a consistency which would be typical for low-volume sprayed application.

The mortars were mixed using a 0.043 m^3 capacity forced-action paddle mixer according to the manufacturers instructions, with 3.3 to 4.0 litres of water per 25 kg bag of dry material for approximately 4 minutes. The mortars were pumped with a Putzmeister TS3/EVR variable-speed worm (i.e. screw-type) pump with a 25 mm diameter rubber hose, an air pressure of 300 kPa and an output of approximately 6 L/min. The mortars were sprayed into 500×500×100 mm deep panels whilst trying to minimize both voids and rebound. Mix P2 was also sprayed with a large diameter worm pump (termed P2W) and mix P1 with a piston pump (termed P1p).

3.1 Fresh property testing

The workability was measured by the slump test (BS1881: Part102: 1983) and by a modified form of the shear vane test for soils (BS 1377: Part 9: 1990). The shear vane consisted of a torque measuring device at the head of the instrument together with a set of vanes to provide sufficient shear resistance to register on the torque scale. The shear strength for the mortar (in kPa) was then calculated from the maximum torque. The pressure bleed apparatus (Browne & Bamforth 1977) consisted of a 125 mm diameter steel cylinder lined with a 75 μm mesh on the inside of the base and a bleed hole with a stop tap located beneath the mesh. The apparatus was filled with approximately 1700 cm^3 of mortar and subjected to a load of 12.2 kN, equivalent to 10 bar (1000 kPa), which was the highest pumping pressure recorded with the TS3EVR worm pump. The valve was opened after 10 seconds and the liquid emitted collected on a digital balance and the change in weight was data-logged for 30 minutes. Sprayability was assessed both qualitatively (did the material pass through the nozzle) and quantitatively (in terms of the amount of material that could be built up on a vertical grit-blasted concrete substrate).

Table 1. Composition of mortars (P:Pre-blended, D:Designed mix, w/W: small/large diameter worm pump, p: piston pump)

Mix	Water/ cementitious Ratio	Aggregate/ cementitious Ratio	Polymer Modified	Polypropylene Fibres	Shrinkage Compensators	Light-weight Filler	Description
P1w	0.59	2.3	N	N	Some	N	Basic repair mortar
P2w	0.41	1.45	Y	Y	Y	Y	High build repair mortar
P3w	-	1.58	Y	Y	Y	Y	2-part re-profiling mortar
P4w	0.47	2.31	Y	Y	N	Y	Basic repair mortar
P5w	0.39	1.33	Y	Y	N	Y	Repair mortar
P6w	0.45	1.62	Y	Y	Y	N	Repair mortar
P7w	0.90	3.42	Y	Y	Y	Y	Lightweight mortar
P1p	-	2.3	N	N	Some	N	Piston pumped
P2W	-	1.45	Y	Y	Y	Y	Large diameter worm pump
D1p	-	3:1	Y	N	N	N	Designed (piston pumped)

The mortar was sprayed horizontally onto a 300×300 mm target area in order to obtain as large an amount of material as possible on the substrate whilst keeping within the 'target'. The mortar would then fail under its own weight either cohesively, adhesively or a combination of both. The sprayability was also measured in terms of the reinforcement encasement.

This test consisted of a 500×500×100 mm deep panel fitted with steel reinforcement of 8 and 12 mm diameters at 100 mm centres with some of the bars placed in pairs to produce an effective maximum thickness of 24 mm. The panel was sprayed to obtain as complete encasement as possible. At 28 days the intersections of the bars were cored and a 5 mm disc was cut from the bottom (i.e. moulded face) of the 55 mm diameter core and discarded. A 20 mm thick disc was sawn from the same end and a sorptivity test conducted on both the disc and the remainder of the core The sorptivity was then related to the density of the reinforcement at the bar intersection.

3.2 Hardened property testing

All material within 50 mm of the panel edge was discarded to avoid the effects of rebound entrapment around the edges of the moulds. All samples were sawn approximately 24 hours after spraying and cured under water at 20±2°C. The specimens sprayed and cast into steel moulds were struck and cured in the same manner.

Cores were capped with a sulphur compound, whilst sawn cubes were capped with a 2-3 mm layer of high-strength plaster between steel plates due to the imperfect orientation and texture of the sawn sides. Compressive cube tests were carried out at 28 days in accordance with BS1881: Part116: 1983 and the compressive core tests were conducted at 28 days in accordance with BS1881: Part120: 1983.

The tensile bond strength was measured by a core pull-off test (using the Limpet apparatus (McLeish 1993)) at 7 and 28 days. Five 55 mm diameter partial cores were cut through the repair material and into the concrete substrate (which had been grit-blasted) to a depth of approximately 10 mm and a 50 mm diameter steel dolly was then glued to the top of the core and an axial tensile load applied at a rate of 2 kN/min to failure.

The saturated hardened densities of the cubes were calculated by weighing in air and determining their volume from measured dimensions.

Prisms to monitor drying shrinkage vary according to different standards, although even the largest are too small to spray directly into a mould. 75×75×229 mm specimens were therefore cast to BS1881: Part5: 1970 and also sawn from the sprayed panels. Pairs of demec pips were glued to

three of the longitudinal faces on a 200 mm gauge length and the specimens were stored in a climatic cabinet at 20°C and 50% RH.

4 TEST RESULTS

4.1 Fresh properties

The shear vane provides a basic measure of the shear strength (in kPa) of a mortar and this is plotted against slump (in mm) in Figure 1. As would be expected, this shows a decrease in shear strength for an increase in slump. The shear vane test can provide an instantaneous result exactly where the rheological properties of the mortar needs to be measured, e.g. in the hopper of the pump.

Figure 2 shows the flow resistance (g) and plastic viscosity (h) for the mortar P2 after it has been mixed, pumped and sprayed.

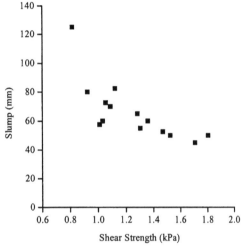

Figure 1. Shear vane strength compared with slump.

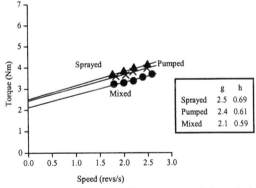

Figure 2. Two-point test: effect of mix P2 being mixed, pumped and sprayed.

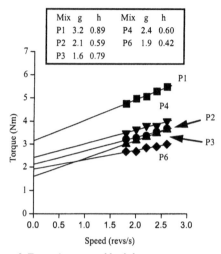

Figure 3. Two-point test: pre-blended mortars.

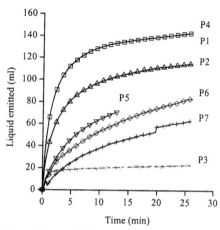

Figure 4. Pressure bleed test.

The increase in the values of both g and h as the mortar is pumped and then sprayed would be expected as the excess air is forced out of the mortar. The two-point test results for all the mortars in this study are shown in Figure 3.

The pre-blended mortar with both the highest g and highest h values was mix P1 which had the most 'basic' mix design of all the pre-blended mortars tested, and contained no polymers, fibres or lightweight fillers. The mix with the next highest value of g, P4, also had a relatively basic mix design and together with P1 were the cheapest commercially of all the pre-blended mortars that were tested. The two highly polymer-modified mixes (P6 and P3) had the lowest values of g, although their corresponding values of h were very different. The mix P3 is a two-part (powder and liquid) re-profiling mortar which had been formulated to enable it to be applied in thin layers without it separating or being too 'sticky', which may explain why it had the smallest value of g.

Figure 4 shows that the total liquid emitted from the pre-blended repair mortars in the first 30 minutes ranged from 20 to 140 mL. This liquid was a combination of water, SBR, Portland cement, silica fume and very fine (<75 µm) sand particles. The relatively basic mortars (P1 and P4) that contain little or no polymers emitted the largest total amount of liquid at the fastest rate and the highly polymer-modified mixes (P6 and P7) emitted a smaller total amount of liquid at a slower rate. The two-part re-profiling mortar (P3) emitted a small amount of liquid (20 mL) very quickly in the first 2.5 minutes but then the rate of bleeding decreased rapidly. The resistance of a mix to bleeding is dependant upon the mix composition, especially the grading of the constituents and the mixes examined here with the lowest proportion

of fine material emitted the most liquid and vice versa.

The build compared with the slump before pumping is shown in Figure 5. This indicates an increase in build as the slump increases from zero until a slump of approximately 60 mm is reached, at which point the build begins to decrease. The relationship is not distinct due to the variability in the mix designs of the mortars. Beaupré (1994) found no correlation between slump before pumping and build, although he did conclude that a correlation could exist between concretes of the same mix design. As the slump decreases the mortars change from a cohesive failure (within the fresh material) to an adhesive failure (at the substrate), these failure modes being indicated on the Figure. This change in failure mode could possibly account for the build of

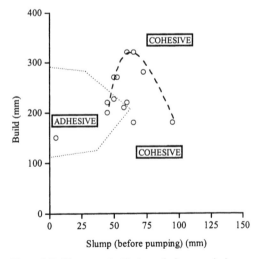

Figure 5. Build compared with slump (before pumping).

the mortars increasing as the slump decreases until a slump of approximately 60 mm when the build appears to fall for a decrease in slump.

Figure 6 suggests that the build at first increases with increasing vane shear strength, and then begins to decrease at a vane shear strength of approximately 1.4 kN/m^2, as would be expected due to the relationship between the slump and the vane shear strength (Fig. 1).

The influence of the density of reinforcement on the sorptivity (of the top of the core, i.e. the material just behind the bars) is shown in Figure 7. Note that the sorptivity of all the mortars except P2W increases between bar overlaps of zero and 96 mm^2, and then levels off and only increases slightly as the bar overlap area increases. Trend lines are shown on the first two points of each mortar for clarity. The bar overlap area was taken as the cross-sectional

area on plan of the intersection of the bars (e.g. 8×12 mm = 96 mm^2). In general, the sorptivity of the pre-blended mortars does not increase greatly as the density of reinforcement increases. A reinforcement encasement test was conducted on mortar P4w using bars with overlap areas up to 576 mm^2 (equivalent to 2×12 mm bars overlapping 2×12 mm bars). This showed a much steeper increase in sorptivity with bar area due to the increased voidage caused by the additional reinforcement (Fig. 8).

4.2 Hardened properties

Figure 9 shows the equivalent cube strengths of the worm-pumped mortars obtained from in-situ cores, cubes cut from panels, and the cast and sprayed cubes. The mortars with the lowest strengths of 26-34 MPa were, as expected, obtained with the render/profiling and lightweight repair mortars (P3w and P7w). The simple laboratory designed mix D1w produced the highest strengths compared to the more sophisticated (and therefore expensive) pre-blended mortars.

The in-situ cube strengths are generally higher than the corresponding cast cubes, due mainly to the greater compaction obtained with the spraying proc-

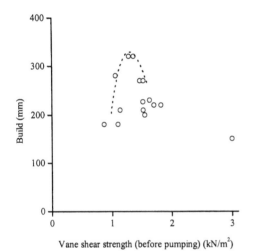

Figure 6. Build compared with vane shear strength (before pumping).

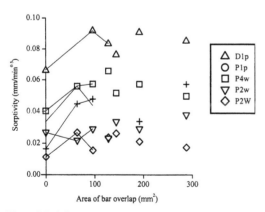

Figure 7. Reinforcement encasement: sorptivity compared with area of bar overlap.

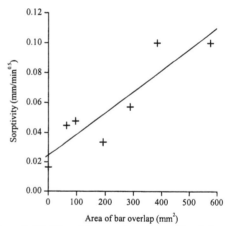

Figure 8. Mix P4w reinforcement encasement: Sorptivity compared with area of bar overlap.

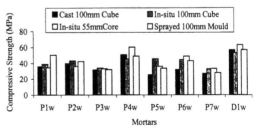

Figure 9. Compressive strengths of mortars.

ess. This greater compaction can be seen in the greater densities of the in-situ cubes compared with the corresponding cast cubes (Figure 10). It is generally agreed that in-situ sprayed concretes produce higher strengths than for similarly cast mixes although the opposite has also been observed (Banthia et al., 1994). P5w, P6w and P7w have low cast cube strengths (and correspondingly low cast densities) as these specimens contained a large number of air voids, even after considerable vibration. There is a good correlation between the in-situ cube strengths and the cubes sprayed in moulds, despite the difficulty in obtaining a sample with no voids and low rebound (samples with excessive voidage being discarded).

The vertical and overhead bond strengths of the small worm pump mortars are shown in Figure 11. All the pre-blended mortars achieved at least 1.7 MPa at 28 days (with the exception of the lightweight mortar P7) which comfortably exceeds the Concrete Society minimum bond strength of 0.8 MPa. The mortars in this study (except P7) possess a relatively narrow range of vertical bond strengths (1.7-2.25 MPa), despite having a broad range of in-situ compressive strengths (32-57 MPa).

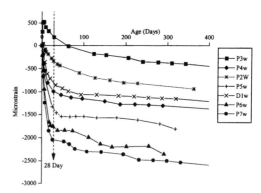

Figure 12. Drying shrinkage of prisms taken from in situ material.

The drying shrinkage results for the 75×75×229 mm in-situ prisms are shown in Figure 12. A wide range of results were obtained, despite all the pre-blended mixes being described as low shrinkage. P2W and P3w expanded initially due to the presence of shrinkage compensators. However, the inclusion of these admixtures appeared to have little affect on the other mortars containing shrinkage compensators (P6w and P7w). The mortar which shrank the greatest at the fastest rate was the lightweight mortar P7, which would be expected due to the very high water/cementitious ratio.

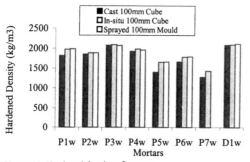

Figure 10. Hardened density of mortars.

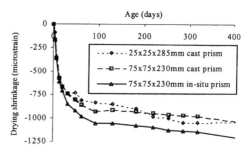

Figure 13. Drying shrinkage: mix D1.

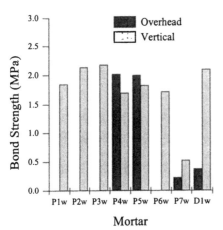

Mortar

Figure 11. Bond Strength Overhead and Vertical at 28 days

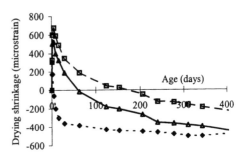

Figure 14. Drying shrinkage: mix P3w.

The rates of drying shrinkage for three types of prism (from an average of two samples) are shown for mix D1 in Figure 13 and for mix P3w in Figure 14. The larger (75×75×229 mm) in-situ prisms had the highest total drying shrinkage for the designed-mix D1w (approximately 1200 microstrain at 1 year), possibly due to the greater compaction and slightly lower aggregate content (due to differential rebound) compared with the cast samples. However, the shrinkage-compensated pre-blended mix P3w expanded before it began to shrink. This demonstrates that care should always be taken when quoting drying shrinkage values, especially when shrinkage-compensators are present, as the dimensional change of the sample can be very different depending on the age and size of the sample.

5 CONCLUSIONS

This paper has presented and discussed a variety of data on the rheological and hardened performance of wet-process sprayed mortars and shows that proprietary pre-blended repair mortars designed for hand application are suitable for wet-process application with a small-diameter worm pump.

A shear vane test has been developed and a good correlation with the slump has been found. The two-point apparatus was successful with low-workability mortars and their flow resistance and torque viscosities were determined. The pressure bleed test demonstrated that the presence of an SBR significantly influences both the rate and total emission of liquid from the mix under pressure. The proportion of fine material and the water content of the mix were also crucial factors in the amount and rate of liquid emitted. A test was devised to quantify the degree of reinforcement encasement and in general the sorptivity (which was related to the amount of voidage behind the bars) did not increase greatly as the density of reinforcement increased.

The relatively simple laboratory-designed mortar possessed high compressive strengths compared with the proprietary pre-blended mortars. There was a good correlation between the in-situ and the sprayed mould compressive cube strengths, providing that no large voids or excessive rebound is present. Except for the lightweight mortar P7w, all the mortars possessed a relatively narrow range of bond strengths compared with their compressive strengths. The cast and the in-situ prisms exhibited very similar rates of drying shrinkage, suggesting that cast prisms could be used for quality control purposes to measure and monitor in-situ drying shrinkage.

ACKNOWLEDGEMENTS

The authors are grateful for: the financial support of the EPSRC (Grant number GR/K52829); the assistance of the industrial collaborators Balvac Whitley Moran, Fibre Technology, Fosroc International, Gunform International Ltd and Putzmeister UK Ltd; and the supply of additional materials by CMS Pozament, Flexcrete Ltd., and Ronacrete Ltd.

REFERENCES

Austin, S.A, 1997, Repair with sprayed concrete, *Concrete*, 31 (1): 18-26. (Invited).

Austin, S.A. 1995a, Production and Installation, *Sprayed Concrete: Properties, Design and Application* (Eds. Austin, S.A. and Robins, P.J.), Whittles Publishing, Latheronwheel, UK, pp.31-51.

Austin, S.A. 1995b, Materials selection, specification and quality control, *Sprayed Concrete: Properties, Design and Application* (Eds. Austin, S.A. and Robins, P.J.), Whittles Publishing, Latheronwheel, UK, pp.87-112.

Austin, S.A., Robins, P.J. & Goodier, C.I., 1999a, The Rheological Performance of Wet-Process Sprayed Mortars, *Magazine of Concrete Research*, 51 (5): 341-352.

Austin, S.A., Robins, P.J. & Goodier, C.I., 1999b, Workability, Shear Strength and Build of Wet-Process Sprayed Mortars, *Proc. Int. Congress on Creating with Concrete*, Dundee, September 1999.

Austin, S.A., Robins, P.J. & Goodier, C.I., 2000a, The Performance of Hardened Wet-Process Sprayed Mortars, *Magazine of Concrete Research*, 52 (3), June.

Austin, S.A., Robins, P.J. & Goodier, C.I., 2000b, *Construction and repair with wet-process sprayed concrete and mortar - good concrete guide*, (in press), Concrete Society, UK.

Banthia, N., Trottier, J.F. & Beaupre, D., 1994, Steel-Fiber-Reinforced Wet-Mix Shotcrete: Comparisons with Cast Concrete, *Journal of Materials in Civil Engineering*, 6 (3), August.

Beaupre, D., 1994 *Rheology of high performance shotcrete.* PhD Thesis, University of British Colombia, February.

British Standards Institution, 1983, *Method for determination of slump*, BS1881:Part102:1983, London.

British Standards Institution, 1990, *Specification for soils for civil engineering purposes, Part 9. In-situ tests*, BS 1377:Part 9:1990, London.

British Standards Institution, 1983, *Testing Concrete: Method for Determination of the Compressive Strength of Concrete Cubes*, BS1881: Part116: 1983, London.

British Standards Institution, 1983, *Testing Concrete: Method for Determination of the Compressive Strength of Concrete Cores*, BS1881: Part120: 1983, London.

British Standards Institution, 1970, *Determination of Changes in Length on Drying and Wetting (Initial Drying Shrinkage, Drying Shrinkage and Wetting Expansion)*, BS1881: Part5: 1970, London.

Browne, R.D. & Bamforth, P.B., 1977, Tests to establish concrete pumpability. *ACI Journal*, May: 193-203.

Ede, A.N., 1957, The resistance of concrete pumped through pipelines. *Magazine of Concrete Research*, November: 129-140.

Gordon, K., 1991, Wet-spraying of pre-packaged mortars for concrete repair, *Proc. of European seminar and workshop on sprayed concrete*, Loughborough University, Loughborough, UK, 3 October, pp.9.

Hills, D.L., 1982, Site-produced sprayed concrete, *Concrete,* 16 (12): 44-50.

Loadwick, F., 1970, Some factors affecting the flow of concrete through pipelines. *Proceedings of 1st Int. Conf. Hydraulic Transport of Solids in Pipes.* British Hydromechanics Research Association, Bedford, UK.

McLeish, A., 1993, *Standard tests for repair materials and coatings for concrete,* CIRIA Technical note 139, CIRIA, London.

Tattersall, G.H., 1976, Relationships between the British Standard tests for workability and the two-point test, *Magazine of Concrete Research,* 28 (96) September: 143-147.

Shotcrete: Engineering Developments, Bernard (ed.) © 2001 Swets & Zeitlinger, Lisse, ISBN 90 5809 176 7

Design guidelines for the use of SFRS in ground support

N.B.Grant, R. Ratcliffe & F. Papworth
Scancem Materials Pty Ltd, Australia

ABSTRACT: There are presently no design guidelines based on toughness for the use of Steel Fibre Reinforced Shotcrete (SFRS) in ground support for underground mine development. Typically, in the Australian mining environment, the approach to the use of SFRS in ground support has been one of borrowing experiences from other mines and a "trial and error" method of design, installation, and assessment. There is a need, particularly in the mining industry, for a design guide that can be simply applied by "front line" personnel as an aid to their ground support decision making.

This paper provides an overview of the performance characteristics of SFRS and how the various shotcrete guides specify its use. Practical experiences of the use of SFRS in Australia and Canada, in various applications and ground conditions, are combined with existing empirically-based ground support design methods in order to develop a ground support guideline that incorporates the concept of toughness. An assessment of structural synthetic fibres shows that their low modulus makes their performance characteristics different to steel fibres and hence the recommendations made in this paper should not be used to guide their usage.

1 INTRODUCTION

Steel Fibre Reinforced Shotcrete has been used successfully for ground support for more than twenty years. However, although its use today is widespread globally, the understanding of how it works is limited and application assessment is subjective.

The performance of SFRS can be characterised using a variety of test methods taken from European, Japanese, and American Standards and, more recently, by a method developed in Australia. These tests characterise the performance of SFRS by measuring the ability of this composite material to carry load in flexure beyond the flexural capacity of the concrete itself, i.e. ductility or "toughness". Extensive use of these tests to assess the ever increasing range of steel fibres available on international markets shows that the performance of different fibres varies enormously, that many of the test methods give poor repeatability, and that many tests are undertaken erroneously. The resultant confusion may explain why most current specifications for SFRS linings typically specify the compressive or flexural strength of the shotcrete and a minimum fibre dose rate with no mention of the performance of the resultant SFRS.

In Australia the approach to the use of SFRS in ground support within mines is typically based on shared experiences and trial and error rather than an engineered approach. A potential consequence of this is that a lack of success leads to poor perceptions of the real capability of SFRS. Because there is no starting point to design a SFRS lining for a given set of ground conditions, other methods of ground support are often sought as an easier and more conservative solution.

Field experience has shown that SFRS is a safe, efficient, and economic ground support method. To promote its adoption, a performance-based design guide that can be simply applied by "front line" personnel is required. This paper reviews testing methods and application assessment in the industry to develop such a performance-based design guide.

2 CHARACTERISATION AND SPECIFICATION OF TOUGHNESS

The post-crack capacity of SFRS can be determined through a variety of internationally recognized methods. Some of these methods are described below.

2.1 Norwegian Concrete Association "Sprayed Concrete for Rock Support" (1993)

Increasing toughness requirements for SFRS are given in three classes within this guide:

Class 1 - Type and dosage of fibre in shotcrete placed is specified.

Class 2 & 3 - Residual flexural stress is specified at 1mm and 3mm deflection, as determined by testing a 75 × 125 mm beam in third-point loading.

2.2 EFNARC "European Specification for Sprayed Concrete" (1996)

This document specified toughness in two ways:
1. Residual strength based on the shape of the beam stress/deflection curve and defined in terms of the five different classes.
2. Energy absorption is determined from tests on shotcrete plates supported on four sides and loaded in the centre. Absorbed energy at a central displacement of 25 mm provides three toughness classes.

2.3 Template Method by Morgan et al. (1990)

ASTM C1018 beam test results define five Toughness Performance Levels (TPL) over deflections from 0.5 mm to 3 mm.

2.4 Austrian Concrete Society - Sprayed Concrete Guideline (1999)

This requires that steel fibre reinforced shotcrete contain a minimum fibre content of 30 kg/m^3 in situ. SFRS toughness is specified in two ways:
1. "Toughness range class" draws on the template method by Morgan et al. (1990), as noted in section 2.3. The method suggests that the evaluations should be performed after 1, 7 or 28 days according to the situation requirements.
2. "Energy absorption" based on limits of 500, 700 and 1000 Joules, are specified for three classes based on EFNARC panel tests as outlined in section 2.2.

Early strength classes for fresh sprayed shotcrete are required. Melbye (1997) suggests that the flexural strength required from shotcrete to provide support to the rock mass must exceed 4 MPa. This requirement for early strength is important for NATM tunnels and any application where total ground deformations need to be limited and support provided in the shortest possible time frame.

2.5 Preferred test method

Beam test results show poor repeatability and reproducibility. The high variability requires taking the average of at least five samples. The test is also too complex to set up and is not available in many labs, nor does not represent how shotcrete fails. The EFNARC panel test is undertaken on representative samples, is easier to undertake, but has problems arising from uniform seating of samples. A better plate test, in the form of the Round Determinate panel test (Bernard 2000), has recently been developed and once established as a standard should become the most commonly used assessment method.

3 UTILIZATION OF SFRS IN GROUND SUPPORT

The major problem in designing support to underground openings is in determining the strength and deformation properties of the ground and matching it with the chosen support structure. Though a great deal of resources are utilised in trying to quantify the strength and deformation properties of the ground, and sophisticated modeling programs have been developed for analysing ground behaviour, there is presently no link between the behaviour of the ground and that of thin SFRS linings. As a decision regarding the SFRS lining must be made quickly as the ground is exposed, a design method that can be applied with relative ease by suitably qualified personnel at the development face is needed. Standards and guidelines noted in section 2 are reviewed below to see how they incorporate design elements for SFRS.

3.1 Norwegian Concrete Association "Sprayed Concrete for Rock Support" (1993)

This guide states: "...in major weak zones containing clay, in weak rock and rock subjected to high pressure, concrete toughness and flexural tensile strength are of major importance. Under conditions of low or no deformation, bond, compressive strength and shear strength are the most important." However, it acknowledges that no documented design models exist incorporating the parameters of flexural tensile strength and toughness.

The general design approach is based on the widely recognised empirical rock stability classification, the Q-System developed by Barton et al. (1974). The relationship between rock mass quality, Q, and the associated rock reinforcement measure is summarised in a single chart, often referred to as the "Barton chart".

The "Barton chart" relates rock mass quality, Q, excavation dimensions and end use, to recommend bolt length, spacing and shotcrete thickness (plain or steel fibre reinforced). However, it makes no recommendation as to SFRS toughness requirements.

3.2 EFNARC "European Specification for Sprayed Concrete" (1996)

The EFNARC specification does not provide any guide or reference to the design of shotcrete for SFRS linings.

3.3 Template Method by Morgan et al. (1990)

Morgan et al. (1990) does not provide any guide to the use of SFRS or toughness characteristics required for tunnel or mine drive linings. However, Morgan (1998) does provide some insight into the use of SFRS according to his toughness performance template for certain applications using Toughness Performance Levels (TPL), as follows:

TPL IV – Appropriate for situations involving severe ground movement, with an expectation for cracking of the SFRS lining, squeezing ground in tunnels and mines, where additional support in the form of rock bolts and/or cable bolts may be required.

TPL III – Suitable for relatively stable rock in hard rock mines or tunnels where relatively low rock stress and movement is expected and the potential for cracking of the SFRS lining is expected to be minor.

TPL II – Should be used where the potential for stress and movement induced cracking is considered low (or the consequences of such cracking are not severe) and where the fibre is providing mainly thermal and shrinkage crack control and perhaps some enhanced impact resistance.

3.4 Austrian Concrete Society - Sprayed Concrete Guideline (1999)

The Austrian guideline does not provide recommendations on the design of SFRS linings with respect to thickness, toughness, or rock reinforcement measures.

3.5 Summary of existing guidelines

Only the Norwegian guideline and Morgan's template with application examples provide some direction on how toughness is used as a parameter in lining design.

4 METHOD FOR INCORPORATING TOUGHNESS IN DESIGN

The advice by Morgan (1998) on the toughness requirements for shotcrete in various applications can be incorporated into the Norwegian Q-system approach to rock mass classification and support recommendations to derive an approach to ground support design that includes toughness criteria. This approach is described below.

4.1 Methodology

The Norwegian design approach is based on allocating the ground a "Q" value (which can change many times and considerably over the length of a tunnel) by an appropriately qualified geotechnical engineer.

Once a Q value, or range of Q values, is established, the required ground support is determined using the "Rock Mass Classification, Q Chart" or "Barton Chart". A modified version of this chart is presented in Figure 1. It should be noted that the modifications evident on this chart (which are an outcome of the analysis described in sections 4 & 5) are intended to provide guidance on the required toughness of SFRS and do not alter the original format or support recommendations in any way.

Example: Suppose the ground has an average Q value of 0.2, and the mine decline is 4 metres high at the crown and 5 metres wide at the base.

- The span of 5 metres is critical, being larger than the height,
- From Table 1 select a value of 1.3 for Excavation Support Ratio (ESR)
- Hence, Span/ESR is 5/1.3 = 3.1.
- On Figure 1, plot 3.1 on the vertical axis against a Q value of 0.2 to give the support requirements as 9 cm of SFRS with bolts 1.4 m apart. The right hand side vertical axis indicates the bolts should be around 1.8 metres long.

The deficiency with this design approach is that although the thickness of the SFRS is given there is no toughness requirement indicated.

Table 1. Recommended values for ESR from Barton & Grimstad (1994).

Type of Excavation	ESR
Temporary mine opening	2 - 5
Permanent mine openings, water tunnels for Hydropower (excluding high pressure penstocks), pilot tunnels, drifts and headings for large openings, surge chambers.	1.6 - 2.0
Storage caverns, water treatment plants, minor road and railway tunnels, access tunnels	1.2-1.3
Power stations, major road and railway tunnels, civil defence chambers, portals, intersections	0.9-1.1
Underground nuclear power stations, railway stations, sports and public facilities, factories, major gas pipeline tunnels	0.5-0.8

With the wide range in performance for different fibres (Clements 1996, Bernard 1999) the SFRS generically expressed in the Barton chart could range in toughness from 400 to 1400+ Joule energy absorption based on the EFNARC panel test (1996). Given the structural requirements of the SFRS, this is highly susceptible to misuse. The Norwegian

113

specification suggests that the choice of fibre should be based on anticipated rock deformation and bond potential. It states that in cases of good bond potential, and where deformations are small, toughness is less important. In cases where large deformations are expected and where bond potential is poor, toughness should be given priority along with stabilisation with rock bolts, i.e. as ground conditions and rock mass quality deteriorates, the toughness level of the SFRS should increase.

Based on the description of the ground conditions applicable to the different TPL's given in section 3.3 above by Morgan and their own experience, the authors initially proposed a correlation between the given description of ground conditions and the different rock classes given across the top of the Q-system Design Chart (Figure 1). The result is summarised in Table 2. A Toughness Performance Level (TPL) of IV is required in our example, i.e for a Q value of 0.2, "very poor" or class E rock mass.

Table 2. Correlating Morgan's TPL's to Q-system rock classes.

Toughness Performance Level (TPL)	Rock Class
IV	E & F
III	D
II	C
I	B
0	A

Morgan's TPL's are based on ASTM C1018 beam tests but as outlined in section 2, results from panel tests are preferred for shotcrete assessment.

For this reason, the authors went on to develop the EFNARC panel-based toughness performance recommendations in Table 3, derived from Morgan's values of TPL based on published data. For these EFNARC toughness ranges, the most suitable fibre type and dosage can be estimated taking into account an appropriate fibre rebound of, say, maximum 20% for wet process and possibly 40% for dry process.

Table 3. EFNARC Panel Toughness Related to Rock Class

Rock Class	EFNARC Panel Toughness (Joule)
F	1400+
E	1000 - 1400
D	700 - 1000
C	500 - 700
B	500
A	0

4.2 Limitations of this design approach

The design approach described above is based on tests on beams and panels undergoing limited deflections. For this reason, and the fact that fibres will pull out and become ineffective over a certain crack width, this approach may not be adequate in ground experiencing excessive movement.

Where excessive movement is considered inevitable it will be necessary to give special consideration to the method of ground support, possibly incorporating one of the following options:-

• Sets or reinforced shotcrete ribs, as detailed in Norwegian Concrete Association Publication No. 7 (1993)
• Conventional mesh reinforcement
• Secondary/permanent lining

Rock Class G should be given special consideration along with any ground where the deflection between bolts is expected to be greater than 0.05 times the bolt spacing.

5 PRACTICAL EXPERIENCES WITH THE USE OF SFRS

Having established a broad relationship between rock class, rock quality value, Q, and SFRS toughness according to EFNARC (Table 3), the authors checked this relationship by collecting input from Australian mines regarding their use of SFRS.

A total of fourteen metaliferous mines in Australia were visited and consented to input basic information regarding their operation, and answered questions about the following;

• Rock mass conditions and classification
• Type of excavations and end use
• Dimensions of excavations
• Shotcrete thickness and properties
• Type and design of rock reinforcement
• Development sequence / cycle
• General comments

The objectives were to:

• Establish the extent of use of the rock quality classification, Q,
• Review the application and performance characteristics of SFRS used,
• Overlay this information onto the existing Barton chart.

In this way a comparison between the toughness of real shotcrete used in the field, and the relationship between toughness and rock class Q, could be established.

5.1 The extent of use of shotcrete and SFRS

All of the mines either presently or previously used shotcrete or SFRS within their operations, with use varying from full production cycle shotcrete to random campaigns. The range of applications was broad, and included:

• Plain shotcrete for rock sealing and scat control,
• Essential application in development cycle,

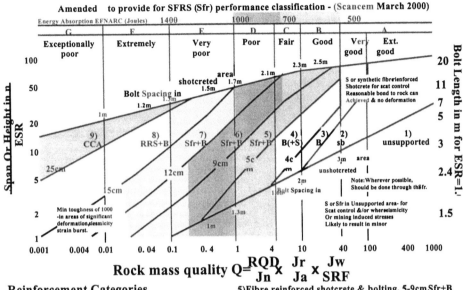

ROCK MASS CLASSIFFICATION

Amended   to provide for SFRS (Sfr) performance classification - (Scancem **March 2000**)

Rock mass quality $Q = \dfrac{RQD}{Jn} \times \dfrac{Jr}{Ja} \times \dfrac{Jw}{SRF}$

Reinforcement Categories

1) Unsupported
2) Spot bolting, Sb
3) Systematic bolting, B
4) Systematic Bolting,
   (and unreinforced shotcrete, 4-10cm), B(+S)

5) Fibre reinforced shotcrete & bolting, 5-9cm,Sfr+B
6) Fibre rienforced shotcrete and bolting9-12cm,Sfr+B
7) Fibre rienforced shotcrete and bolting12-15cm,Sfr+B
8) Fibre rienforced shotcrete >15cm,Sfr,RRS+B
(rienforced ribs of shotcrete&bolting, Sfr,RRS+B)
9) Cast concrete lining,CCA

Figure 1. Q-System Rock Mass classification and support recommendations modified to include SFRS toughness.

- Low maintenance permanent structures such as conveyor drives, pump stations, workshops, etc.
- Draw points for impact and abrasion resistance,
- Rehabilitation of deteriorated rock or previously meshed areas,
- Contingency treatment for fault or weakness zones,
- Large span intersections that may be subjected to intermittent blasting or mining induced stresses,
- Multi-layer treatments in areas of extreme deformation.

5.2 *What toughness has been used?*

The toughness of the SFRS had not been specified in any of the cases analysed. However, in operations where a large volume of SFRS was used, the type, (or general description) and dosage of the steel fibre was generally specified based on previous testing programs and/or experiences. The authors determined the toughness of the SFRS used by relating the characteristic of the concrete mix, fibre type and dosage to test results in the public domain.

The estimated EFNARC energy absorption ranged from 500 Joule for minor weakness zones and for sealing of sound rock in areas unlikely to experience deformation, to 1400+ Joule in rock subject to high stresses, potential strain bursting, and areas likely to experience large deformation.

5.3 *Shotcrete thickness*

Shotcrete thickness were generally specified for the various applications and ranged from a low of 30 mm up to 125 mm, with the typical range being from 50 mm to 75 mm. The thickness was normally deemed to be a "nominal" thickness.

For less demanding, low toughness shotcrete, thickness was usually 50 mm minimum. For high toughness shotcrete, 75 mm was typical, but in one case a multi-layer treatment of 75 mm + 50 mm was used.

5.4 *The extent of use of "Q-value" and the Barton chart*

Of the fourteen mines, all used some form of rock mass classification, ranging from the determination of RQD to estimate Q, intermittent determination of Q, formal determinations of Q, RMR (Bieniawski, 1999) and MRMR (Laubscher, 1990). Eleven of the fourteen mines were able to provide some measure of Q or a range of Q's for their rock types.

In May 1999, the Department of Mines and Energy Western Australia (1999) issued a technical note titled "The Q-System Geotechnical Design Method Was Updated In 1994" in which it outlines the updates made to the Q-system determination and the potential impact on support design. The major changes involved ESR values and significantly increased the SRF (Stress Reduction Factor) values used in the estimation of Q. It was not possible to ascertain whether the Q-value calculations were based on the original 1974 system or the updated 1994 system. Even though Q-values were commonly determined for the various rock masses, the Barton chart was rarely used for support determination. Some mines perceived that it inadequately catered for "mining induced stresses", while in conjunction others considered it too conservative.

In all cases, the span or height/ESR value on the left axis of the Barton chart was less than 3, and higher toughness shotcrete was used as the value of Rock Mass Q reduced. Numerous SFRS applications were in area 1 of the chart), i.e. no support necessary. Whether this is a result of "mining induced stresses" or superseded calculations of Q is not known.

Analyses of these results were used to verify the toughness levels in the "modified Barton chat" (Figure 1) but also led to the following conclusions:

- In areas of anticipated "significant" deformation, seismicity or potential strain burst, a minimum Energy Absorption capacity of 1000 Joule should be used based on EFNARC panel tests (1996). In extreme cases this should be 1400 Joule.
- The application of SFRS is not a "fix-all". It must be applied prior to failure of the rock mass in order to work effectively.
- Wherever possible, always bolt through the SFRS.
- Shotcrete or SFRS may be required in areas designated as "Unsupported" in the Barton chart due to "mining induced stresses".
- Unreinforced shotcrete is an effective measure for controlling scats and replacing mesh used for this purpose. However, the bond strength should be considered and if likely to be very weak, or the ground is subject to minor deformation, postbolted SFRS should be used.

6 STRUCTURAL SYNTHETIC FIBRES AND USE OF THE MODIFIED BARTON CHART

Large diameter (0.5-1.0 mm) Structural Synthetic Fibres (SSF) are typically manufactured from polypropylene and, while quite similar in size to steel fi-

bres, tend to vary significantly in other regards (Table 4).

With millions of cubic metres of steel fibre reinforced shotcrete performing satisfactorily, the principle questions for synthetic fibres which exhibit such markedly different properties are:

- Will shotcrete reinforced with structural synthetic fibres work to support ground pressures in the same way as currently available steel fibres?
- Can current test methods, developed for steel fibres, be used to quantify the performance of synthetic fibres without modification?

Table 4. Fibre Properties

Property	Steel	SSF
Specific Gravity	7.85	0.9-0.91
Strength (MPa)	300-1800	130-690
Elastic Modulus (10^3 MPa)	200	3.4-4.8

Analysis of the shapes of load-deflection graphs derived from beam tests for the best performing steel and synthetic fibres from 57 different mix designs tested by Bernard (2000), shown in Figure 2, shows:

1. Steel fibres provide superior resistance to crack opening at narrower crack widths than synthetic fibres.
2. Structural synthetic fibres maintain load better across wide flexural cracks while the load carrying capacity of steel fibre reinforced specimens diminishes.

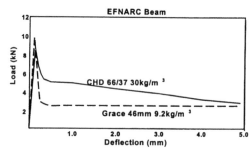

Figure 2. Steel versus synthetic fibre reinforced shotcrete beam test results (Bernard 1999, 2000).

Specifying performance in terms of the load capacity at a defined deflection in a beam test would typically mean that uneconomically high dosages of polypropylene fibres would be required to match the performance provided by steel fibres at normally specified deflections. This supports the traditional theory as to how steel fibres and synthetic fibres carry load.

Steel fibres are primarily anchored at each end. As a crack opens, the strain is supported along the entire length of the fibre and, because the fibre has a high elastic modulus, significant load is supported for a relatively small crack opening. A critical factor in obtaining high toughness with steel fibres is the requirement to design the anchorage to pull through as the steel reaches ultimate capacity.

Polypropylene fibres generally behave very differently. If only anchored at each end the low elastic modulus of these fibres means that cracks would have to open 50-100 times more than steel fibres to carry the same load (i.e. a crack in SFRC of 0.1 mm width is equivalent to a crack of 5-10 mm width in synthetic fibre reinforced shotcrete). This can be overcome by:
1. Increasing the synthetic fibre dosage to reduce the stress induced and hence extension experienced by each fibre at a given load,
2. Reducing the length of fibre between anchorages to increase the relative extension and hence stress due to opening of the crack.

Traditionally, microfine synthetic fibres have been used at dosages of less than 1 kg/m^3. Changing to large diameter (0.5-1.0 mm) has enabled workable dosage levels to increase by about a factor of 10 times. To reduce the length between anchorages, deformed fibres and/or improved chemical bond is required.

This raises a significant question about the long term performance of structural synthetic fibres. The mechanical anchorage of steel fibre provides a high degree of security regarding long term performance. The creep of synthetic fibres and use of chemical bonding means long term test results need to be available wherever support is required for more than a short period of time. Even though synthetic fibres are definitely improving, they still cannot compete with steel fibre to support loads at low deflections. The significance of this is shown in Figure 3 which overlays load deflection curves obtained by Bernard (2000) using Round Determinate panel tests for steel and synthetic fibres of similar toughness levels onto a hypothetical support pressure versus radial displacement curve. In this figure, d_r is the radial displacement prior to support going in and d_s is the radial displacement of the supported ground required to reach equilibrium.

Although purely illustrative, it indicates the interaction between ground movement and a supporting shotcrete layer, highlighting the ability of steel fibre to provide stable support to a rock mass at a lower total radial displacement.

Where a support regime has been determined on the basis of steel fibre shotcrete of a specific toughness level, the substitution of structural synthetic fibre reinforced shotcrete of equivalent toughness (from a panel test) will result in support being achieved at a greater rock mass displacement. Hence, the modified Barton chart (Figure 1) cannot be used for synthetic fibres as it presently stands.

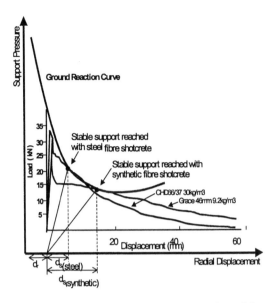

Figure 3. Hypothetical ground reaction and support for steel fibre and synthetic fibre reinforced shotcrete.

7 CONCLUSIONS

Toughness is the defining characteristic of fibre reinforced shotcrete. There are many toughness test methods available internationally, but the EFNARC panel test is currently the most appropriate for shotcrete assessment. The Round Determinate panel test overcomes problems with the EFNARC panel test and is likely to supersede it.

The Barton chart is widely used to assess ground conditions but its support recommendations do not include a toughness requirement. The method suggested in this paper is proposed as a link between EFNARC plate test values and ground conditions in the Barton chart.

The toughness values suggested here are considered appropriate for SFRS based on experience with this high modulus material. Specifying the same toughness values when using structural synthetic fibres may not provide the same degree of support due to their different anchorage mechanism.

ACKNOWLEDGEMENTS

The Authors would like to thank the following mines, their geotechnical and mining services personnel, and their companies, for their data input and time contributed to this analysis: Northparkes Mine, NSW; Henty Gold Mine, Tasmania; WMC Leinster Nickel Operations, WA; Jundee Gold Mine, WA; KCGM Mt Charlotte Mine, WA; WMC Kambalda Nickel Operations, WA; Big Bell Gold Mine, WA;

Forrestania Nickel Mine, WA; Bronzewing Gold Mine, WA; Kanowna Belle Gold Mine, WA; Kundana Gold Mine, WA; Black Swan Nickel Mine, WA; Yilgarn Star Gold Mine, WA; Beaconsfield Gold Mine, Tasmania

REFERENCES

Austrian Concrete Society, 1999. Sprayed Concrete Guideline, Karlsgasse: Osterreichischer Betonverein.

Bernard E.S, 1999. Correlations in the Performance of Fibre Reinforced Shotcrete Beams and Panels. *Engineering Report No. CE9*, School of Civil Engineering and Environment. University of Western Sydney.

Bernard E.S. 2000. Correlations in the Performance of Fibre Reinforced Shotcrete Beams and Panels: Part 2. *Engineering Report No. CE15*, School of Civil Engineering and Environment. University of Western Sydney.

Clements M. J. K. 1996. Measuring the Performance of Steel Fibre Reinforced Shotcrete. *IX Australian Tunneling Conference*. Sydney.

Department of Minerals & Energy Western Australia, 1999. The Q-System Geotechnical Design Method Was Updated In 1994.

EFNARC. 1996. European Specification for Sprayed Concrete.

Melbye, T. 1997. Sprayed Concrete for Rock Support. MBT. 6th edition.

Morgan D. R. 1998. Agra Earth & Environmental Communication to Bekaert NV.

Morgan, D.R., Chen, L. & Beaupre, D. 1990. Toughness of Fibre Reinforced Shotcrete.

Sprayed Concrete for Rock Support 1993. Technical Specification and Guidelines, Norwegian Concrete Association, *Publication Number 7*, Committee Sprayed Concrete.

Bienaiawski Z.T. 1981 Engineering Rock Mass Classifications. New York: Wiley Interscience.

Laubscher D.H. 1990. A geometrics classification system for the rating of rock mass in mine design. *J. S. Afr. Inst. Min. Metall.* 90 (10): 257-273.

Shotcrete: Engineering Developments, Bernard (ed.) © 2001 Swets & Zeitlinger, Lisse, ISBN 90 5809 176 7

Behaviour of steel fibre reinforced shotcrete during large deformations in squeezing rock

E.Grimstad
Norwegian Geotechnical Institute, Oslo, Norway

ABSTRACT: In a major fault zone under more than 1000 m of overburden in the 24.5 km long Laerdal Tunnel, Norway, large deformations have occurred in squeezing rock a long time after excavation. Heavy support including rock bolts, many layers of Fibre Reinforced Shotcrete (FRS), and reinforced ribs of sprayed concrete (RRS), were installed during construction. Cracks started to develop in the FRS and the RRS due to large deformations. Several layers of FRS were applied during the next 6 weeks as the cracking continued. The total thickness of FRS applied was 95 cm in the crown and 60 cm in the walls. In spite of this, the convergence continued, and sheared and crushed both the FRS and the rock to a depth of at least 150cm behind the shotcrete. In this case, the FRS behaved very similarly to brittle rock and the fibres had only a minor effect on behaviour during development of cracks. However, the fibres prevented release of concrete slabs from the crown. During shearing a large number of the fibres were cut off, while most of the fibres were pulled out during bending failure.

1 DEFORMATION OF SPRAYED CONCRETE DURING CONVERGENCE IN A TUNNEL

1.1 *General description of the 24.5 km long Laerdal road tunnel*

The 24.5km long Laerdal road tunnel broke through on September 3, 1999 after approximately four years of excavation. The excavation was carried out on three faces, two of which were constructed from a 2.1km long access tunnel. Three safety turning bays for lorries have been excavated with a 30m span inside the tunnel. The tunnel is excavated through Precambrian gneisses with a uni-axial unconfined compressive strength of 100-210 MPa in unweathered rock, and had a maximum overburden of 1450 m, which induces a vertical stress of 39 MPa. About 20 km of the total length has an overburden greater than 800 m. The effect of the stress has been variable along the tunnel, and partly independent of the overburden due to changes in jointing and elastic modulus (25-55GPa) along the route. In spite of the large overburden the major principle stress is sub-horizontal in most of the tunnel.

In the majority of the tunnel length all levels of spalling and rock burst have been noticed. In two of the caverns with a 30 m span, and at some other places where heavy rock burst occurred, shear failure deep into the tunnel periphery has been observed. This induces shearing of the shotcrete instead of flexural failure which is the more normal result of buckling of a shotcrete layer caused by ordinary spalling and slabbing of the rock induced by general convergence of a tunnel under high stress (see Figure 1).

1.2 *Cave in and large deformations in squeezing rock in the Laerdal tunnel*

Squeezing rock in a 50-60m wide fault zone 11 km from the entrance caused a cave-in in which 1200-1500 m^3 of rock fell from the crown during two days. A total length of 17 m of the supported tunnel collapsed (see Figure 2).

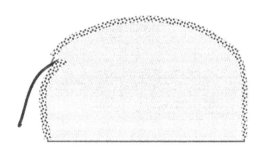

Figure 1. Shear failure in the rock mass also causes shear failure in the shotcrete.

During construction, only ordinary rock support was installed as is usual for moderate rock stress with minor spalling, despite the extremely high drilling rate. Drilling through weathered rock with swelling clay in a dry condition is a cause for concern. The cave-in area had an overburden of 1000 m with higher mountains on both sides, giving a theoretical vertical stress of approximately 30 MPa. The rock mass was partly weathered and contained several irregular joints filled with swelling clay that resulted in a maximum swelling pressure of 0.7 MPa. Fortunately no water leakage was evident. The rock mass quality, Q, in the fault zone was about:

$$Q = \frac{RQD}{J_n} \times \frac{J_r}{J_a} \times \frac{J_w}{SRF} = \frac{10-60}{12} \times \frac{1}{10-15} \times \frac{1,0}{10} = 0,0055 - 0,05$$

which corresponds to "exceptionally to extremely poor" ground (Barton et al. 1974). According to Singh et al. (1992), the compressive strength of the rock mass may be found from the rock mass quality, Q, and the specific weight using the equation:

$$\sigma_{cmass} = 7 \gamma Q^{1/3} (MPa) = 7 \times 2.7 \times 0.01^{1/3} = 4 \text{ MPa}$$

The estimated ratio between tangential stress, σ_θ and compressive strength of the rock mass, σ_{cmass}, $\sigma_\theta / \sigma_{cmass} = 55/4 \approx 14$. This is more than double the limit for heavy squeezing according to empirical equations (Bhasin & Grimstad 1996).

The cave in was initiated during drilling for rock bolts when a drilling rod got stuck. The swelling clay probably absorbed water from the drilling jumbo. The combination of water supply to a dry swelling clay and a overstressed soft fault zone gave a progressive failure which continued for two days, and extended 15 m metres into the already supported tunnel.

After the cave-in, the last 20m of tunnel behind the cave-in was supported by supplementary reinforced ribs of shotcrete in order to stop further cave-ins. Following this, 750 m³ of concrete was pumped into the void behind a wall of shotcrete between the

debris and the crown. After curing of this 3-5 m thick concrete slab, the debris was gradually mucked out simultaneously with anchoring and spraying of the walls below the large concrete slab.

Ten days after the cave-in the old face was made accessible again. From this point onward the length of the rounds was reduced from 5 to about 2.5 m. Moreover, 6m long grouted spilling bolts were installed in front of each round.

After blasting, each round was supported with 30-40 cm thick FRS and 5 m long grouted rock bolts at a spacing of 1 m, and completed with a steel bar reinforced rib of shotcrete in the area excavated after the cave-in.

This procedure was continued for the next 45 metres of the tunnel. All together, 22 reinforced ribs of shotcrete were placed in this fault zone. During a summer holiday break in construction lasting 3 weeks, the un-reinforced invert close to the face (17 m from the edge of the cave-in) was heaved approximately 3 cm in 17 days following the last round.

After some weeks the shotcrete and the ribs began to crack. Additional layers of shotcrete were sprayed combined with supplementary 5m long grouted rock

Figure 2. Cave-in in a fault zone in the Laerdal Tunnel.

Table 1. Distribution of pulled out and sheared off fibres in tensional and shear failure.

Slab	Type of failure	Length (cm)	Width (cm)	Area (cm²)	Pulled out fibres	Pulled out fibres (/cm²)	Sheared fibres	Sheared fibres (/cm²)
1	Shearing	28	4.3	120.4	18	0.150	23	0.191
1	Bending	37	5	185	66	0.357	9	0.049
2	Bending	38.4	4.7	180.48	54	0.299	13	0.072
2	Bending + shearing	45.9	3.9	179.01	59	0.330	19	0.106
3	Bending	29.5	4.1	120.95	63	0.521	16	0.132
3	Bending + shearing	29	3.4	98.6	44	0.446	17	0.172
4	Bending + shearing	47	4.7	220.9	77	0.349	24	0.109
4	Bending	28	4.5	126	54	0.429	14	0.111
5	Bending	36	3.8	136.8	71	0.519	12	0.088
5	Shearing	36	19	684	76	0.111	157	0.230

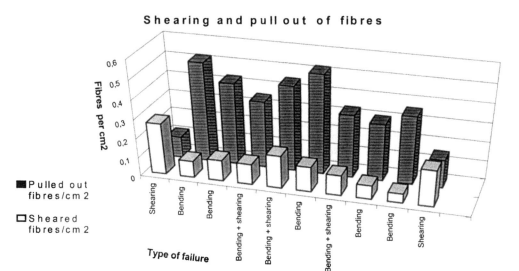

Shearing and pull out of fibres

Fibres per cm2

■ Pulled out
fibres/cm2

□ Sheared
fibres/cm2

Type of failure

Shearing · Bending · Bending · Bending + shearing · Bending + shearing · Bending · Bending + shearing · Bending · Bending · Shearing

Figure 3 Distribution of pulled out and sheared off fibres

bolts and some 3-4 m long resin anchored rock bolts in the deforming areas. Six weeks after excavation in the worst area, the total thickness of the shotcrete was equal to 95 cm in the crown and 60 cm in the walls, but the deformations nevertheless continued. It was observed with an endoscope that all the shotcrete and rock mass up to 150 cm behind the shotcrete lining was completely sheared or crushed (Grimstad & Kvaale 1999). A few of the rock bolts were torn off and/or the anchorage was pulled out. A majority of the visible triangular steel plates on the bolts were also deformed. The steel fibre reinforced shotcrete behaved very much like massive rock under the influence of high stress. Between the rock bolts, shear failure and spalling in the massive shotcrete went on near the surface. Several pieces of shotcrete were loosened in the crown of the lining, but almost all of these were restrained from falling by the steel fibres until they were taken down with scaling bars. The deformation (convergence) of the crown after installation of rock bolts was estimated to be 20 cm.

The deformation in the crown and spring line occurred at a rate of approximately 2 cm per week until the invert was reinforced with a cast concrete lining combined with rock bolts. The rock under the invert was more or less transformed into clay because it had been saturated by water from the drilling jumbo. Following reinforcement of the invert about 2 months after blasting, the rate of deformation was slower but still continued. Nearly 5 months after excavation in the fault zone the owner was persuaded to carry out convergence measurements. The rate of convergence was at that time equal to a maximum 2.0 mm per week, and slowed down to between 0.3 and 1.0 mm per week 8 months after excavation. As

a temporary support for the cracked shotcrete, rock bolts anchored with resin capsules were installed with large triangular plates outside the spalling shotcrete.

As a part of the permanent support, a 15 cm thick layer of FRS incorporating Dramix RC 65/35 fibres at a dosage rate of 60 kg/m^3, which according to tests carried out by Bernard (1999) absorbs 1700 Joules in an EFNARC panel test, combined with 4 m long rock bolts anchored with resin capsules, were placed in May/June 2000. The average bolt spacing was approximately 0.6 m following installation of all the various bolts. At the spring line and the crown, deformation slots were filled with polyethylene foam parallel to the tunnel axis in order to absorb the deformation and prevent shearing of the last layer of shotcrete. Due to widening of the tunnel during blasting, there was sufficient space for all the shotcrete applied in the efforts to restrain convergence.

2 COUNTING OF STEEL FIBRES IN SLABS RELEASED FROM SHEAR ZONES IN THE LAERDAL TUNNEL

Some sprayed concrete slabs from the Laerdal Tunnel were collected after failure as a result of the deformations described above. Most of the slabs exhibited tension failure surfaces, but some showed a combined flexural-tensile failure and shear failure or pure shear failure. The crack surfaces formed under pure shear showed a small majority of sheared fibres compared to pulled out fibres. On the other hand, the crack surfaces which were formed by bending or tensile failure had a large majority of pulled out fibres. Table 1 and Figure 3 show the distribution of the two fibre failure modes observed in the collected lining segments.

3 DISCUSSION AND CONCLUSION

No lining is strong enough to stop convergence in squeezing or spalling rock under high stress, but it should be able to control the deformation and prevent cave-in. Hence it should be accepted that rock support is deformed or partly destroyed in a controlled way. There is no doubt regarding the importance of a flexible support in such cases.

Almost all commonly performed laboratory tests for the mechanical performance of shotcrete result in an estimate of either the uniaxial unconfined compressive strength (cubes or cylinders) or bending flexural strength (beams or panels). These tests induce mainly a compressive, bending or tensile failure. In tunnels affected by significant convergence, shearing of the sprayed concrete often occurs. This results in a different mode of failure in the fibres, where shearing of the fibres is more likely than during tensile deformation in which pull-out of the fibres is common. High steel quality will probably provide better resistance to shearing compared to low steel quality with lower tensile and flexural strength. During shearing the steel fibre reinforced shotcrete in thick layers behaves like a cast concrete lining or massive hard rock in that it develops spalling and slabbing under high stress. Tough fibres prevent immediate release of slab fragment from the lining after formation of cracks. The dosage and type of fibres are important to the residual strength of the concrete following large deformations. Laboratory tests should be carried out in order to evaluate the behaviour of fibres during shearing.

REFERENCES

Barton, N. et al. 1974, Engineering Classification of Rock Masses for the Design of Rock Support. *Rock Mech.* 6 (4): 189-239.

Bhasin, R. And Grimstad, E. 1996, The use of Stress-Strength Relationships in the Assessment of Tunnel Stability, *Tunnelling and Underground Space Technology,* 11: 93-98.

Singh, B. et al. 1992, Correlation between Observed Support Pressure and Rock Mass Quality, *Tunnelling and Underground Space Technology,* 7 (1): 59-74.

Grimstad, E. & Kvaale, J. 1999. The Influence of rock stress and support on the depth of the disturbed zone in the Laerdal Tunnel. A Key to differentiate the Rock Support. Proceedings, *World Tunnel Congress,* Oslo, Norway.

Bernard, E.S, 1999. Correlation in the Performance of Fibre Reinforced Shotcrete Beams and Panels, *Engineering Report No. CE9,* University of Western Sydney, Nepean.

Shotcrete: Engineering Developments, Bernard (ed.) © 2001 Swets & Zeitlinger, Lisse, ISBN 90 5809 176 7

Active design in civil tunnelling with sprayed concrete as a permanent lining

E.Grøv
O.T. Blindheim AS, Trondheim, Norway

ABSTRACT: Sprayed concrete in civil tunnelling has been acknowledged world-wide as a legitimate rock support measure. The advantages of the wide range of properties that can be achieved and high application rates allow tunnelling to proceed with minimum delay. In parallel with this, the technology associated with shotcrete is rapidly developing to allow usage as a permanent lining for rock tunnels. In many projects, cast-in-place concrete or similar rigid support measures are the only approved methods for the permanent lining, whilst sprayed concrete is limited to providing temporary support. Modern tunnelling demands cost-effective techniques, long-term durable products, and a safe working environment. These demands require comprehensive documentation to be prepared. In Scandinavia, temporary rock support is generally approved as permanent when conforming to relevant technical specifications. Through project references, this article aims to present and describe the elements which together form a tunnelling method including active design with permanent sprayed concrete lining as an integral part. Two solutions for application of sprayed concrete in adverse rock mass conditions will also be described.

1 INTRODUCTION

Sprayed concrete as rock support has achieved world-wide recognition in civil tunnelling. Practical advantages, such as the wide ranges of structural/physical properties possible, and the high rate of application, allow tunnelling to proceed with minimum delay. The technology involved is rapidly developing. Modern tunnelling depends on: cost effective tunnelling techniques, long term durability of products used, and the provision of a safe working environment. Sprayed concrete as a permanent lining fulfils these requirements, leading to the prospect of a demand for increased use in the future.

Continuous development of the technology enables improvements in the physical properties of sprayed concrete. These improvements offer a variety of possibilities for the designer to customize particular properties for any given project. This includes high early and final strength, resulting in higher stiffness and increased impermeability, and the introduction of fibre reinforcement increases ductility (Grøv et al. 1997). The sprayed concrete mix design can be determined on the basis of project specific requirements with respect to properties of the sprayed concrete. The challenge for the designer is to identify relevant and applicable quality criteria for the sprayed concrete based on, e.g. defined functional requirements.

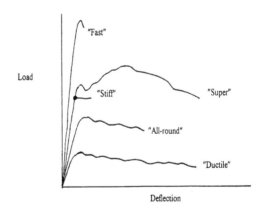

Figure 1. Illustration of "typical mix designs" for shotcrete.

Primary rock support is applied in order to fulfil the technical specifications as set forth for a permanent lining. For the contractor, increased tunnelling progress can be achieved as improved workability allows faster application. The possibility of spraying thick layers is obtained using alkali-free accelerators, which is particularly useful in poor rock mass conditions. The owner will thereby get a final product that ultimately fulfils his requirements.

In Scandinavia, tunnelling is dominated by the application of "sprayed concrete-based tunnelling". Garshol (1997). The support structure commonly associated with this tunnelling method is the single shell sprayed concrete lining. The basis for the method is the application of sprayed concrete both for initial rock support and as the permanent final lining. As an alternative to massive, cast-in-place concrete structures, or pre-cast segments often placed inside a temporary lining consisting of competent sprayed concrete, the capability of the initial primary shotcrete lining in rock tunnels can be utilised to the extent the quality of the materials allow. Such primary linings consist of sprayed concrete with a thickness ranging from 50-200 mm, combined with fully grouted and corrosion protected rock bolts.

This article will present a tunnelling method based on active design, and describe the fundamental elements for its application, in which the use of sprayed concrete is an integral part. Through a selection of examples, this article outlines the application of sprayed concrete as a permanent lining in a variety of underground projects. The article describes how permanent sprayed concrete linings can also be used in adverse rock mass conditions.

2 TUNNELLING METHODOLOGY

2.1 Introduction

Sprayed concrete linings as permanent rock support, combined with other tunnelling techniques, constitute a tunnelling method. It is important for the quality of the work, and the success of the application, that the entire structure of the tunnelling method is familiarised. The tunnelling method involves measures as described in the sections below.

2.2 Ground water control

Ground water control is achieved by the use of probe-drilling ahead of the face and pre-grouting of the rock mass. The primary purpose of a pre-grouting scheme is to establish an impervious zone around the tunnel. This impervious zone will reduce the water gradient to avoid water of high pressure reaching the tunnel contour. In addition, pre-grouting will have the effect of improving the stability situation in the grouted zone within the rock mass. The experience obtained with respect to probe-drilling and pre-grouting has also been adopted for tunnels in urban areas in Scandinavia with maximum allowable permanent water ingress rates as low as 2-10 litres per minute per 100 m.

2.3 Cautious blasting

New blasting technology can be applied to limit the damage to the tunnel periphery. Techniques and materials are at hand to perform blasting operations so that the tunnel advance can be made efficient, and at the same time without imposing adverse effects on the tunnel periphery. Tunnel excavation by drill and blast often induces secondary cracking of the rock mass in the periphery of the tunnel opening requiring additional support. The extent of this phenomenon can be reduced significantly, if cautious blasting is focused appropriately. Cautious blasting is particularly important where a pre-grouted zone has been created ahead of the tunnel face, the result being that the rock mass, to a greater extent, keeps its impermeable capabilities intact.

Further, the blasting technology available today provides a smooth and even rock surface with little deviation compared to the theoretical rock contour. This improves the possibility of constructing a support structure with close to an ideal arch shape.

2.4 Empirical guidelines and analytical/numerical calculations

Together with empirical guidelines such as the Q-system (Løset et al. 1997), RMR (Bienawski 1984) or similar, analytical and numerical calculations assist in rock support determination. Depending on the complexity of any given tunnel project, the need for analytical and/or numerical modelling must be judged. In many cases a support design based on empirical guidelines will do. In rock engineering the application of numerical modelling, both as a basis for decision making and documentation of design tasks, has grown in popularity. In addition, numerical modelling has been widely used as a tool for design verification and for assessments and follow-up during tunnelling works.

Software codes are capable of taking into account complicated geometrical conditions as well as support measures most commonly used. Further, these codes can simulate different rock mass conditions with some degree of realism.

2.5 Observations and monitoring

Currently the tunnelling industry utilises a variety of different methods for monitoring of deformations for stability control and verification of tunnel integrity. The NATM method introduced the systematic use of such follow-up measurements. The increased application of sprayed concrete as a permanent lining also calls for monitoring. Visual observations, however, can be a first step. Installation of convergence sections (tape extensometers), rod extensometers, measuring rock bolts or similar devices can be the next. The purpose is primarily to detect any occurrence of deformations, the results from observations and monitoring will then form the basis for additional support installations if required.

2.6 *Working procedures*

Project specific working procedures based on general codes and standards are available to ensure the desired quality of the work. Quality control is an essential factor to ensure that both grouting and support work meet the intended goals. Furthermore, it is essential that critical work, described in the procedures, are carefully recorded and properly documented. All work relating to probe drilling, grouting, and support, must be monitored by experienced professionals who are capable of incorporating modifications during the actual execution of the work. Materials, methods, and results must be documented for continuous evaluation.

Pre-construction trials are required to identify the best suited mix-design for the sprayed concrete and to ensure the final product will meet the given technical specifications. During construction, applied quality is documented by testing schedules as outlined in the working procedures.

3 BASIS FOR IMPLEMENTATION OF SPRAYED CONCRETE BASED TUNNELLING

3.1 *Self supporting capacity*

Any rock mass has a certain self-supporting capacity, although this capacity may vary within a wide range (Bienawski 1984). The fact that there is some "stand-up" time implies that the rock mass for a certain time is not a dead load, thus it shall not be treated as if it was (Figure 2). An appropriate engineering approach is to take this capacity into account when designing permanent support. Rock strengthening may, however, be needed to secure certain properties/specified capacities, in the same way as is the case for any other construction material. The fact that the rock mass is not a homogenous material shall not disqualify the utilization of its self supporting and load bearing capacity.

3.2 *Drained structure*

Another important aspect of the sprayed concrete based tunnelling method is the concept of a drained structure. The rock mass in combination with the rock support constitute a drained structure. This means that the support measures installed have not been constructed to support external water pressure. Excessive water must therefore not be allowed to build up behind the rock support measures.

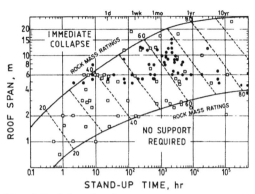

Figure 2. Typical stand up times.

However, even in a tunnel which has been subject to an extensive, customized pre-grouting schedule, some seepage may occur. The tunnelling method includes a controlled handling of excessive water at the tunnel periphery and behind the sprayed concrete lining. Excessive water must either be piped to the water collection system in the tunnel or taken care of by water shielding. Drainage can be achieved by means of installing, for example, local collection devices to confine the water and transfer the water via pipes to the drainage system in the tunnel. Alternatively, a structure can be installed at the rock surface or as an interlayer between two subsequent layers of sprayed concrete. A water and frost protection system including sprayed concrete may also be required, see principles in Figure 3.

3.3 *Primary lining approved as permanent*

Approved primary rock support constitutes a part of a permanent support structure. In most tunnel proj-

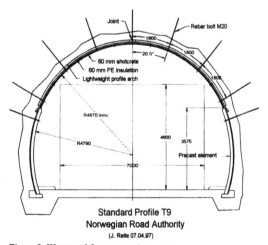

Figure 3. Water and frost protection system

125

ects, an initial primary lining will be established. One objective of such a lining is to temporarily serve as rock support to maintain safe working conditions during construction. It may consist of sprayed concrete and rock bolts. In Scandinavia, it has become a common practice that this initial and primary lining constitutes a part of the permanent lining.

This conversion can be effected if the support measures applied in the primary lining conform to the technical requirements applicable for the permanent lining. Consequently, it is widely believed that as much as possible of the permanent lining should be placed at an early stage producing a completed tunnel shortly behind the tunnel face.

In this respect, the technical specifications and quality of the sprayed concrete is of utmost importance. In addition, with the concrete technology at hand today, there is no difficulty involved in producing a sprayed concrete which can satisfy a wide range of requirements. It may, though, be an expensive routine to specify at all times a "super" mixture that can cope with every requirement. However, when the designer knows the purpose of the sprayed concrete and the functional requirements he intends to designate it for, the contractor can prepare the mix design accordingly.

The permanent lining can be established either by approving the primary lining as it is, or, if necessary, including a supplementary support to fulfil specific requirements, for example, related to the desired final sprayed concrete thickness, surface exposure, or any other aspect that may require an outer layer.

3.4 Active design, adaptability

The predicted rock mass conditions will vary along a tunnelling route. This fact implies that the engineering staff in any tunnel project taking advantage of sprayed concrete-based tunnelling shall adapt the tunnelling method to the actual rock mass conditions. In this way the documentation available together with the evaluation of the rock mass forms an important basis for a continuous modification of the procedure. A design which, in a given location, works to the designer's satisfaction can be quite ineffective only a few tens of metres away.

Recording the rock mass conditions and implementing the grouting and support procedure to produce a flexible tunnelling scheme is a crucial element in this method. Active design is characterised by the following main elements:

- An established geological model based on information at hand prior to excavation works.
- A predefined set of rock support classes based on, for example, empirical guidelines.
- A sound verification of these support classes by utilisation of analytical and/or numerical models.
- A quantitative rock mass classification.
- A confirmed procedure for the application of

support classes, combined with rock mass classification, and rules to handle occurrences beyond the coverage of the system.

- A continuous evaluation of the geological model and the predefined rock support classes based on experiences gained with modifications if needed.
- An immediate classification of the rock mass quality at the tunnel face.

Today's tunnelling industry sets forth a number of pre-requisites that must be fulfilled for any tunnel project. These can be summarised as: flexibility, adaptability, experience, cost efficiency, and decision making at the work face. At the same time the tunnelling method shall allow reliability, predictability, planning, cost control and documentation. The system described by the term "active design" provides a flexible tunnelling approach to adapt the support and grouting efforts to the actual rock mass encountered while still maintaining the requirement on documentation.

4 EXAMPLES OF IMPLEMENTATION

The implementation of permanent sprayed concrete in civil tunnelling has been reported widely over recent years. In the following, some examples of the application of shotcrete-based tunnelling in a number of projects will be presented. Technical requirements for sprayed concrete applied in the following examples are summarised in Table 2.

4.1 The Nathpa Jakry Hydroelectric Project, India

The Nathpa Jakry Hydroelectric Power Project is being built by the Nathpa Jakry Power Corporation Ltd., and is the largest tunnel project in India (Maharjan 1999). The Himalayan rivers carry heavy loads of silt. As the project is a run of river schemes, the removal of silt was important. This is being done through the use of underground de-silting chambers. The de-silting chambers were designed with a length of 525 m, width 17 m and height 27.5 m.

The geological conditions encountered included a variety of rock types, such as: closely jointed and weathered gneiss and augen-gneiss, amphibolite schistose in nature, slightly weathered along joints, biotite schist, and moderately jointed quartz mica schist.

As per the original design, the de-silting chambers were planned to be provided with a 300 mm thick reinforced concrete lining along with grouting and consolidation grouting. During excavation, rock bolts and 100 mm thick sprayed concrete with welded wire mesh were proposed.

The Nathpa Jakry Power Corporation Ltd. saw the potential of substituting the original support concept with fibre reinforced, permanent sprayed concrete. An analysis schedule was established including numerical modelling, seepage analysis, effective stress analysis, cost analysis, etc. to document the suitability of permanent sprayed concrete lining.

The conclusion yielded a sprayed concrete lining being chosen as the permanent support, accompanied by rock bolts and shallow drainage holes of 1-1.5 m to minimise pore pressure building up behind the lining. The sprayed concrete lining consisted of a 50 mm initial layer, and a second layer varying between 100 and 150 mm, all fibre reinforced, including 60 kg/m³ Dramix steel fibres, and wet-mix sprayed concrete. Weak zones and locations with permeability beyond 5 Lugeon (L/min/m at 10 bars pressure) were treated with grouting.

This project was the first of its kind in India and produced a permanent support structure which gave a 15% saving on the support costs, but more importantly, reduced the construction time by 8 to 10 months. A safe working environment was achieved avoiding the erection of a 22.5 m high gantry.

4.2 The Hvalfjördur Sub-sea Road Tunnel, Iceland

The Hvalfjördur Sub-sea Road Tunnel in Iceland breaks new ground for a tunnelling concept which to a great extent has been previously applied solely in hard rock environments in Scandinavian countries. The tunnel is the first ever sub-sea road tunnel built in a young basaltic rock mass which is active, both from a geothermal and seismic point of view.

The tunnel excavations encountered lava flows which varied in thickness from 2 to 10 m. Due to the thin layered nature of this material, and the gentle dip of the lava succession, a total of 80% of the tunnel was excavated under mixed face conditions. The rock type distribution encountered in the tunnel yielded 62% basalt and 30% scoria, the remaining 8% was mainly sediments. As many as 4 different rock units occurred simultaneously at the same working face. The bedrock was found to be highly affected by intrusive basaltic dykes and faults. A total of 135 dykes and 105 faults, mostly striking perpendicular to the tunnel route, were encountered during the tunnel excavation (Grøv et al. 1999).

The Q-system, applied for rock mass classification, included a modification to reduce the effect of the joint number. The encountered rock mass was classified as a total of 90% "good" and "fair" rock, whilst 10% was of "poor" quality. Rock support by means of rock bolts and wet-mix sprayed concrete became the only measure that was used. An average of 3.25 rock bolts per metre of tunnel, and an overall permanent sprayed concrete lining, 60-80 mm thick, has been applied throughout the tunnel. Rock mass with quality "worse than poor" was only recorded in limited zones. Zones of heavily jointed and clay filled faults and a section of sedimentary rock in the crown were negotiated with the additional support of 100-150 mm fibre reinforced (50 kg/m³ steel fibres)

wet-mix sprayed concrete combined with systematic bolting at 1.5 m spacing. All surfaces with fibre reinforcement have been covered by 20 mm plain sprayed concrete.

The geothermal activity in the tunnel area yielded a rock surface temperature of approximately 30°C. A maximum of 56°C was recorded in leaking ground water. A comprehensive procedure for surface wetting was conducted after spraying to avoid an accelerated hydration of the fresh concrete and to maintain its long-term durability.

The tunnel cross-sections varied, but were in general between 55 and 65 m², with enlarged sections in niches etc. A predefined rock support system was applied throughout the tunnelling work for cross-sections of 55 and 65 m² respectively (see Table 1 for detailed description of the rock support classes). A support structure consisting of fibre reinforced sprayed concrete, in combination with rock bolts, was found also to fulfil the requirements for flexibility and ductility resulting from the moderate seismic activity in Iceland.

4.3 The Stockholm Ring Road System, Sweden

The Southern Link is a major tunnelling project currently being undertaken to guide traffic around the central parts of Stockholm. The tunnel system includes a two lane, dual-tube system with a rock pillar of 10 m between the parallel tubes (Widing 1998) being constructed at a shallow depth of 5-25 metres in an area of densely populated residential and business centres. A maximum allowable water ingress of 2 to 4 litres per minute per 100 m is specified. Pregrouting is required to maintain the groundwater regime above the tunnel where gneiss, gneiss granite, and greenstone dominate the geology. Rock support measures will have a design life of 120 years.

Table 1. Predefined rock support classes, Hvalfjördur

Cross Section	Class 1 Q > 4	Class 2 1 < Q < 4	Class 3 0.1 < Q < 1
55 m²	S40 mm (r) S30 mm (w) B spot L=3 m	S60 mm B spot L=3 m Bcc 2.2 L=3 m	Srf80S20 B cc 1.7(r) B cc 2.0(w) L=3 m
65 m²	S40 mm (r) S30 mm (w) B spot L=3 m	S80 mm (r) S60 mm (w) B spot L=3 m B cc 2.2 L=3 m	Srf80S20 B cc 1.7(r) B cc 2.0(w) L=3 m

Cross Section	Class 4a Q < 0.1	Class 4b Q < 0.1	Class 5 Q < 0.1
55 m²	Srf130S20 B cc 1.5 L=4 m	Srf130S20 Ribs cc 2 m B cc 1.5 L=4 m	Cast-in-Place concrete lining
65 m²	Srf130S20 B cc 1.5 L=4 m	Srf130S20 Ribs cc 2m B cc 1.5 L=4 m	Cast-in-Place concrete lining

The following abbreviations have been applied in Table 1: S = Sprayed concrete (plain), Srf = Fibre reinforced sprayed concrete, B spot = Rock bolts (spot application), L = Length of rock bolts, r/w = roof/walls, respectively.

The rock support design involved calculations by both numerical and analytical methods including rock support measures such as rock bolts, sprayed concrete, and fibre reinforcement. A series of rock support classes have been established to guide the determination of permanent rock support with Q-values as low as 0.1. These classes include variable parameters such as: rock mass quality, rock cover, and tunnel width. As an example, in very poor rock (1<Q<0.1) and shallow rock cover (< 5m) the fibre reinforced sprayed concrete will amount to a thickness of 150 mm combined with rock bolts (of 25 mm diameter) in a pattern of 0.6 bolts per m².

The project has established a procedure for determination of permanent rock support based on pre-defined rock support classes, and rock mass conditions as encountered. Conditions outside the pre-defined system trigger particular actions to be taken, such as additional modelling and consequently new support classes to be issued, if required.

4.4 The Romeriksporten Railroad Tunnel, Norway

The Romeriksporten Railroad Tunnel accommodates one section of the railroad system that connects the City of Oslo, in Norway, with the new airport at Gardermoen. The double track tunnel, which has a cross section of almost 115 m², had a requirement for a maximum average water ingress rate of less than 20 litres per minute per 100 m. The project design life is a minimum of 50 years, the design travel speed is 210 km/hour.

The installed permanent support consists of sprayed concrete and fully grouted rock bolts. The contractor was responsible for the working environment and installed primary support close to the tunnel face. Rock mass classification and determination of permanent rock support were the responsibilities of the tunnel owner. Upon the instructions of the owner, the contractor installed all support measures before the ventilation tube, producing a supported tunnel some 50 m behind the tunnel face. Fibre reinforced sprayed concrete was applied with a thickness of 50 mm in rock mass of fair to good quality, which dominated the rock mass classification.

In distinct zones of worse rock mass quality, several thin layers building up from 100 to 220 mm wet-mix sprayed concrete lining, in combination with spiling bolts and radial bolts, were applied.

The tunnel has been designed as a drained structure where the water and frost protection system consists of PE-foam and sprayed concrete applied in both the roof and walls, as shown in Figure 3.

4.5 The Vereina Railroad Tunnel, Switzerland

The Vereina Railroad Tunnel in the east of Switzerland is approximately 19 km long. The tunnel was excavated by the use of TBM in one section and conventional drill & blast in another section. The tunnel is mainly single track, with a cross-section of 40 to 46 m². The overburden is as great as 1500 m. Some two-thirds of the tunnel length is in an old crystalline complex only slightly affected by the alpine folding and comprises gneiss and amphibolite.

The Vereina tunnel is designed and built with a single shell sprayed concrete layer. The term implies that sprayed concrete is used for both primary support and final lining. The main aspects of the single shell sprayed concrete construction are:
1. A composite structure,
2. Combined immediate rock support and lining,
3. Monolithic sprayed concrete lining, and
4. No sealing membranes separating the lining.

For the design of the support measures the following hazardous scenarios were defined: spalling of small rock fragments from the roof, rock failure from the roof, brittle failure (rock burst, separation of bedding layers), plastic deformations, fault zones, and leakage of water. The rock support needed to be designed to allow adjustments according to local changes of the geological conditions.

Assisted by rock mechanics considerations, the single shell sprayed concrete lining for the Vereina project was applied to take into account the hazardous scenarios described above, and included installed rock support measures in the final permanent lining. A typical overall lining thickness of 200-300 mm was applied. The method mobilised the flexibility of the sprayed concrete technology, but required a working process that adapted the rock mass as encountered during tunnelling (Amberg 1999).

5 SPRAYED CONCRETE LININGS IN ADVERSE ROCK MASS CONDITIONS

5.1 Introduction

For construction purposes the rock mass must be

Table 2. Summary of sprayed concrete requirements.

Project	Early strength MPa	Final strength MPa	Flexural strength MPa	Toughness MPa	Rebound %
Naptha Jakry	< 10* (3 days)	< 30 *	> 3.8	> 3.8	<12/8 (r/w)
Hvalfjördur	Not specified	40/45**	50 kg/m³ ***	50 kg/m³ ***	< 10
Vereina	No information	43/56	No information	No information	< 10
Romeriksporten	Not specified	40	40 kg/m³ ***	40 kg/m³ ***	< 10

The following footnotes apply in Table 2: * Indicates cast cylindrical tests; ** Indicates sub-sea part of the tunnel; *** Specification directs type of fibre and dosage; r/w roof and walls, respectively considered as a construction material. Throughout Scandinavia, general rock mass conditions are favourable for such utilisation. The geological setting is dominated by igneous rock types such as granite, together with metamorphic rocks of various types and origins like gneiss, shale, etc. The host rock is more or less intersected by weak zones, which may have an intense tectonic jointing, hydro-thermal alteration, or be faulted and sheared, constituting significant weaknesses in the rock and making the rock mass far from homogenous. These conditions may require rock strengthening measures.

The host rock in Scandinavia general varies from poor to extremely good rock quality according to the Q-system. The zones of weakness can exhibit great variation in quality, their Q-classification ranging from "extremely poor" rock mass at the lower end of the scale, to "good", with width extending from only a few centimetres to tens of metres. The stand-up time of many of these zones may be limited to only a few hours.

The tunnelling industry has acknowledged that the interface between existing infrastructure and planned future underground utilisation may be as decisive on the site selection of underground projects as the actual rock mass conditions. In this context, a need arises to provide cost effective support measures in situations with adverse rock mass conditions to secure appropriate performance of such projects.

One measure suitable for use in adverse rock mass conditions is sprayed concrete with alkali-free accelerators. Another is a support system based on reinforced ribs of sprayed concrete, particularly developed to be applied in adverse rock mass conditions, defined as Q < 1 or "very poor or worse" rock. For both systems, examples will be provided.

5.2 Alkali-free accelerators, specifications and application

The term "alkali-free" accelerator for sprayed concrete should only have one meaning. The alkali-free accelerator has no (or below 1%) alkali cations, reducing the risk of alkali cations reacting with sensitive minerals that occasionally are associated with aggregates used in concrete. Such reactions may have a detrimental effect on the sprayed concrete matrix due to the fracturing of aggregate grains resulting from their expansion.

The alkali-free accelerators for sprayed concrete have been developed as a result of market demands. These demands, according to (Garshol et al. 1999), are related to one or more of the following issues:

- Alkali-aggregate reaction reduction; by removal of the alkali content arising from common use of caustic, aluminate based accelerators.

- Work safety improvement; by reduced aggressiveness of the accelerators, avoiding skin burns, loss of eyesight, and respiratory health problems.
- Environmental protection improvement; by reducing the amount of aggressive and harmful components being released to the ground water.
- Final strength compensation; by forming a homogenous and compact concrete matrix without internal tension and micro-cracking. Traditional accelerators produce a loss of 15-50% in compressive strength.

The various types of alkali-free accelerators on the market can provide the following selection of characteristic properties for sprayed concrete:

- Early strength of 1 MPa after 1 hour of curing.
- Final strength reaching as a minimum the same level as without accelerator (some produce a 20% increase according to suppliers' brochures).
- Low rebound.
- 300 mm thickness sprayed in one operation.
- Low corrosiveness.
- Reduced permeability.

The Norwegian Public Roads Administration has performed a full scale test on a number of alkali-free accelerators currently available in the market (Storås et al. 1999). The results can be summarised as follows:

- No difference in personal dust exposure between the alkali-free and silicate based accelerators.
- Improved early strength development for the alkali-free accelerators compared to water glass.
- Wet conditions at spraying surfaces delay the early strength development for some accelerators.
- The tests indicate a durable, homogenous final product.

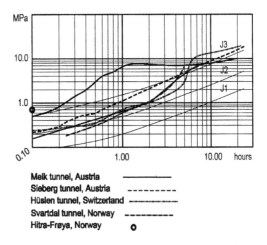

Melk tunnel, Austria ——————
Sieberg tunnel, Austria - - - - - - - -
Hüslen tunnel, Switzerland ·······
Svartdal tunnel, Norway — · — · —
Hitra-Frøya, Norway o

Figure 4. Strength developments using 7% Meyco SA 160

In northern Norway the Public Roads Administration (NPRA) has recently completed the construction of the FATIMA-tunnel. Veidekke AS was the civil works contractor (Hovland et al. 1999). The tunnel, a sub-sea crossing with a length of almost 7 km, encountered severe rock conditions from the very beginning of excavation at one tunnel face. This forced the implementation of a cast-in-place concrete lining. A sedimentary rock type consisting of shale, siltstone, schist and sandstone hampered progress. Clay-fillings, slickensided joint surfaces, and a rock consisting of "chips" in general less than 5 cm long (and occasionally up to 20 cm) was common. After a few kilometres of good progress on the other face the same sedimentary package was encountered. The rock mass was in general classified as $0.05<Q<0.1$, equivalent to "extremely poor" to "very poor" rock mass according to the Q-system. The situation called for an investigation of methods to replace the cast-in-place concrete lining.

The contractor looked at a spraying method that could produce a lining up to 500 mm thick in one spraying operation, high early strength, and a final strength fulfilling C45 criteria. A test program including alkali-free liquid accelerators commenced to prove that wet-mix sprayed concrete could replace cast-in-place concrete lining. Tests were successful and the NPRA approved the application involving a fibre reinforced (40 kg/m^3 steel fibres) wet-mix design including an alkali-free liquid accelerator in combination with systematic rock bolting.

A sprayed concrete lining, with a thickness of 200 to 250 mm, was established emphasising that the effect of a best possible arch-shape should be achieved. Half the thickness was sprayed immediately after mucking out, the remainder after the next blast round, to allow some deflection to take place. Rock bolts (0.25 bolts per m^2) were installed some 3-5 days after the completion of concrete spraying.

A monitoring program has been implemented consisting of convergence measurements. Almost six months after the support installation, deformations in the range of 1 mm/month took place. However, the best indicator of the behaviour of the support measures is the presence of visual cracking or damage on the sprayed concrete, of which there has been none reported.

The sprayed concrete design replaced the general application of cast-in-place concrete as permanent lining and the contractor experienced an increased progress from an average of 17.5 m per week per face to 30 m per face per week. However, some zones of exceptionally poor quality required additional support in the form of a cast-in-place concrete lining.

For the purpose of documenting that sprayed concrete linings with alkali-free accelerators can replace cast-in-place concrete linings as the main support solution, a full scale test was recently undertaken for a metro project in Portugal. The project specifications for the concrete lining were to be fulfilled by the sprayed concrete lining (Proenca 1999), including, amongst other aspects, the following:

- Design life of 100 years.
- Compressive strength of C30/35.
- Thickness of 400 mm.
- High impermeability, 5 mm after 28 days (DIN 1048).

The project, Lisbon Metro, was excavated through soft soil and geologically complex formations. A mix design including alkali-free accelerator in powder form was applied.

The aggressive environment required low permeability and a special cement, SR/MR ($C_3A\approx1\%$). The wet-mix design chosen included 1% superplasticiser to reduce the water/cement ratio and maintain workability. This produced a sprayed concrete lining with a rebound of 4%, and spraying was possible as long as 5 hours after mixing. Even below the water table (at 7 m), no membrane waterproofing system was used. A check two years after application showed a further increase of 35% over the 28 day compressive strength.

The primary support consisting of sprayed concrete reinforced with steel ribs was covered by a final lining of plain sprayed concrete.

Permanent support in "good" to "very poor" rock mass is predominantly handled with a combination of sprayed concrete and rock bolts in Norway. A number of options to replace cast-in-place concrete linings in adverse rock mass conditions have been tested. This has included combinations of spiling bolts, radial bolts, and rebar reinforced sprayed concrete, as well as pre-grouting depending on the actual conditions. One such successful option, termed "reinforced ribs of sprayed concrete" will be described in the following.

When the Q-value falls below 1, bolting as a support measure may not be adequate on its own. The rock mass between the bolts must be stabilised by means such as sprayed concrete. An increasing number of tunnelling projects in Norway face the situation of encountering adverse rock mass conditions. Therefore, alternative solutions have been considered. Reinforced ribs of sprayed concrete is one solution which has become a useful application in adverse rock mass conditions. It consists of a sandwich type construction, based on fibre reinforced (and also plain) sprayed concrete, radial bolts, and rebars. Figure 5 shows a principal solution of reinforced ribs of sprayed concrete (Aagaard et al. 1997)

The system has the following advantages:

- Materials to be used are normally available on most construction sites.
- Convenient construction, easy to handle materials, and on-site production.
- Flexible installation and wide span in capacity.
- Cost effective.
- Ductile, allowing rock deformations without imposing load concentration on support.
- Allows tunnel progress shortly after installation.
- Easy to repair and custom design by spraying thicker concrete or adding new ribs.

For the Bjorøy sub-sea road tunnel outside Bergen in western Norway, the tunnel encountered a zone of loose silty sand over a distance of 25 m (Aagaard et al. 1999). The Q-values in the critical section ranged from 0.08 to 0.003. The zone was successfully passed with a design including pre-grouting, pre-bolting (spiling), fibre reinforced sprayed concrete, and radial bolts. The invert was supported with a cast-in-place concrete lining. A monitoring program was established to follow-up potential deformations. The monitoring confirmed a stable situation without long term deformations. Despite the stability of the situation, the tunnel owner, the NPRA, required the rock support to withstand full hydrostatic pressure, which at this particular point was approximately 80 m, requiring a concrete cast-in-place lining to be installed.

The application of a similar structure was installed in Chile, in a tunnel prone to squeezing and swelling rock (Stefanussen 1998). In an area with lutite (rock consisting entirely of particles in silt/clay

fractions) and some 800 m rock cover, the criteria for squeezing was met. On this basis, a design criteria for the support pressure of maximum 2.5 MPa was established. Using circular reinforced ribs of sprayed concrete a capacity of 3 MPa per rib was calculated.

The Frøya-tunnel outside Trondheim faced a number of weak zones during the course of excavation. Rock support covered a wide range of measures from rock bolts and sprayed concrete to cast-in-place concrete. The effectiveness of various support solutions has been demonstrated by the use of numerical modelling (UDEC) for a relatively wide zone with an average Q-value of 0.0013, i.e. extremely poor rock (Bhasin et al. 1999).

The most favourable combinations (Bhasin et al. 1999) for support in the Frøya-tunnel were found to be: fibre reinforced sprayed concrete (Sfr), thickness 250 mm, combined with concrete lined invert and rock bolts (B) in roof and walls; reinforced ribs of sprayed concrete with 2 m spacing (RRS); and finally cast-in-place concrete (CCA), thickness 0.6 m in invert and 0.4 m in roof and walls. The modelling showed only marginal differences between these 3 alternatives. In Table 3 below, a summary of the results from numerical modelling is given.

A Master of Science thesis (Rødseth 2000) at the University of Technology in Trondheim applied FLAC 3D numerical modelling to the analysis of reinforced ribs for the same reinforcement schemes and rock mass conditions as described above. The thesis attempted to compare the UDEC results with a 3-dimensional approach.

The results confirm earlier calculations (Bhasin et al. 1999), indicating that the two methods (reinforced ribs with sprayed concrete and cast-in-place concrete lining) produced deformations with insignificant difference where a concrete invert was applied in both cases. FLAC 3D calculations, however, resulted in lower deformations than were observed for similar UDEC-calculations, but these were in the same order of magnitude as the measured conver-

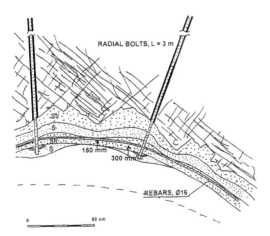

Figure 5. Reinforced ribs of sprayed concrete.

Table 3. Results from numerical modelling (UDEC)

Type of Support	Sprayed concrete 250 mm, concrete invert, rock bolts	Reinforced ribs and sprayed concrete (RRS)	Cast-in-place concrete lining (CCA)
Max. Displacement after equilibr.	14.4 mm	17.1 mm	17.3 mm
Max. axial loading on bolts	3.3 tonnes	11.6 tonnes	-
Max. axial load on the structure	1.96 MN (roof)	0.88 MN (roof)	1.4 MN (roof)
Max. joint aperture	3.3 mm	3.3 mm	3.5 mm
Max. shear displacement	10.7 mm	10.7 mm	11.7 mm

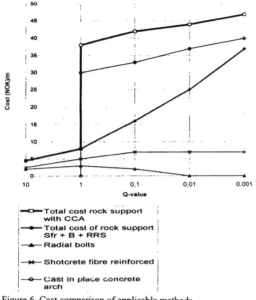

- Total cost rock support with CCA
- Total cost of rock support Sfr + B + RRS
- Radial bolts
- Shotcrete fibre reinforced
- Cast in place concrete arch

Figure 6. Cost comparison of applicable methods

gence. A cost comparison of various support measures for adverse rock mass conditions is provided in Figure 6, including reinforced ribs and sprayed concrete compared to cast-in-place concrete lining (Aagaard et al. 1997). One US dollar is equivalent to approximately 9 NOK.

Making a comparison of the costs associated with the solution involving reinforced ribs of sprayed concrete with traditional cast-in-place concrete it can be seen that for rock mass classified as 1>Q>0.001, the application involving reinforced ribs is the most cost-effective.

6 CONCLUSIONS

Sprayed concrete is a technology which is developing quickly. The solutions for permanent linings described herein involve a tunnelling method including sprayed concrete as an integral part of civil projects developed through active design.

The examples given demonstrate that the sprayed concrete-based tunnelling method has a successful record, including application in adverse rock mass conditions. For each application it is important that the technique and properties specified are adapted to the actual geological quality. This tunnelling method includes a number of elements that must be taken into account when applying permanent sprayed concrete lining. On this basis the method may also be applicable in adverse rock mass conditions. Calculations and experiences have proved that the sprayed concrete alternative can be applied in adverse rock mass conditions as a valid alternative to cast-in-

place concrete linings. Typical soft rock/soil tunnelling may, however, be outside the scope of the method, but alkali-free accelerators yield promising possibilities in this respect.

Sprayed concrete linings applied together with a range of tunnelling techniques, constitute a competitive and cost effective, flexible tunnelling method fulfilling rigorous demands on adaptability, predictability, forward prediction, and documentation.

REFERENCES

Aagaard, B. & Blindheim, O. T. 1999. Crossing of exceptionally poor weakness zones in three subsea tunnels, Norway. World tunnel congress '99. Norway. Balkema.

Aagaard, B., Grøv, E. & Blindheim, O. T. 1997. Sprayed concrete as part of rock support systems for adverse rock mass conditions. International symposium on rock support. Norway. Norwegian Society of Chartered Engineers.

Amberg, F. 1999. Single Shale Lining in the Vereina Tunnel. Third international symposium on sprayed concrete. Norway. Norwegian Society of Chartered Engineers.

Bhasin, R., Løset, F. & Lillevik, S. 1999. Rock support performance prediction of a sub-sea tunnel in western Norway. Third international symposium on sprayed concrete. Norway. Norwegian Society of Chartered Engineers.

Bieniawski, Z. T. 1984. Rock Mechanics in Mining and Tunnelling. Balkema. 1984.

Ericsson, S., Grøv, E., Hardarson, B. & Kröyer, J. 1998. The Hvalfjordur sub-sea tunnel, Iceland. A hard rock tunnelling concept in young basaltic rock mass. Underground construction in modern infrastructure. Sweden. Balkema.

Garshol, K. 1997. Single shell sprayed concrete linings, why and how. International symposium on rock support. Norway. Norwegian Society of Chartered Engineers.

Garshol, K. & Melbye, T. 1999. Practical experiences with alkali-free, non-caustic liquid accelerators for sprayed concrete. World tunnel congress '99. Norway. Balkema.

Grøv, E. & Blindheim, O. T. 1997. Selecting properties of sprayed concrete for different rock support applications. International symposium on rock support. Norway. Norwegian Society of Chartered Engineers.

Grøv, E. & Haraldsson, H. 1999. Hvalfjordur sub-sea road tunnel Iceland. Pre-construction forecasts and construction experiences. World tunnel congress '99. Oslo, Norway. Balkema.

Hovland, I. D., Holtmon, J. P. & Hauck, C. 1999. FATIMA – increased tunnelling results with replacement of fully casted lining at the North Cape Tunnel. Third international symposium on sprayed concrete. Norway. Norwegian Society of Chartered Engineers.

Løset, F., Kveldsvik, V. & Grimstad, E. 1997. Practical experience with the Q-system. Norwegian Geotechnical Institute. Norway. Published by NGI.

Maharjan, S. C. 1999. Practical application of steel fibre reinforced shotcrete in de-silting chambers of Nahpta Jakry Hydro Electric Project. Third international symposium on sprayed concrete. Norway. Norwegian Society of Chartered Engineers.

Proenca, A. M. 1999. Sprayed concrete-wet mix-replacing reinforced cast "in situ" concrete. Third international symposium on sprayed concrete. Norway. Norwegian Society of Chartered Engineers.

Rødseth, S. V. 2000. Numerical analysis of rock support. Master of Science Thesis, NTNU. Norway.

Storås, I., Bakke, B., Hauck, C. & Davik, K. I. 1999. Full scale testing of alkali-free accelerators – with special emphasis on working environment, safety and quality. Third international symposium on sprayed concrete. Norway. Norwegian Society of Chartered Engineers.

Stefanussen, W. 1998. Rock support by use of shotcrete reinforced ribs. Workshop on squeezing rock conditions in tunnelling. Underground construction in modern infrastructure. Sweden. Unpublished.

Widing, E. 1998. Planning of underground. Rock support design and construction for Sodra Lanken in Stockholm. Underground construction in modern infrastructure. Sweden. Balkema.

Shotcrete: Engineering Developments, Bernard (ed.) © 2001 Swets & Zeitlinger, Lisse, ISBN 90 5809 176 7

Monocoque lining design for the Gotthard Base Tunnel

H.Hagedorn & Z.Q.Wei
Amberg Consulting Engineers Ltd., Switzerland

ABSTRACT: Based on experience gained in the construction of two long tunnels (Furka, 15.4 km; Vereina, 19 km) and large caverns, we have formulated the steps involved in monocoque lining design with regard to rock mechanics. With this formulation, we hope that the single shell lining method of tunnel construction, which offers a potential total cost reduction up to 25% compared to a double shell lining, will encourage people to take advantage of this technology to develop cost-effective, environment-friendly and safe methods for underground construction to meet the increasing demands for underground space, traffic, cost reduction, quality and safety.

1 INSTRUCTIONS

The underground construction method involving a single shell lining reduces the required amount of materials and labour and the structure comes into service earlier than with alternative methods of tunnel construction. With the introduction of steel fibre reinforced sprayed concrete, the labour-intensive work of steel-mesh fixing is eliminated, and the lining thickness is further reduced. Generally speaking, it is possible to save up to 25% (Melbye & Garshol 1999) of the total cost compared to a double shell lining and even more if the financial aspects of early operation of the structure are taken into account. For these reasons, single shell linings are being applied increasingly in underground construction (Brux 1998).

Support is defined as all the measures needed to stabilize the underground opening permanently, to limit deformations, and give the structure its required shape and other desired properties. These properties include, for example, durability, impermeability, and the appearance of the underground structure. The support then comprises the immediate excavation support measures and the subsequent lining. The former involves the technical measures to be implemented directly after the excavation process. These guarantee the stability of the opening, prevent unacceptable deformations and also provide safety for the workforce and the equipment. The lining represents the completion of the rock support

and also serves the purpose of providing the structure with its required shape and properties.

In the case of a single shell lining, the immediate rock support measures and the lining are combined in such a way that they provide force transmission to act as a permanent composite structure and fulfill shape requirements, or it is simply a single pre-cast concrete segment lining. Monocoque linings are a special case in that shotcrete is used both for the immediate support and the permanent lining combined with other support measures. In the case of a single layer support system the means of providing rock support must be flexible and must be adjusted to changing geotechnical conditions and geometry. This flexibility relates to the required support action and the time and place of installation of different excavation support measures.

For the construction of a double shell lining, the primary support is required to maintain stability at least up to the time the final lining comes into effect, and the long term stability generally must be guaranteed by the final lining. Due to the large safety margin, there are fewer requirements for the primary support than for the single shell lining. The composite action of the monocoque lining must be guaranteed so that it can act as both the primary support and the permanent lining. This paper will examine the rock mechanics aspects of monocoque lining design for a long deep railway tunnel (the Gotthard Base Tunnel, Switzerland) based on experience and analysis.

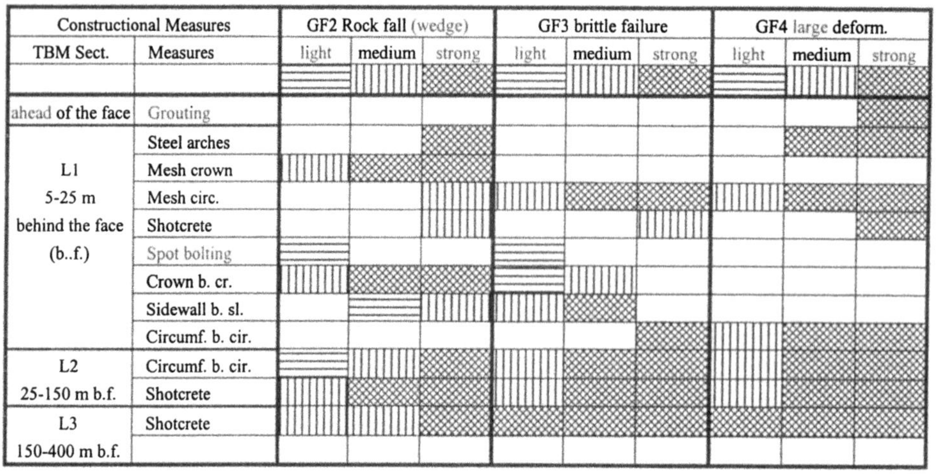

Constructional Measures		GF2 Rock fall (wedge)			GF3 brittle failure			GF4 large deform.		
TBM Sect.	Measures	light	medium	strong	light	medium	strong	light	medium	strong
ahead of the face	Grouting									
L1 5-25 m behind the face (b..f.)	Steel arches									
	Mesh crown									
	Mesh circ.									
	Shotcrete									
	Spot bolting									
	Crown b. cr.									
	Sidewall b. sl.									
	Circumf. b. cir.									
L2 25-150 m b.f.	Circumf. b. cir.									
	Shotcrete									
L3 150-400 m b.f.	Shotcrete									

Figure 1. Assignment of support elements to the hazard scenarios with weighting

2 POTENTIAL HAZARDS, CONSEQUENCES AND RISKS

The first step in the design of support for tunnels or underground caverns is to determine potential hazard scenarios from the geological and geotechnical conditions, and the geometrical properties of the proposed structure. For the preliminary design of the support measures and the lining of the proposed deep tunnel the following hazards have been defined:

GF1 Small rock fall
GF2 Large unstable rock wedges
GF3 Brittle failure (rock bursts, spalling and violent buckling of layers)
GF4 Large deformation and high ground pressure
GF5 Instability of the face
GF6 Fault zones
GF7 Ingress of high pressure groundwater

The combination of the above hazards and some other scenarios during construction can also occur. Therefore, the rock support must be planned in such a way that it can be adjusted to the changing rock conditions.

The second and third steps of the support design are to consider the consequences and risks of each potential hazards. The scenario GF1, "small rock fall", although not of a large scale, can have serious consequences because it can damage equipment and endanger personnel without any warning. Thus it is considered high risk.

Scenario GF2, "large unstable rock wedges", will only occur if there is a lack of ground confinement or if an incorrect method of excavation and construction has been chosen, or the work has been carelessly carried out. The risk can be high or low depending on several aspects, but generally, unstable wedges in the roof have high risk, because there is no warning before they fall down.

Scenario GF3, "brittle failure", has different forms and consequences, and therefore risks. The condition for rock burst and spalling is brittle rock under high ground pressure. Their occurrence can happen at any time for a long period of time. The dynamic behaviour of brittle failure, as in the scenario GF1, can also damage equipment and endanger personnel without any warning. They thus have a high risk potential. The most dangerous case of scenario GF3 is the violent buckling of layers without any warning, which usually leads to catastrophes. This is also considered high risk.

Scenario GF4, "large deformations and high ground pressure", occurs in soft ground and also in hard rock under high ground pressure. Without groundwater and for small excavation areas, this scenario takes place slowly compared to the advance rate, and is therefore considered a low risk. However, when combined with groundwater or a large excavation in soft ground, it becomes very dangerous and risky.

Scenario GF5, "instability of the face", usually occurs in loose ground or in weak rocks under high ground pressure. In most cases it is accompanied by warnings but generally people do not recognize them since they are on the face and roof, thus it should be classified as high risk, in particular, if combined with groundwater. If the face is left open too long, the violent buckling of scenario GF3 can also occur at the face.

Scenario GF6, "fault zones", has various consequences and risk levels depending on the width, orientation to the structure axis, mechanical properties, and hydro-geological conditions of the fault zones. Narrow fault zones (< 3 m) do not have many problems and risks. Wide weak fault zones with water always pose big problems and high risks. For fault zones (both narrow and wide) under the strong influence of tectonic activities one can only reduce the damage and risks and not eliminate them.

Scenario GF7, "high groundwater pressure and ingress of water", occurs in jointed rocks or saturated ground or fault zones and always poses a high risk. As an example, in an accident in the investigation tunnel, measuring 5 m in diameter, max. 1400 m depth, and 5.5 km length, for the Gotthard Base Tunnel, sands under 900 m of water pressure flooded the tunnel on March 31, 1996.

According to the consequences and risks of each potential hazard, different intensities can be specified, e.g., low, medium or high. The span or height of the structure plays an important role in the classification of the intensities. Figure 1 shows a part of the classifications devised for the TBM driven sections of a deep tunnel. Different hazard scenarios are given in the table headings. The different intensities of shading in the cells of the headings denote the severity of the hazard.

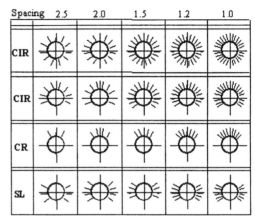

Tunnel diameter: 9.6 m
Anchor length: 3-8 m

Figure 2. Pattern bolting according to the hazard scenarios.

3 COUNTER MEASURES AGAINST THE POTENTIAL HAZARDS

The fourth step in support design is to select counter measures against different grades of potential hazard under consideration, taking into account the feasibility of techniques, costs, environment, and work conditions (personnel health and safety). For the support of the Gotthard Base Tunnel the following means, which may be combined, were chosen:
- Reinforcement mesh or steel fibres
- Lattice girders or steel profiles in the roof
- Systematic grid of anchors in the roof (CR, Figure 2)
- Systematic grid of anchors partially distributed such as in the side walls (SL) in case of high vertical tangential stresses giving rise to rock bursts and spalling (Figure 2).
- Systematic grid of anchors in the tunnel circumference (CIR, Figure 2)
- Pre grouting and pre-reinforcing
- Steel arches (girders, TH-profiles, HP-profiles etc.)
- Shotcrete or sprayed concrete

For the TBM driven sections of the deep tunnel, the possible support measures are given (Figure 1). The different intensities of shading in the cells of the matrix denote the necessity of the different measures. L1, L2 and L3 denote the working areas 5-25m, 25-150m and 150-400m behind the face.

It should be pointed out that, for each intensity of a hazard scenario, the specification of support measures must consider when, where, and how they should be put in place, in addition to the feasibilities of techniques, costs, environment, and work conditions (personnel health and safety). Thus quantified analysis methods must be used.

Different technically feasible support measures have different qualities, effects, and risks. In the quantitative analysis, various safety factors should be set according to the importance (and requirements) of the structure, intensity of hazard, and quality of support measure. For each intensity of a hazard scenario there may be a few alternative support measures possible. Before going into the detailed quantified analysis and in order to reduce the amount of work, the selection of alternatives should be done after rough comparison.

4 METHODS OF ANALYSIS FOR SUPPORT DESIGN

There are many methods available for the analysis of support design. A few of them will be presented below. The problem is to properly select them. The general criterion is that the principal mechanisms of deformation and failure and the interaction between ground and support should be considered both on the large scale as well as on the local scale. Unlike surface structures, underground structures have the advantage of making use of the self-supporting capacity of the ground. Underground construction is still expensive, which is partly due to the fact that the self-supporting effects of the ground have not been properly exploited in most cases. With a cost reduction of up to 25%, the single shell lining method will encourage people to take full advantage of this technology to develop cost-effective, environment-

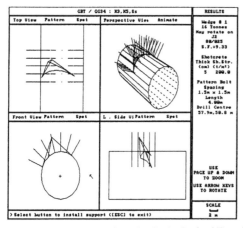

Figure 3. Different views of a wedge for the Gotthard Tunnel.

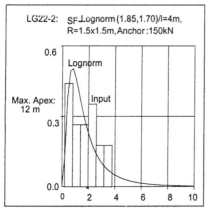

Figure 4. Distribution of safety factors for different wedges.

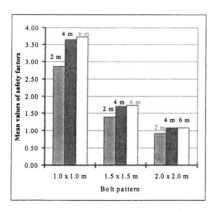

Figure 5. Mean values of safety factors of different bolt patterns and bolt lengths.

friendly, and safe methods for underground construction.

4.1 *Large unstable rock wedges*

The first computational model refers to the hazard scenario GF2, "large unstable rock wedges". These instabilities occur due to unfavorable orientations of joint systems. They jeopardize the workforce and machinery in the area of excavation. Three joints form a wedge which can detach from the rock mass and fall into the opening. According to the number of discontinuity planes there is a given number of wedge combinations. The positions of these wedges can be determined and a safety factor for a given bolt length and bolt pattern can be computed. Figure 3 shows one wedge formed by 3 of a total of 6 discontinuity planes in a gneissic rock formation of the deep tunnel. For a given bolt system (pattern, bolt length and bolt capacity), the lognormal distribution of the safety factors for all wedge combinations is depicted in Figure 4. The mean values of all the safety factors for different bolt pattern densities and bolt lengths are shown in Figure 5. For this investigation the program Unwedge was used (Carvalho et al 1992).

4.2 *Roof collapse*

A further computational model refers to the investigation of roof collapse. This hazard is present when there is an unfavorable orientation of joint planes with respect to the tunnel and cavern axes. If these are known in advance, then the potentially dangerous blocks of rock will be recognized and technical measures necessary to remedy the situation can be planned in the design stage. For a shallow tunnel in

Surface

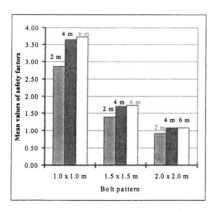

65 m

13 m

No convergence of deformations

Figure 6. Without a lining the roof is unstable.

massive steeply jointed rock, computational investigations for stability were carried out using a computer program. This is based on a distinct element code where the discontinuity formed by the joints can be simulated. Figure 6 shows the deformation vectors in the roof area of the excavation without a lining or rock support. The size and development of the deformations point to a collapse of the system which would involve a collapse up to the ground surface. In this way, a rock loosening zone is formed, which acts as a loading on the lining. In this particular case, systematic anchoring in the roof area combined with a reinforced shotcrete lining and, in some cases, steel girders, are required to ensure tunnel stability in the excavation area. The reinforcing mesh of the first shotcrete layer should be fixed to the rock anchors. This connection prevents the detachment of the shotcrete from the rock in case the bond strength between them is exceeded (see Figure 7).

4.3 Large deformation and high ground pressure

4.3.1 Axisymmetric model computation

For a deep tunnel with an overburden of up to 2500 m, extensive feasibility studies for a TBM heading together with the use of a monocoque lining based on shotcrete were made. In order to confirm the structural integrity of the lining it was necessary to quantify the deformations that occur after installation of each layer. The extent to which the deformations can be reduced by means of additional support measures such as grouting or pattern anchoring must be calculated. The deformations of the first shotcrete layer therefore depend, among other factors, on the distance to the face at which it is installed. This influence was investigated using an axi-symmetric model for a circular tunnel section of 9.6 m diameter and an overburden of 2000 m (see Figure 8). The modulus of the rock is 20 GPa, the coefficient of cohesion c is equal to 0.8 MPa and the friction angle ϕ is taken to be 34°. The first strengthening was through an increase of rock cohesion by 0.3 MPa in a grouted annulus of 3 m thickness over a distance of 10 m between 5-15 m behind the face. At a distance of 15-25m, a further increase in strength to a total increase of 0.50 MPa followed due to grout hardening or rock bolting. Simultaneously a shotcrete lining 30 cm thick was installed in this latter interval. The increasing load bearing capacity of the lining is numerically accounted for by increasing the modulus

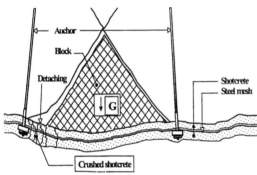

Figure 7. Preventing the detachment between rock and shotcrete.

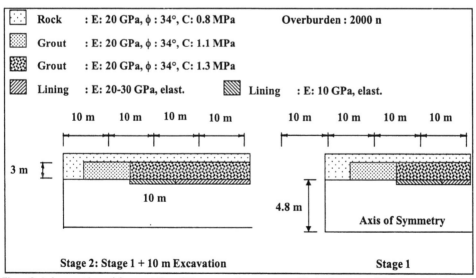

Figure 8. Axisymmetric model computation of 10 m excavation segments with simultaneous stabilisation of the rock (cohesion increase) and placing of the lining.

of elasticity of the shotcrete at each excavation stage. The variations of the tangential normal stress Szz in the concrete, shown in Figure 9 for points at 15, 20 and 25 m distance from the face, indicate the varying loading of the lining. The highest increase of deformation occurs close to the face and the stresses in the shotcrete layer are therefore considerably higher at a distance of 15 m from the face compared to 25 m.

4.3.2 Characteristic line method

Increasing the strength of rock around the opening is a way to reduce deformations for low strength rock masses or high overburdens to an admissible value and ensure stability until the first shotcrete lining is installed. By strengthening the rock mass, the final deformations of the underground opening are controlled and the shotcrete lining can be placed closer to the working face to prevent rock fall or to enhance the preliminary rock support, if necessary. Strengthening can be accomplished by grouting or systematic anchoring. Both are measures that can be implemented immediately behind the working face. The influence on the deformations of the underground opening of rock strengthening and of successively applied shotcrete layers can be estimated using the characteristic lines of the structural system "rock mass – lining" (Brown et al. 1983). Every increase in rock strength, or every additional layer of shotcrete, changes the system, i.e. the characteristic line must be re-determined for each stage of support. The computational procedure for the determination of the characteristic line includes the excavation, rock strengthening, and three successively applied 10 cm thick layers of shotcrete for a monocoque lining. For each constructional measure such as rock strengthening and the application of the shotcrete layers, a new characteristic line for the corresponding statical system was computed. In Figure 10, curve R represents the characteristic line of the rock mass, RG that of the strengthened (grouted or anchored) rock ring and the rock mass, and RGSL1 that of the strengthened rock ring, the rock mass, and the first shotcrete layer installed at an internal pressure of 1 MPa (Point 1 in Figure 10) near to the working face. The characteristic line for the first shotcrete layer is labeled SL1 and would intersect curve RG at an absolute radial deformation of approximately 100 mm. At point 2 the characteristic line of the system with the strengthened ring plus 2 shotcrete layers start so the bearing capacity of the second layer was not added up to that of the first layer. The intersection point EQ of this curve with the characteristic line of the third shotcrete layer marks the point of equilibrium of the system. From the characteristic lines for the three shotcrete layers SL1 - SL3 (only SL1 is marked in Figure 10) it is evident that the third layer behaves elastically up to the equilibrium point EQ. For a hydrostatic primary stress state, this method is suitable to estimate the expected radial deformations of the individual shotcrete layers and the necessary increase of rock cohesion, such that the deformations are limited to minimize damage of the first shotcrete layer applied. In a similar way the method also shows which rock support measures have to be carried out using flexible steel arches (TH-profiles with frictional connections) or lattice girders to keep deformations of the shotcrete linings to a minimum.

Rock (R) + Stabilized rock ring (G) + Shotcrete Layers (SL)

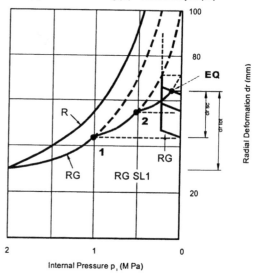

Figure 10. Characteristic line method for a system with rock strengthening and monocoque shotcrete lining applied in 3 layers.

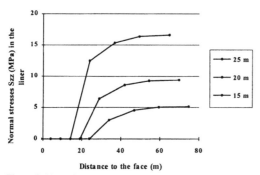

Figure 9. Normal stress Szz in the lining as a function of the distance to the tunnel face.

5 EXPERIENCE OF THE AMBERG ENGINEERING GROUP IN MONOCOQUE LINING

5.1 *Furka Tunnel, 15.4 km, Switzerland*

The Furka Railway Tunnel was constructed in the Swiss Alpine region at up to 2000 m above sea level, where winter produced arctic-like conditions. This project involved a long railway tunnel at a maximum depth of 1500 m that was put into service on June 25, 1982. The span of the double track railway tunnel is 8.84 m, the maximum span of the branching section is 13 m. Amberg Engineering Group did the detailed design and construction supervision: drill-blasting excavation method, tunnel support and final shotcrete lining (moncoque), construction planning, cost estimation and control, quality assurance and geotechnical monitoring. The encountered problems were ingress of water (flooding of tunnel), top heading and face collapse, spalling, buckling of layers and large deformations of steel arches. Excavation was divided into seven types. The support and lining consisted of shotcrete, mesh, anchors and, in some sections, also steel arches. For the details of the monocoque lining design, see Amberg & Sala (1984) and Hagedorn (1982). The rock bolts were corrosion-protected (Hagedorn 1983).

5.2 *Vereina Tunnel, 19 km, Switzerland*

The Vereina Tunnel is a 19 km long single track railway tunnel at a maximum depth of 1450 m that was put into operation on November 19, 1999, mainly for car transportation between Klosters and Lavin in the Alps of the Canton Grison. The TBM normal profile had a diameter of 7.64 m and the conventional one 6.5 m. Amberg Engineering Group also did the detailed design and construction supervision. At both portals and in the middle of the tunnel the profile was enlarged to a double track cross section for operational purposes. The portals, each with a length of approximately 2000 m, were excavated by blasting. On the Klosters side, this double track section had to be driven through a tectonically strongly sheared Serpentinite formation of length 500 metres. Already during the top excavation, radial deformations of a magnitude of 20 cm were measured and the profile had to be enlarged correspondingly in order to keep the required clearance. Due to the poor rock conditions this part was excavated mechanically by means of a pick hammer. TH-steel profiles 36/58 as well as the first shotcrete layer of 10 - 15 cm were installed immediately after each excavation length of 80 cm on the top. A similar procedure was used for the bench and invert. The full action of the primary support could therefore be provided after setting the steel arches in the invert.

The TH-profiles are connected by means of steel clamps in the overlapping section of the profile segments, thus enabling the arch to sustain large peripheral deformations without reducing the support capacity. Figure 11 shows this connection in the area of the bench excavation. In the footings of the top and bench excavation section, longitudinal steel reinforcement fixed with anchors were applied in order to stabilize the abutments and to distribute differential radial arch loadings in the direction of the tunnel axis.

The design of the final lining required a knowledge of the deformation behaviour of the above primary support and convergence measurement sections were therefore installed. The horizontal convergence of all measuring sections is shown in Figure 12. The line marked "Broken" represents a measuring section placed immediately behind the top excavation.

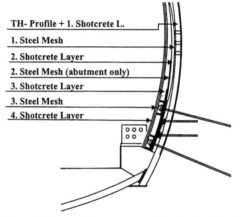

TH- Profile + 1. Shotcrete L.

1. Steel Mesh
2. Shotcrete Layer
2. Steel Mesh (abutment only)
3. Shotcrete Layer
3. Steel Mesh
4. Shotcrete Layer

Figure 11. Set up of the monocoque lining in the abutment of Vereina tunnel.

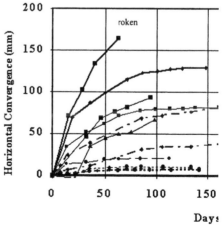

Figure 12. Horizontal convergence in the measuring sections in Vereina tunnel.

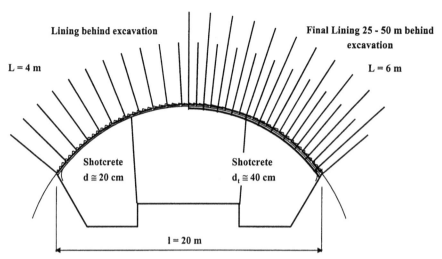

Lining behind excavation

Final Lining 25 - 50 m behind excavation

L = 4 m

L = 6 m

Shotcrete
d ≅ 20 cm

Shotcrete
d_t ≅ 40 cm

l = 20 m

Figure 13. Primary support and final lining of a large cavern in limestone.

By means of a back analysis it was possible to assess the amount of relaxation which took place before placing the final lining consisting of an additional reinforced shotcrete layer. The design of the final lining therefore considered this relaxation.

Before placing the final lining the rock around the tunnel was grouted with a cement suspension on both sides of each steel arch. This measure was necessary to prevent corrosion of the steel and a corresponding reduction of the primary support bearing capacity.

The constructional set-up in the change from the arch to the invert lining is shown in Figure 11. The results of the computational investigations showed a concentration of shear forces in this abutment. For this reason two additional rows of fully grouted anchors were installed and connected by means of steel profiles in the longitudinal direction for the distribution of the abutment forces.

5.3 System of large caverns, Switzerland

Another example of the successful application of a monocoque lining is a large cavern system where Amberg Engineering Group did the detailed design and construction supervision. A cavern with a span greater than 20 m in a limestone formation at a depth of approximately 1000 m will be presented. The cavern had a horizontal cylindrical shape and was constructed top-down in 4 stages. Only when an entire stage was completed, including the final lining of the whole cavern length, was the next stage below commenced.

The lining setup consisted of a primary support with 4 m long fully grouted (cartridges) epoxy reinforced glass fibre anchors (according to the pattern depicted in Figure 13) and a 20 cm thick shotcrete

lining reinforced with steel mesh. This support was placed immediately after each excavation step of 4 m length with a safety distance of 75 cm to the rock excavated with the next blasting. This offset was sufficient to prevent destruction of the primary support due to blasting effects.

The final lining was placed at a distance of 25 to max. 50 metres behind the excavation. It consisted of an additional shotcrete layer 20 cm thick reinforced with steel mesh as well as 6 m long fully grouted (epoxy resin) reinforced glass fibre tube anchors. With this lining procedure, each stage was fully supported at a maximum distance of 25 m behind the face.

The reinforcing mesh for the primary support and those for the final lining were connected with the heads of the anchors according to Figure 7.

6 MONOCOQUE LINING DESIGN FOR THE GOTTHARD BASE TUNNEL

This twin tube railway tunnel is 57 km long, has an excavation diameter of 9.3 m in the normal profile, and was designed for high speed rail transit. About 2×26 km of the tunnel was located at a depth of 1000 to 2500 m. Already in the preliminary design stage, a detailed determination of the support elements has been made for the different geotechnical conditions in order to arrive at a better cost estimate. Thus, different anchor patterns had to be defined (see Figure 2), which had to be adjusted to the different hazard scenarios, e.g., those in Figure 1. The circumferential patterns (CIR) of the anchors are applied in cases where failure of the rock is expected around the opening and the pattern and anchor length can be adapted to the intensity of the failure mechanisms. The assessment for the required pattern densities and bolt lengths is described in Section 4.3. To assess the increase in cohesion achieved through pattern bolting, see Egger & Pellet (1991). The crown bolting (CR) is especially necessary to prevent rock fall from the roof and a corresponding design method is described in Section 4.1.

Bolting of the side walls (SL) is recommended in case of spalling, buckling, and rock burst effects as well as for yielding due to overstressing. It is emphasized that the precise location of the bolts in the cross section depends on the direction of the principal stresses. The depicted arrangement is therefore valid for a vertical major principal stress direction.

The thicknesses of shotcrete layers for the monocoque lining were determined for the working sections L1, L2, L3 (see Figure 1), using various rock mechanics design methods, where the individual shotcrete layers had to be applied. An example of forms of different shotcrete linings is given in Figure 14. These were combined with different anchor patterns and steel sets for a section of Gotthard Massif. Three lining types, light, medium and heavy, were defined for selection.

The initial decision of the Swiss Federal Railways favoured the monocoque lining largely due to the positive influence and cost-efficiency of the two long Swiss railway tunnels mentioned above. All initial design was based on the construction method of a monocoque tunnel lining. Because of the required high train frequency (up to 200 trains per day, or 4 trains per half an hour) and high train speeds over 200 km/h, a 100 year design life without any major interference or repair work is required. Therefore, the decision was later changed to a double shell lining. But one can imagine how much this decision costs, considering that the inner lining must withstand the various loadings by itself without taking the supporting effects of the primary support into account. The inner lining can be installed only when rock deformations cannot damage the inner lining. Because it can take months or even more than a year for rock deformation to stabilize, and because there may be delayed lining construction, the initial monocoque design documents were used for the double lining design with minor changes regarding material types and permeability requirements.

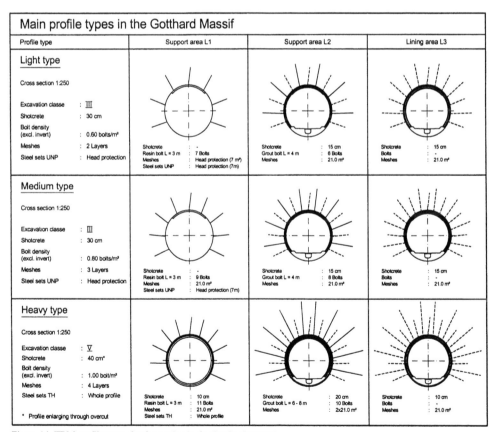

Figure 14. TBM profile types for the section in the Gotthard Massif.

143

7 CONCLUSIONS

From experience gained on several projects, the following steps for monocoque lining design were formulated with respect to rock mechanics issues:

- Identify potential hazard scenarios
- Consider the consequences of hazards
- Evaluate the risks of each potential hazard
- Assign intensities, low, medium, and high, to each potential hazard if refined classification is required
- Select counter measures against different (grades of) potential hazards
- Set safety factors according to the importance (requirements) of the structure, intensity of hazard, and quality of support measures
- Select the methods for support design analysis, taking into account the principal mechanisms of deformation and failure, and the interaction between ground and support
- Detailed quantitative analysis
- Decisions for design

Using a single shell lining can reduce the total tunnel cost by up to 25% compared to a double shell lining. With this formulation we hope to encourage people to take full advantage of this technology to develop cost-effective, environment-friendly and safe methods of underground construction to meet increasing demands for underground space, traffic, cost reduction, quality, and safety.

REFERENCES

Amberg, R. and Sala, A. 1984. Shotcrete as Permanent Lining for the Furka Base Tunnel. *Rock Mechanics and Rock Engineering* 17(1): 1-14.

Brown, E.T., Bray, J.W., Ladanyi, B. & Hoek, E. 1983. Characteristic line calculations for rock tunnels. *J. Geotech. Engng Div., ASCE* 109, 15-39.

Brux, G. 1998. Einschaliger Tunnelausbau mit Spritzbeton. *Tunnelbau:* 172-224. Essen: Glückauf.

Carvalho, J., Hoek, E. & Li, B. 1992. Unwedge – a program for the 3D visualization of potentially unstable wedges in the rock surrounding underground excavations and calculation of factors of safety and support requirements for these wedges. Rock Engineering Group, University of Toronto.

Egger, P. & Pellet, F. 1991. Strength and deformation properties of reinforced jointed media under true triaxial conditions. In *Proc. 7th ISRM Congr., Aachen:* 215-220.

Hagedorn H. 1982. The stretch of double tracks in the section 61 of Furka Tunnel. *Journal of Swiss Engineers and Architects:* June, 535-542 (in German).

Hagedorn H. 1983. Synthetic resin grouted tube anchors. *Rock Mechanics and Rock Engineering* 16: 143-149.

Melbye T. & Garshol, K. 1999. Concrete with all the advantages. *Tunnels and Tunnelling International*, Aug., 52-53.

Shotcrete: Engineering Developments, Bernard (ed.) © 2001 Swets & Zeitlinger, Lisse, ISBN 90 5809 176 7

The M5 motorway tunnel: an education in Quality Assurance for fibre reinforced shotcrete

S.A.Hanke & A.Collis
Baulderstone Hornibrook and Bilfinger + Berger Joint Venture (BHBB JV), Sydney, Australia

E.S.Bernard
University of Western Sydney, Nepean, Australia

ABSTRACT: The underground extension of the M5 Motorway through the southern suburbs of Sydney, Australia, resulted in a two-lane dual carriage way of 3.8 km length under sensitive urban areas. Fibre Reinforced Shotcrete (FRS) was used in combination with permanent rock bolts to support the roof of these tunnels. The specified design life of 100 years for the major support elements required Quality Assurance (QA) measures that were extra-ordinary in their frequency and breadth. Persistent problems in achieving post-crack load carrying specifications using beams led to the adoption of Round Determinate panels as the basis for performance assessment. However, correlations in performance between the beams and panels had to be established, and obstacles to effective specimen production, handling, and testing overcome, before this test could be fully implemented as the basis for performance assessment.

1 INTRODUCTION

1.1 Project overview

The M5 Motorway is an important transport thoroughfare through the south-western suburbs of Sydney. The first stage of this motorway was constructed in 1985, and extended from Beverly Hills to Casula in Sydney's southern suburbs. The new M5 East Motorway will connect the existing M5 to the General Holmes Drive adjacent to Sydney's Kingsford Smith Airport (Fig. 1). The project is due to be completed in early 2002.

The total length of the M5 East Motorway Project is 10 km. The principal elements of the project include 4 km of open dual two-lane carriage-way, 3.8 km of dual two-lane tunnel, a 700 m long under-river crossing constructed by a cut & cover method, and two viaduct structures 300 m and 700 m in length. The viaducts pass over sensitive wetlands at the eastern end of the thoroughfare.

The main tunnels comprise two parallel drives with cross sections of between 49 m^2 and 56 m^2. There are three entry/exit ramps running parallel and connecting with the driven tunnels at the eastern end. The cross section of these ramps varies between 50 m^2 and 80 m^2 with the span at the intersection between the main tunnels and the ramps widening up to 20 m. Pedestrian cross-passages are provided every 120 m, and a vehicular cross-passage is located at the tunnel midpoint.

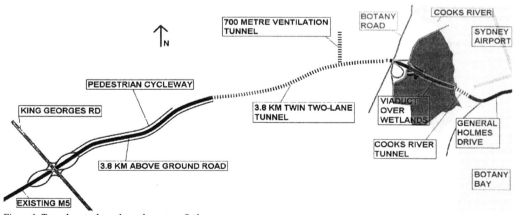

Figure 1. Tunnel route through south-western Sydney.

1.2 Tunnel design and construction

The majority of each driven tunnel is in moderately weathered to fresh Hawkesbury Sandstone common to the Sydney region. Hawkesbury Sandstone comprises massive, laminated and cross-bedded quartz sandstone in near horizontal beddings with some siltstone layers generally less than 4m in thickness. The unconfined compressive strength (UCS) ranges from 10 to 65 MPa.

Excavation of the main tunnels was carried out by road-headers, with a total of 5 machines in service (Fig. 2). Excavation rates varied between 4 and 8 m/day depending upon rock conditions. The properties of the Hawkesbury Sandstone allowed the tunnel to be excavated to an almost rectangular shape with a nearly flat tunnel roof in ground exhibiting good conditions. This is illustrated by profile Types 1 and 2, shown in Figs. 3 and 4. At the eastern end of the tunnel around the Princes Highway exit ramps, excavation was required under an existing rail line through loose sand, with only 5 m of cover. This required the adoption of a different type of ground support that included the installation of Alwag AT Casing System canopy tubes as a pre-excavation support measure in combination with Alwag lattice

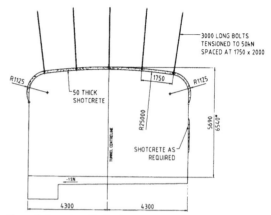

Figure 3. Tunnel profile for Type 1 support.

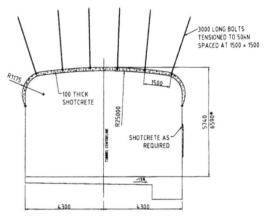

Figure 4. Tunnel profile for Type 2 support.

Figure 2. Excavation with road header and parallel bolt installation.

Table 1. Standard ground support details.

Support	Details
Type 1 Very good conditions	Sandstone with siltstone laminations less than 5 mm thick, profile almost rectangular. 50 mm thick steel FRS 3 m CT bolts @ 1.75×2.00 m
Type 2 Good conditions	Inter-bedded sandstone and siltstone, or siltstone, profile almost rectangular. 100 mm thick steel FRS 3 m CT bolts @ 1.50×1.50 m
Type 3 Poor conditions	Dyke rock and/or fault zone within sandstone, horseshoe shaped profile. 250 mm thick steel FRS 3 m CT bolts @ 1.0×1.0 m

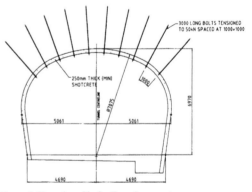

Figure 5. Tunnel profile for Type 3 support.

girders and mesh reinforced FRS (Fig. 5). Details of the ground support measures employed in the main-line tunnels are given in Table 1.

Rock bolts were used together with the FRS for all three ground support profiles. Sheathed solid bar rock bolts (Strata Control Systems CT bolts) of between 3 m and 5.5 m length were used for tunnel widths less than 15 m. Sheathed Megabolt cable bolts of 7 m length were used for sections where the tunnel width exceeded 15 m. The bolts were pretensioned to 50kN for the CT bolts, and 100 kN for the cable bolts, respectively. Full encapsulation with shrinkage compensated cement grout was used to prevent corrosion by groundwater. Installation of the rockbolts was carried out using automated rock bolting equipment (shown in Fig. 2).

The main purpose of the shotcrete lining in the hardrock tunnels was to seal the weathered rock and stop further deterioration, prevent fall out of rock wedges, and help prevent groundwater leaking onto the roadways. The objective of the rock support was to generate a reinforced rock beam spanning across the tunnel width. The steel FRS lining for the main tunnel drives was designed as a final lining with a design life of 100 years. It was applied using remotely controlled spraying manipulators. No secondary lining was employed.

During extensive field trials prior to commencement of the excavation works, an appropriate shotcrete mix containing 60kg/m³ of Dramix RC 65/35 steel fibres was developed which appeared to meet the high quality demands required for this project. The main characteristics of the mix design adopted were a total cementitious content of 520 kg/m³ including 60 kg/m³ fly ash, and 40 kg/m³ silica fume, and a w/b ratio of 0.38. The concrete was produced as wet-mix shotcrete in a local ready-mixed plant, and transported in 6 m³ agitator trucks. The dosage of alkali-free accelerator was set at 4.0M% of total cementitious content, and the slump was 90 mm at the pump.

2 QUALITY ASSURANCE PROCEDURES

2.1 Original QA procedures

The original Quality Assurance (QA) procedures selected for this project included a number of standard tests intended to assess compressive strength, adhesion, and durability, and beam tests to assess post-crack performance. The beam test was the only procedure adopted to assess the unique property of FRS, which is post-crack residual strength, or *toughness*. Each of the procedures specified are listed in Table 2.

The testing procedures adopted to determine the Un-confined Compressive Strength (UCS) of cast cylinders and cores were the normal specifications common to Australia and most other countries. The specification adopted for the determination of adhesive strength between the rock and shotcrete was initially based on a pull-off test (EN 1542 1999), but

Table 2. Quality Assurance procedures used to assess FRS.

Parameter	Procedure & Minimum Requirement	Frequency of Testing
Un-confined Compressive Strength (UCS) of cast cylinders (28 days)	AS1012 Part 9 Min. 40 MPa	2 cylinders per 25 m³
Un-confined Compressive Strength (UCS) of cores (28 days)	AS1012 Part 14 Min. 40 MPa	2 cores per day's production
Adhesion between rock and shotcrete (28 days)	EN 1542 Min. 0.5 MPa, or Hammer test	One core test per 250 m², or one hammer test per 10 m²
Flexural Strength at 28 days (MOR)	EFNARC/ C-1018 Min. 4.5 MPa	One set of 3 beams per 250 m²
Residual Flexural Strength at 3.0 mm deflection (28 days)	EFNARC/ C-1018 Min. 3.0 MPa	One set of 3 beams per 250 m²
Steel fibre content	Wash out with seive, 60 kg/m³	One test per 5 batches
Shotcrete Thickness	Cores 50/100 mm	One core per 50 m²
Water Permeability	DIN 1048 Part 5 Max. 30 mm	One test per 250 m²

Figure 6. Production of Round Determinate panels and the panel from which beams and cores were later cut.

later reverted to a hammer test for de-laminations (Norwegian Concrete Association 1993). The procedure adopted for the determination of water permeability of the shotcrete was a German specification (DIN 1048 1991). Most tests were carried out on panels sprayed as part of daily production (see Fig. 6). The procedure adopted for beam testing was the only problematic area in that no single specification could be found that satisfied all parties involved.

2.2 Beam testing

Beams were tested in accordance with the EFNARC specification (1996) with the test machine operated in displacement-control. Performance was reported as a Modulus of Rupture for the concrete matrix, and post-crack residual strength at 3.0 mm central deflection. However, since the procedures required for displacement measurement in the EFNARC specification are relatively vague, the more explicitly described procedures for displacement measurement detailed in ASTM C-1018 (1997) were instead employed. This 'mixed' approach was adopted in the belief that tests could be undertaken in a more stable manner using the slender EFNARC beam (measuring 75×125×550 mm, tested on a 450 mm span) compared to the stocky 100×100×350 mm ASTM beam, but that results could be re-produced more reliably using the detailed procedures described in ASTM C-1018.

The Universal Test Machine used was an electromechanical Instron 6027-5500R (Fig. 7); the LVDT output was digitally recorded by an Instron 5500 controller (with *Merlin* software) and later used to plot the load-deflection curve for the specimen. Although the machine had the capacity to operate in closed-loop control (meaning that the displacement of the beam as measured by an LVDT could be used to control crosshead movement), the tests were undertaken in displacement control (such that the crosshead was advanced at a constant rate) because this is a more reliable method of testing. The difference between these two modes of control was explained in more detail by Bernard (1998).

3 FIELD RESULTS

3.1 Un-confined compressive strength

Un-confined Compressive Strength tests were carried out on pairs of cast cylinders and cores taken from numerous batches of FRS used in the project. The mean within-batch and between-batch variation for the many samples tested are listed in Table 3.

The Coefficient of Variation (C.O.V.) data listed in this table was obtained as the standard deviation divided by the mean and expressed as a percentage.

The data for UCS indicate that the between-batch variation exceeded the within-batch variation for both cast cylinders and cores. Moreover, no significant relationship existed between the two parameters when cast and cored samples were compared for the same batch (see Fig. 8). This is likely to have been due to the limited strength range for concrete produced during the project, but nevertheless suggests that cast cylinders have limited value as QA specimens for FRS. Although they may be useful for batch plant QA, they indicate little about the in-place shotcrete. In this and the following figures, the crosses and diamonds represent the results obtained from the two contractors involved in the project.

Table 3. Variation in performance parameters obtained from tests on FRS. The beam results are not corrected for offset.

Parameter	Within-batch C.O.V. (%)	Between-batch C.O.V. (%)
Un-confined Compressive Strength (UCS) of cast cylinders at 28 days	1.8[a]	10.4
Un-confined Compressive Strength (UCS) of cored samples at 28 days	4.9	10.9
Modulus of Rupture (MOR) at 28 days	6.4	18.8
Residual Flexural Strength at 0.5 mm deflection, 28 d (without offset correction)	13.4	19.7
Residual Flexural Strength at 3.0 mm deflection, 28 d (without offset correction)	18.7	19.5
Density of cast cylinders		1.80
Density of cores	0.35	1.69
Water Permeability	13.3	35.8

[a] Limited by the requirement to exclude results that differed by more than 2.0 MPa from the higher value.

Figure 7. Testing of an EFNARC beam.

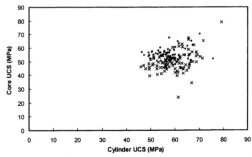

Figure 8. Correlation between UCS obtained from cores and UCS obtained from cast cylinders for batches of concrete in which both were available.

3.2 Beam results

A set of three beams was tested for every 250 m² of FRS sprayed in the tunnel. The within-batch and between-batch variation evident in the Modulus of Rupture (MOR) and residual strength parameters obtained from these sets is shown in Table 3. The within-batch variation in the Modulus of Rupture was relatively modest and has therefore been plotted against the UCS of cores obtained for the same batches of concrete in Fig. 9. The UCS of cores has been used in preference to cylinders because the beams were cut from the same samples of as-sprayed FRS. This figure appears to indicate a weak relation, but the data do not agree with published relationships between compressive strength and MOR obtained for cast concrete in laboratories (Rapheal, 1984).

Although only required at 3.0 mm central deflection, the post-cracking residual strength was measured at both 0.5 and 3.0 mm central deflection to obtain a more detailed understanding of performance at varying levels of deformation. The mean within-batch and between-batch variation in both these parameters, listed in Table 3, indicate the very high variability typical of post-crack performance as-sessment using beams. This is also evident in the data shown in Fig. 10, which shows the difference in load-deflection curves typical of a set of three beams displaying average within-batch variability.

4 OFFSET CORRECTIONS TO BEAM RESULTS

Following a period of production and testing during which repeated numbers of beams failed to satisfy minimum post-crack performance requirements, an attempt was made to determine the cause of the failures. Examination of the results indicated that very low performance in isolated beams had the effect of depressing the mean residual strength results in a number of sets. The cause of this appeared to be an excessive offset between the location of the crack and the centre of the beam in many specimens.

Holmgren (1993) observed that the position of the crack, even when it occurs between the rollers under third-point loading, has a significant effect on the angle of rotation experienced at the centre for a given central deflection. Since the majority of fibres show decreased load capacity as crack widths increase, this can have a substantial effect on apparent performance. Holmgren proposed a method of correcting for this by adjusting the level of deflection at which residual strength is measured. This method of correction was therefore examined as a possible means of reducing the high variability of beam results in this project.

To be admissible as a method of performance correction, it was necessary to consider whether the proposed adjustments conflicted with design assumptions about behaviour of the FRS in the lining. Selection of 3.0 MPa as the minimum post-crack residual strength required at 3.0 mm central deflection in the beams was based on a set of assumptions about the load-carrying capacity of the FRS lining at anticipated ground deformations. Cracks were expected to occur half way between rock bolts, giving rise to a continuous beam mechanism that was required to support displaced rock strata in the immediate vicinity. Following discussions between all the parties involved, the designers of the lining and Roads and Traffic Authority of NSW (RTA) representatives agreed that it was more rational to specify minimum levels of performance at a deflection equivalent to the crack rotation angle that would occur in a beam failing with a central crack. The adjustment procedure was therefore trialled.

Holmgren suggested that the residual strength be measured at a central deflection that is adjusted to account for the distance between the location of the crack and the centre of the beam (known as the 'offset'). Performance is thereby measured at a consistent crack rotation angle. Since an off-centre crack results in an increased crack rotation relative to that

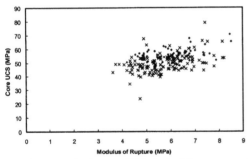

Figure 9. Correlation between UCS obtained from cores and Modulus of Rupture obtained from EFNARC beams.

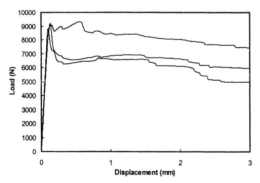

Figure 10. Load-displacement curves for three EFNARC beams made with the same batch of FRS.

which occurs when the crack is centrally located (Fig. 11), the magnitude of the central deflection at which performance is assessed is always adjusted downwards according to this procedure. For strain-softening FRC, the average residual strength is therefore increased. The procedure for conducting an off-set correction following a test is as follows:

1. Measure the offset between the location of the crack and centre of the beam,
2. Find the central deflection that results in the same crack rotation angle that would occur had the beam cracked at the centre. This is known as the 'adjusted' central deflection.
3. Determine the load capacity of the beam at the adjusted central deflection and calculate the 'corrected' residual strength of the beam on the basis of this load capacity.

To carry out this procedure, it is essential that the crack offset be measured for each and every beam, and that a continuous load-deflection response be recorded for the beam between the start of loading and the nominated maximum central deflection.

Based on Figure 11, it is clear that the crack rotation angle is 50% greater for a crack located under a third-point roller than for a central crack. The maximum adjustment to the magnitude of the central deflection that is therefore possible for a third-point

loaded beam is a 33% reduction. This will be necessary for beams in which the crack occurs directly under one of the load points.

For the beams examined in this project, the residual strength was assessed at 3.0 mm central deflection so the adjusted deflection at which performance was assessed was 2.0 mm for an offset of 75 mm (half the span between the third-point rollers). For offsets between zero and 75 mm, the adjusted central deflection at which residual strength was assessed, d', was calculated according to the following equation

$$d' = \frac{x}{(1 + y/150)} \qquad (1)$$

where y is the offset between the crack and the centre of the beam, and x is the central deflection at which residual strength is determined.

4.1 Offset-adjusted beam results

Correcting the performance of the beams for crack offsets resulted in an average 0.3 MPa increase in residual strength at 3.0 mm central deflection. However, the within-batch variation in post-crack performance did not change (compare Tables 3 and 4). This is perhaps explained by the fact that variability in residual strengths was high through most of the load-deflection envelope for each beam set. The maximum reduction in deflection at which performance was assessed was 33%, which corresponded to a drop from 3.0 to 2.0 mm. Examination

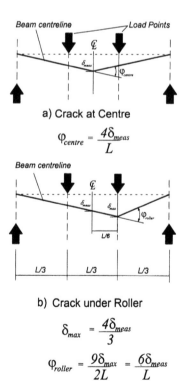

a) Crack at Centre

$$\varphi_{centre} = \frac{4\delta_{meas}}{L}$$

b) Crack under Roller

$$\delta_{max} = \frac{4\delta_{meas}}{3}$$

$$\varphi_{roller} = \frac{9\delta_{max}}{2L} = \frac{6\delta_{meas}}{L}$$

Figure 11. Increase in crack rotation angle due to crack offset from centre of beam.

Table 4. Variation in performance parameters obtained from beam tests on FRS, with correction for offset.

Parameter	Within-batch C.O.V. (%)	Between-batch C.O.V. (%)
Residual Flexural Strength at 0.5 mm deflection, 28 d	13.1	19.0
Residual Flexural Strength at 3.0 mm deflection, 28 d	17.8	18.9

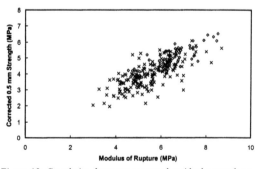

Figure 12. Correlation between corrected residual strength at 0.5 mm central deflection and Modulus of Rupture in the EFNARC beams. Each point is the average of three beams.

of many sets of load-deflection curves indicated that substantial reductions in variability only occurred for reductions greater than about 66%. This is evident in the difference between results at 3.0 and 0.5 mm deflection (Table 4). The Holmgren offset adjustment procedure therefore did not offer relief from the problem of excessive within-batch variability that appears to be an inherent feature of beam tests.

Following offset adjustments to all the beam results, the relationship between the performance of the concrete matrix and the post-cracking performance of the beams was examined by plotting the Modulus of Rupture against the corrected residual strength at 0.5 mm central deflection in the beams (Fig. 12) and at 3.0 mm central deflection (Fig. 13). These two figures show the mean performance for sets of three beams, and indicate that while post-cracking performance at low levels of deformation, represented by 0.5 mm central deflection, correlate relatively well with concrete strength, no relationship exists at greater levels of deformation.

The data for individual beams has been plotted in Figures 14. This reveals a slightly different picture, indicating a less defined relationship between matrix strength and residual strength at 0.5 mm deflection. However, Fig. 15 shows that an optimum Modulus of Rupture exists at around 6.0-6.5 MPa at which residual strength at 3.0 mm central deflection is at a maximum. The difference between Figures 12 and 14, and Figures 13 and 15, demonstrate that minor variations in the method of assessment can alter the perceived outcome.

5 ROUND DETERMINATE PANELS

Failure of the Holmgren offset adjustment method to reduce the high variability in beam results led the contractors to consider alternative methods of post-crack performance assessment. Based on favorable laboratory results (Bernard 1999), the procedure described in RTA T373 (2000) was selected as a possible method of post-crack performance assessment for Round Determinate panels produced for this project.

5.1 Panel testing procedure

The T373 specification (RTA 2000) is similar to the draft ASTM Standard Test Method (2000), but differs in details regarding measurement and strain rates. The tests undertaken as part of this project complied with T373 as this was a requirement upon the contractors.

The test machine used for all tests was an MTS 250 kN servo-hydraulic actuator controlled by an MTS TestStar II controller operating with a 5000 Hz feedback signal. The frame supporting the panel during testing consisted of three symmetrically ar-

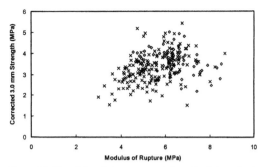

Figure 13. Correlation between corrected residual strength at 3.0 mm central deflection and Modulus of Rupture in the EFNARC beams. Each point is the average of three beams.

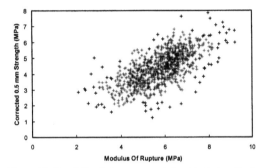

Figure 14. Correlation between corrected residual strength at 0.5 mm central deflection and Modulus of Rupture in the EFNARC beams. Each point represents a single beam.

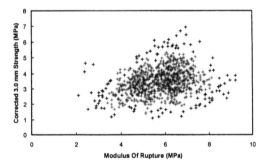

Figure 15. Correlation between corrected residual strength at 3.0 mm central deflection and Modulus of Rupture in the EFNARC beams. Each point represents a single beam.

ranged pivot points on a pitch circle diameter of 750 mm, fixed onto vertical supports (Figure 16). The three supports were fully restrained against lateral translation, but the pivots were free of friction so as not to hinder rotation of the panel fragments after cracking. The support fixture was configured in such a way that the specimen did not come into contact with any portion of the fixture apart from the three pivots during a test. The contact between the specimen and each pivot consisted of a steel transfer plate with plan dimensions of 40×50 mm with a spherical seat machined into one surface to accept a ball pivot. Grease was used to reduce friction in the seat of each pivot, but no rollers were used to lessen friction between the transfer plates and specimen.

The central deflection of the panel was measured relative to the support mounting of the actuator. This resulted in the inclusion of extraneous deflections in the displacement record. However, the test frame was so stiff in the present configuration that extraneous deflections were negligible compared to the deformation of the specimen itself.

To undertake each test, the panel specimens were first mounted in the test apparatus by placing the molded face onto the three transfer plates resting on the pivots. The panel was centred with respect to both the supports and loading piston. Three measures of diameter were made at symmetric angles to each other. The test machine was then operated so that the piston advanced at a constant rate of 10.0 mm/min up to a total displacement of 40.0 mm. For all of the present tests, loading was extended to 100 mm total displacement to examine behaviour at extreme levels of deformation.

Following completion of the test, the failed panel fragments were removed from the test apparatus, and 10 measures of thickness were made along the cracked surfaces. The number of radial cracks occurring between the centre and the perimeter was counted. In most tests, there were three cracks.

Figure 16. Testing of a Round Determinate Panel.

The results for each specimen were processed to correct for excessive deviations in thickness and diameter. This is vitally important to obtain reliable results because the apparent performance of a specimen is very sensitive to thickness. The energy absorption was corrected using the following equation:

$$ W = W' \left(\frac{t_0}{t} \right)^\beta \left(\frac{d_0}{d} \right) \quad \text{where} \quad \beta = 2.0 - \delta/80 \quad (2) $$

where W is the corrected energy absorption, W' is the measured energy absorption, t is the mean thickness, t_0 is the nominal thickness of 75 mm, d is the mean diameter, d_0 is the nominal diameter of 800 mm, and δ is the central deflection at which the load capacity is measured (for performance corrections to residual load capacity after first cracking). This expression was developed by Bernard and Pircher (2000).

5.2 Panel Results

Testing under laboratory conditions had previously indicated a mean within-batch variation in energy absorption at 40 mm central deflection for sets of 3 Round Determinate panels of about 5 percent (Bernard 1999, 2000). Panels were trialled in December 1999 through to April 2000 in parallel with beams to examine whether the excellent reliability in post-crack performance assessment seen in the laboratory could be re-produced in the field. During the trial stage, sets of 3 panels were tested for each batch examined. Following adoption of the Round Determinate panel as the basis of post-crack performance assessment, a set of two panels was tested for every 250 m^2 of FRS sprayed.

The data in Fig. 17 show the within-batch variation in results typical for a set of three Round Determinate panels exhibiting average variability in en-

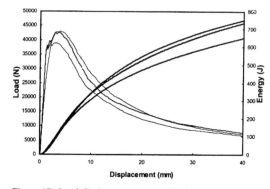

Figure 17. Load-displacement curves for three Round Determinate panels made with the same batch of FRS.

ergy absorption. This represents a considerable improvement in repeatability compared to beams (Fig. 10), with mean within-batch variability dropping from 17.8 percent for the beams to 7.7 percent for energy absorption at 40 mm deflection for the first 60 sets of panels. The between-batch variation remained relatively high at 15.8 percent. Considering that no special procedures were adopted to produce the panels other than the use of a screed, these results confirmed that Round Determinate panels could be used to produce reliable and economical estimates of performance in field-produced set-accelerated FRS.

Since EFNARC beams had previously been used as the basis of performance assessment, it was necessary to establish a correlation in performance between the beams and panels so that the equivalent minimum performance standard in the panels could be determined. The data in Figs. 18 and 19 indicate the relationship obtained between energy absorption at 40 mm central deflection in the Round Determinate panels, and both the corrected and un-corrected residual strength at 3.0 mm central deflection in the beams. The corrected results consisted of residual strength values obtained at deflections that had been adjusted using the Holmgren offset correction. Data from laboratory tests on a large number of FRS

mixes by Bernard (1999, 2000) have also been included.

These figures indicate that a linear relation exists between the two parameters, but that the un-corrected beam data appeared to be more strongly correlated to panel performance. Based on the trend line in Fig. 18, the energy absorption at 40 mm central deflection in the Round Determinate panels deemed equivalent to an uncorrected residual strength of 3.0 MPa at 3.0 mm central deflection in the beams was 570 Joules. When the corrected residual strength is used, the equivalent minimum level of performance in the panels was 496 Joules.

6 DISCUSSION

Beams were found to be a very unreliable means of assessing the post-crack performance of FRS, particularly at relatively high levels of deformation. Although corrections to the deflection at which performance is assessed to account for the offset of the crack was initially believed to offer a solution to this problem, this was not found to be the case. The reason for this appeared to be that variability in residual strength was high across a wide range of deflections. Inherently high variability appears to be due to the small area of the crack at which post-crack residual strength is assessed, which results in poor sampling of the mean performance of the material. Large specimens exhibit lower variability in performance because the effect of a naturally uneven distribution of fibres diminishes as the crack area increases relative to the size of the fibres.

In developing the minimum performance standard for the panels, it was necessary to decide upon an acceptance criterion for the validity of results. Invalid failures are known to occur in Round Determinate panels through the formation of 2 rather than 3 radial cracks. This typically results in a reduction in energy absorption of between 20 and 30 percent compared to a normal failure. Poor results may also occur as a result of cracking during handling, or gross irregularities in the thickness of the specimens. Acceptance criteria for other types of specimen (such as cores) are often based on a maximum allowable variation in performance between two nominally identical specimens equal to three standard deviations. Since the Coefficient of Variation for field-produced Round Determinate panels has been found to be about 7 percent, specimens failing with an energy absorption less than 80 percent of the higher value were therefore taken to be invalid.

Implementation of the Round Determinate panel as the basis of QA for post-crack performance was not without its problems. The most important was the fact that within-batch variability depended strongly on the uniformity with which panel thickness was maintained. While the effect of variations

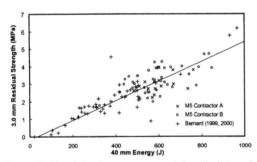

Figure 18. Correlation between un-corrected residual strength at 3.0 mm central deflection and energy absorption at 40 mm central deflection in Round Determinate panels.

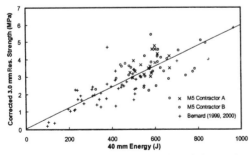

Figure 19. Correlation between corrected residual strength at 3.0 mm central deflection and energy absorption at 40 mm central deflection in Round Determinate panels.

in panel thickness could be corrected using the expressions developed by Bernard and Pircher (2000), these were only applicable to panels of uniform thickness. Irregularly produced specimens remained problematic, and it is believed that these specimens were responsible for the difference in mean within-batch variability between laboratory and field-produced specimens (Bernard 1999, 2000). The problem was particularly noticeable for panels less than 60 mm thick, as these appeared to be more susceptible to the effects of thickness irregularities, and also suffered frequent cracking during handling.

In an attempt to overcome this problem, shift managers were trained to use bars more effectively to screed the surface of freshly prepared panels immediately after spraying. However, set was very rapid in many cases, and screeding did not always result in satisfactory specimens. This aspect of panel production therefore continues to require attention.

Another problem inherent to panels is the large size and mass of the specimens. This makes handling and transport difficult compared to cylinders and cores. Panels also require a dedicated test machine with a peculiar configuration that is not particularly useful for any other type of recognized test. These factors make panels expensive to test, but similar problems exist for beams. The disadvantages of Round Determinate panels are more than compensated for by the fact that concrete cutting is eliminated from the specimen production process. This is the primary reason why QA for post-crack performance based on Round Determinate panels is substantially cheaper than for beams.

7 CONCLUSION

The M5 Motorway Tunnel construction project in Sydney has been a test-bed for the implementation of the Round Determinate panel test as a Quality Assurance procedure for post-crack performance of Fibre Reinforced Shotcrete. Through a large number of field trials under normal conditions of commercial shotcreting, methods have been developed for the effective production, curing, transport, and testing of Round Determinate panels made with accelerated FRS. The process has involved the refinement of techniques, education of personnel, and development of workable performance acceptance criteria that will be transferable to tunnelling projects involving FRS worldwide.

The process of implementing the Round Determinate panel as the basis for QA testing has also demonstrated this procedure to be more reliable and cost effective than beam testing for performance assessment. Beams tested in parallel with Round Determinate panels have established correlations that will allow performance minimums to be selected ra-

tionally. The procedure is therefore attractive to contractors for a number of reasons.

Comparisons between compressive cylinder and core tests have demonstrated no correlation between these two compressive strength parameters. Comparisons of Modulus of Rupture and post-crack performance for FRS beams produced in the field have shown that post-crack performance is related to matrix strength at low levels of deformation, but not at high levels of deformation. However, this observation is only valid for the hooked-end steel fibres used in this project.

ACKNOWLEDGEMENTS

The authors gratefully acknowledge the support and perseverance of BHBB JV, Walters Construction Group (through Alister Rowe), and the Roads and Traffic Authority of NSW in implementing the Round Determinate panel test as a QA procedure in NSW.

REFERENCES

American Society for Testing and Materials Standard C-1018. 1997. Standard Test Method for Flexural Toughness and First-Crack Strength of Fiber-Reinforced Concrete (Using Beam With Third-point Loading). ASTM, West Conshohocken, PA.

American Society for Testing and Materials Draft Standard C-XXXX. 2000. Standard Test Method for Flexural Toughness and of Fiber-Reinforced Concrete (Using Round Determinate Panels). ASTM, West Conshohocken.

Australian Standard AS1012 *Methods of Testing Concrete*, 1986, Part 9. Method for the determination of the compressive strength of concrete. Standards Australia, Sydney.

Australian Standard AS1012 *Methods of Testing Concrete*, 1991, Part 14. Method of securing and testing cores from hardened concrete for compressive strength. Standards Australia, Sydney.

Bernard, E.S. 1998. Measurement of Post-cracking Performance in Fibre Reinforced Shotcrete. *Australian Shotcrete Conference*, Sydney, October 8-9.

Bernard, E.S. 1999. Correlations in the Performance of Fibre Reinforced Shotcrete Beams and Panels. *Civil Engineering Report CE9*, School of Civic Engineering and Environment, UWS Nepean, July.

Bernard, E.S. 2000. Correlations in the Performance of Fibre Reinforced Shotcrete Beams and Panels: Part 2. *Civil Engineering Report CE15*, School of Civic Engineering and Environment, UWS Nepean, June.

Bernard, E.S. and Pircher, M. 2000. The Use of Round Determinate Panels for the Assessment of Flexural Performance of Fiber Reinforced Concrete. *Cement, Concrete, and Aggregates*, ASTM (submitted February 2000).

DIN 1018, 1991. *Testing Concrete*, Part 5. Determination of Water Permeability. DIN.

EN 1542, 1999, *Products and Systems for the protection and repair of concrete structures – Test Methods*, Measurement of bond strength by pull-off.

European Specification for Sprayed Concrete. 1996. European Federation of National Associations of Specialist Contractors and Material Suppliers for the Construction Industry (EFNARC).

Holmgren, J. 1993. The Use of Yield-Line Theory in the Design of Steel Fibre Reinforced Concrete Slabs, In *Shotcrete for Underground Support VI*, Proceedings of the Engineering Foundation Conference, Niagara-on-the-Lake, Canada, May 2-6: 91-98.

Norwegian Concrete Association. 1993. *Sprayed Concrete for Rock Support - Technical Specification and Guidelines*, Publication No. 7.

RTA Specification T373. 2000. Round Determinate Panel Test. in B-82 *Shotcrete Work*.

Rapheal, J.M. 1984. Tensile Strength of Concrete, *ACI Journal*, 81 (2): 158-65.

Shotcrete: Engineering Developments, Bernard (ed.) © 2001 Swets & Zeitlinger, Lisse, ISBN 90 5809 176 7

Use of recycled aggregates in fibre reinforced shotcrete: mix design, properties of fresh concrete, and on-site documentation

C.Hauck
Veidekke ASA, Heavy Construction Division, Oslo, Norway

T.Farstad
Norwegian Research Institute, Oslo, Norway

ABSTRACT: A demonstration project using up to 20% recycled aggregates in fibre-reinforced shotcrete started in the summer of 1999 in Oslo. The project was a full-scale laboratory and on-site test of shotcrete containing recycled aggregate, used to stabilize a permanent lightweight bridge construction. On-site documentation showed that shotcrete with recycled aggregate exhibited excellent spraying and compacting properties, and adheres to the substrate very well, compared to a reference mix. No spraying difficulties occurred due to the use of recycled aggregates and the need for accelerator decreased for all shotcrete-mixes with recycled aggregates. The compressive strength of shotcrete with recycled aggregate was reduced compared to a reference mix without recycled aggregates, but the strength obtained still exceeded 45 MPa at 28 days.

1 PROJECT DESCRIPTION

A demonstration project using recycled aggregates in shotcrete started in the summer of 1999 in Oslo (Lillestøl 1999). The project (Figure 1) is part of the Norwegian research project RESIBA. The objective of RESIBA is to increase the use of recycled aggregates (waste materials of the building industry) in the construction industry in Norway. This research project is the first attempt to use recycled aggregates in shotcrete in Norway.

Shotcrete is in Norway for the most part used to ensure rock stability and fire protection of final PE-linings in tunnels. The project "Gaustadtrikken" was a full-scale laboratory and on-site test of a shotcrete mix containing recycled aggregate, used to stabilize a permanent lightweight bridge construction. The

Figure 1. Gaustadtrikken, a demonstration project using recycled aggregates in shotcrete.

lightweight bridge construction, consisting of 18000 m³ Expanded PolyStyrene (EPS), was stabilized vertically by a 150 mm thick shotcrete layer.

A total area of about 750 m² was covered with shotcrete (approximately 100 m³ shotcrete). An inner layer of steel fibre reinforced shotcrete (40 kg/m³), reinforced with steel mesh, was also covered by an outer layer of shotcrete without steel fibres. The shotcrete layers also served as fire protection of the EPS and as a protection against other EPS-damaging aggressive agents. The project was considered successfully executed.

The recycled aggregates had a grain size of 0-4 mm, the finest parts from the recycling process. This material has a high water demand. For that reason is often not useable as aggregates in cast concrete structures where high consistency and workability is demanded. However, fines containing recycled aggregate grading in the fraction 0-4 mm has been shown to be applicable in sprayed concrete. None of the foreseen difficulties associated with a high fines content occurred: no troublesome concrete mixing, no delivery problems, and no pumping or spraying difficulties.

2 EQUIPMENT

The shotcrete was pumped using a standard worm pump, adding compressed air and accelerator at the nozzle. Worm pumps are more demanding than other pumps with respect to fresh concrete properties, especially consistency. The pump- and spray equipment needs cement or fines rich shotcrete mixes, with a high and stable consistency. Therefore, worm pumps served as a qualifying test for the use of shotcrete with recycled aggregate.

3 AGGREGATES

The purpose when designing the shotcrete composition is to achieve good pumpability, good compression and compactability, good bond, and low rebound. To achieve these rheological properties, the sprayed concrete mix requires an optimized binder or fines content to get a thick, matrix-rich mixture. A matrix-rich mixture does not necessarily imply use of more cement or silica, as long as concrete strength meets the requirements. Therefore, the use of recycled aggregate (0-4 mm) to optimize the grading of fines is a suitable solution for sprayed concrete, both environmentally and economically.

The aggregate grading curve of the control aggregate 0-8 mm consists of two types of natural sand, 72% sand (0-8 mm) from a fluvial sedimentary deposit, and 28% sand (0-8 mm) from a sedimentary lake deposit. The sands were optimized to produce a reference concrete mix with a low rate of rebound and a good bond to the EPS substrate.

Figure 2 shows the control aggregate (0-8 mm) and the recycled aggregate (0-4 mm) grading curves. Recycled aggregate was added to the control aggregates as a percentage by weight of the total aggregate to produce four different mixtures: 0%, 7.5%, 15% and 20% RCA.

Table 1 shows the aggregate compositions for all mixes, both the grain distribution of the control aggregate (0-8 mm) without any recycled aggregate, and the control aggregate with 7.5%, 15% and 20% recycled aggregate.

The fines content measuring less than 125 µm for

Table 1: Grain size distribution of the control aggregate and different mixtures with recycled aggregate.

Grain size distribution: percent accumulated (%)					
	Amount of recycled aggregate				
Sieve (mm)	0 %	7.5%	15%	20%	100%
8	4	3	3	3	0
4	13	12	12	11	6
2	22	23	24	24	32
1	37	38	39	40	53
0.5	65	66	67	67	72
0.25	89	89	89	89	86
0.125	97	96	96	96	93

the four different sprayed concrete mixes (with 0%, 7.5%, 15% and 20% recycled aggregates) were 22.9%, 23.4%, 23.7% and 24%, respectively.

4 SHOTCRETE MIX DESIGN AND LABORATORY TESTS

Mix design is an important process, determining properties such as water/cement ratio, density, and compressive strength. The main objective of the mix design in the present investigation was to achieve a shotcrete mix with good pumpability, compactability, and adhesion to the substrate. In addition, the mix should also have a low rebound rate. These properties all strongly depend on the fresh concrete consistency, which is influenced by the aggregate grading curve, the cement- and fines content, the silica fume content, and the admixtures.

The cement used for all shotcrete mixes was a CEM I-42.5 type, Norcem standard cement, and the microsilica was supplied by Elkem Materials. A lignosulphonate plastizising agent and a melamine-based superplastizising agent were used. Table 2 shows the shotcrete mix design C35 NA.

The liquid accelerator used was sodium silicate (waterglass), at a dosage in the range of 4-5% of cement weight.

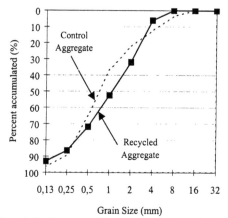

Figure 2. Grading curve of the control and the recycled aggregate.

Table 2. The shotcrete mix design C35.

Ingredient (kg/m³)	0%	7.5%	15%	20%
Cement	485	485	485	485
Silica	30	30	30	30
Aggregate A (0-8 mm)	415	359	303	247
Aggregate B (0-8 mm)	1071	1003	935	866
Recycled Aggregate (0-4 mm)	0	99	198	298
Steel fibres	40	40	40	40
Water	239	239	238	237
Plasticizer P	3.5	3.5	3.5	3.5
Superplasticizer SP	2.7	3.5	4.0	4.5
w/b ready mix	0.44	0.44	0.44	0.44
accelerator	23	20	20	20
W/b shotcrete	0.46	0.46	0.46	0.46

Preliminary laboratory tests at the Veidekke laboratory, and testing at a ready mix plant, showed that it was possible to obtain water-cement ratios as low as 0.44 and a compressive strength of 45 MPa with up to 20% recycled aggregate.

The water absorption of the control aggregate was about 0.3-0.4%, a normal value for Norwegian fluvial sediments of granitite origin. The Norwegian Building Research Institute carried out the water absorption tests of the recycled aggregate, used in this project during 1998 (Mehus et al. 2000). Water absorption was controlled prior to spraying (in the range of approximately 12 %) at the ready mix plant, and this was accounted for in the shotcrete mix design.

5 ON-SITE DOCUMENTATION

On-site documentation consisted of shotcrete mix specification from the ready mix plant, including water cement ratio. Concrete temperature, consistency and air-content were measured on-site for each mixture, see Table 3. In addition, pumping pressure and accelerator dosage was measured on-site during spraying, and specimens were prepared for further testing at the Norwegian Building Research Institute.

On-site spraying showed that shotcrete with recycled aggregate displayed excellent spraying and compacting properties due to the fines content of the recycled material. The fresh shotcrete mixes with recycled aggregates exhibited excellent cohesiveness, thixotropy, and an almost creamy consistency, compared to the reference mix. The nozzleman was surprised by the excellent fresh shotcrete consistency and the ease of spraying shotcrete containing recycled aggregates.

Plastizising admixtures had to be of good quality to achieve a low water/binder ratio and avoid loss of workability during transportation. A combination of lignosulphonate plasticizer and a melamine-based superplasticizer was used at the plant. However, consistency loss during transportation occurred, especially for the 20% recycled aggregate content mixtures. Further superplasticizer (see Table 3) was added to the shotcrete in the agitator truck and blended carefully. The melamine-based superplasticizer could have been exchanged for a polymer-based superplasticizer to obtain a more effective water-reducing effect and a more slump-stable consistency and thereby minimize the total amount of superplasticizer in the mix.

The need for accelerator decreased for all the shotcrete mixes incorporating recycled aggregates. The nozzleman could therefore reduce the accelerator dosage, and still get the same adhesion to the substrate. The pumpability for the mixes with recycled aggregates was very good. The pumping pressure remained unchanged compared to shotcrete without recycled aggregate. During spraying there were no reports of problems with separation or clogging of the hose or nozzle.

The compressive strength of shotcrete with recycled aggregates met the requirements of the specifications for the C35 MPa shotcrete. This was shown by both preliminary laboratory tests and on-site documentation. The characteristics of the hardened shotcrete are shown in Tables 4 and 5.

Preliminary laboratory tests of some shotcrete mixes were carried out according to Norwegian standards at the Veidekke laboratory prior to spraying. The compressive strength required for the project was C35 NA, which means a characteristic compressive strength of 35MPa and a water/binder ratio of maximum 0.60. The tests showed that 7-day compressive strength was 42 MPa for shotcrete with 20 percent recycled aggregate and water/binder ratios of below 0.50.

The specimens from the plant mix were prepared on-site from the shotcrete mixes with 0%, 15%, and 20% recycled aggregates, for strength testing at the Norwegian Building Research Institute. Preparation details are given by Farstad & Hauck (2001).

Compared to the reference mix, the compressive strength was reduced as expected due to the use of recycled aggregate, but still achieved a minimum of 45 MPa at 28 days. The strength reduction due to the replacement by 15% recycled aggregates was about 15-20% at 28 days. At an age of 102 days (3½ months), the strength reduction was still in the range

Table 3. Fresh shotcrete characteristics

Fresh shotcrete	0%	7.5%	15%	20%
Temperature at delivery (°C)	20	21	21	22
Air content (%)	4.5	4.3	4.3	4.8
Superplasticizer (kg/m^3)	0.93	1.75		1.4
Slump (mm)	230	220	220	220
Total aggregate (kg/m^3)	1486	1461	1435	1411
Fines content (<125 µm)	515	520	520	520
Accelerator dosage (kg/m^3)	23	20	20	20
Pumping pressure (bar)	165	165	165	165
Specific Density (calculated)	2.24	2.22	2.19	2.17

Table 4. Hardened shotcrete characteristics: compressive strength

Compressive Strength (MPa)				
Shotcrete with		Lab. mix*	plant mix**	
recycled aggregates %	7 d	28 d	28 d	102 d
casted cubes	0 %		55	58	74
(without	15 %				61
accelerator)	20 %	42	45	50	61
sawn cubes	0 %				65
(with	15 %				56
accelerator)	20 %				54

*Contractor Veidekke ASA
**Norwegian research institute

Table 5: Hardened shotcrete characteristics: Density

Density (kg/m^3) Shotcrete with recycled aggregates (%)		plant mix*	
		28 d	102 d
casted cubes	0 %	2290	2306
(without	15 %		2215
accelerator)	20 %	2185	2213
sawn cubes	0 %		2253
(with	15 %		2234
accelerator)	20 %		2200

*Norwegian research institute

of 15-20%, compared to the reference mix. The increase in recycled aggregate content from 15 to 20% did not change the strength development significantly.

6 CONCLUSION AND FURTHER RESEARCH

Use of recycled aggregates in fibre-reinforced shotcrete produces stable fresh concrete properties, excellent spraying properties, and reduces the need for accelerators. The compressive strength of shotcrete with recycled aggregate is reduced compared to shotcrete without recycled aggregates, but met the requirements of the specifications of the C35 MPa shotcrete.

On-site spraying showed that shotcrete with recycled aggregate exhibited excellent spraying and compacting properties due to the high fines content of the recycled material. Fresh shotcrete mixes with recycled aggregates were of excellent cohesiveness, thixotropy, and an almost creamy consistency, and adheres to the substrate very well, compared to the reference mix.

The need for accelerator decreased for all shotcrete mixed with recycled aggregates. The accelerator demand was reduced by more than 1.5 %.

No spraying difficulties occurred due to the use of recycled aggregates. The pumping pressure for the mixes with recycled aggregates remained unchanged compared to shotcrete without recycled aggregates. During spraying there were no reports of problems with separation or clogging of the hose or nozzle.

The strength requirement of C 35 in sprayed specimens was satisfied by using recycled aggregates with a water/binder ratio of 0.46 and a waterglass accelerator.

On-site documentation showed that the use of waterglass accelerator results in a strength reduction of about 10% compared to a reference mix without accelerator. Alkali free accelerators will normally not have the same negative strength reducing effect.

Polymer-based superplasticizers have a greater water-reducing effect effective and give a more slump stable consistency compared to the melamine-based superplasticizer. To minimize the total amount of superplasticizer in the mix, it could have been exchanged with a polymer-based superplasticizer; especially at high water demanding recycled aggregates content of 20 %.

REFERENCES

Farstad, T. & Hauck, C. 2001. Use of recycled aggregates in fibre reinforced shotcrete: mechanical properties and durability. *International Conference on Engineering Developments in Shotcrete*, April 2-4, Hobart, Tasmania, Balkema.

Lillestøl, B. 1999. Verdens første resirkulerte sprøytebetong. Betong industrien 31 (3). Årgang, 14, Oslo, Norway.

Mehus, J., Lahus, O., Jacobsen, S. & Myhre, Ø. 2000. Bruk av resirkulert tilslag i bygg og anlegg – status 2000.Norges byggforskningsinstitutt. Prosjektrapport 287, Oslo, Norway

Shotcrete: Engineering Developments, Bernard (ed.) © 2001 Swets & Zeitlinger, Lisse, ISBN 90 5809 176 7

Shotcrete use in the Southern Link tunnel, Stockholm

B.I.Karlsson
Swedish National Road Administration, Stockholm, Sweden

T.P.Ellison
Betongsprutnings AB BESAB, Gothenburg, Sweden

ABSTRACT: The Southern Link is a large road tunnel project to the south of Stockholm. This project includes tunnels with imposing dimensions constructed by a drill and blast method through hard rock. Steel Fibre Reinforced Shotcrete (SFRS) and bolts were used as primary rock support. The performance requirements for the shotcrete were very stringent with respect to strength, durability, and environmental considerations. In order to find a mix design that could satisfy these requirements, several pre-construction trials were carried out. To date, approximately 60% of the shotcrete work has successfully been completed. Due to blasting, shotcreting could not take place closer than 50 metres from working face. To determine the effect of peak particle velocities arising from the blasting, a test program was carried out. Results from these tests showed that blasting-induced vibrations decline very quickly with distance from the face.

1 INTRODUCTION

The Southern Link (Södra Länken) is a road tunnel project to the south of Stockholm city. The primary aim of the project is to improve the urban environment for the city's inhabitants by freeing the central urban area of through-traffic. The project will solve a major traffic congestion problem by diverting over 100000 cars daily from the area. According to the plans the project will be completed in 2004.

1.1 *Geology*

The rock tunnels in the project are located in the Precambrian Scandinavian gneiss formation. Rock quality is fairly good, typical Q-values (Barton et al. 1974) are between 1-100. Some fractured zones occur, and cracks are more intense in some areas. Rock support requirements in this project were governed by maintenance considerations rather than geological factors.

1.2 *Design*

The Southern Link consists of a 6 km road of motorway standard. There are several underground grade separated junctions, installation-rooms, shafts, and emergency exits. The total length of rock-tunnel is about 17 kilometres, and the rock volume to be excavated is well over 2 million cubic metres.

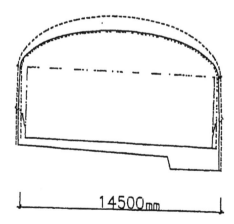

Figure 2. Main tunnel section.

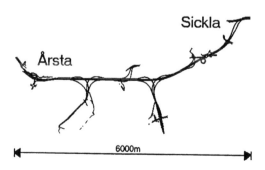

Figure 1. Map of Southern Link route.

Most of the tunnels are located quite close to the ground surface. Soil depth varies from 0 to 20 m, and the groundwater level is near the ground surface. The dimensions of the rock-tunnels are quite imposing with spans up to 32 metres.

The project includes 12 tunnel portals, and due to the rock topography there are underground connections between rock tunnels and concrete-lined tunnels.

One of the tunnel portals has a span of 32 meters and an over burden of less than 5 m. This section has been excavated very carefully by two pilot-tunnels and 28 stope-rounds (see Figures 3 and 4). Longitudinal rock-bolts with a length up to 10 m were installed outside the tunnel profile, and the rock mass excluding tunnel-face were shotcreted before excavation started. Shotcreting and bolting were carried out after each blasting round. The required shotcrete thickness was 0.5 m, and cavities were also filled in places behind the lining. In extreme cases the total thickness of lining was up to roughly one metre. Approximately 500 rock-bolts with a length of 2-4 m were installed over this 20 metre length of tunnel. This method places high demands on the shotcreting process, especially in order to satisfy the requirement for spraying layers up to 200 mm thick that remain in place. This was possible due to the use of alkali-free accelerators. The excavation and rock support in this section took six months to complete. Deformations were checked during the whole operation, and so far no alarming settlements have occurred.

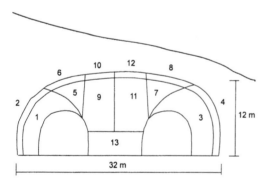

Figure 3. Blasting sequence, cross-sectional view.

1.3 *Environment*

It was required that the groundwater level should not be affected during construction or after the tunnel was completed. This was due to the presence of sensitive structures, streets, railways, pipes, and buildings founded on wooden piles above the tunnels. Considerable efforts were made to limit noise, vibration, and traffic congestion associated with construction. All buildings in the area were inspected

before excavation started, and on-line vibration recorders were used during excavation. Drilling and blasting were not permitted between 10 pm and 7 am.

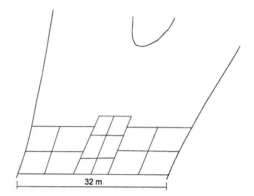

Figure 4. Blasting sequence, plan view.

Providing information about construction activities is an important way to make the people living in the area less concerned about the consequences of construction. Citizens were also invited to visit the tunnel during construction, which has been very popular.

1.4 *Tunnelling*

The excavation method used was traditional drill and blast. All the tunnels were pre-grouted, and some post-grouting is being done in places as well. In some areas with no or very little rock-overburden, the ground was frozen before excavation. In these areas, temporary support was provided by shotcreting, and the final support consists of a pre-cast lining with a thickness of 0.8 m. To prevent frost problems associated with ingressing water in the tunnels, isolated wall drains were used. The rock support mainly consisted of un-tensioned rock-bolts and shotcrete. The crown of all tunnels were supported with fibre-reinforced shotcrete, while most of the tunnel walls were sprayed with unreinforced shotcrete.

2 DEVELOPMENT OF MATERIALS AND EQUIPMENT

2.1 *Requirements*

There were several clauses within the specifications affecting shotcrete properties, and these stipulated the following requirements:
- The tunnels were constructed to last at least 120 years.

- For environmental reasons, use of admixtures that could pollute the ground water was not permitted.
- Adhesion between rock and shotcrete shall not be affected by admixtures.
- The concrete must be resistant to freeze-thaw cycling.
- Strength requirements were considerably higher than normal, both with regard to Modulus of Rupture and post-cracking toughness.

Besides these specifications, there were additional issues to consider:
- The personal health of the men working in the tunnel must be considered.
- It was considered beneficial to be able to place thick layers of shotcrete in one pass.
- Rebound must be low.
- Pumpability and sprayability must be good.
- The shotcrete mix must be cost-competitive to be acceptable.

2.2 Pre-construction Trials

The designers and contractors had no prior experience of any project where the shotcrete properties were as stringent as for these tunnels. For example, frost-durability has usually not been specified in other tunnelling projects in Sweden. It was therefore necessary to conduct pre-construction trials under site conditions to demonstrate that the required FRS properties could be achieved.

An initial mix-design was determined from available literature on materials, see Table 1.

Table 1. Initial mix-design for shotcrete.

Ingredient	Quantity (kg/m³)
Aggregate (0-8 mm)	1600
Portland Cement (SR)	480
Silica Fume	5
Water/cement ratio	0.45

It was also decided that Dramix RC 65/35 hooked-end steel fibres would be used at a dosage rate of 55 kg/m³. Superplasticizers and alkali-free accelerators from Rescon, Sika, and Master Builders were trialed. Test spraying was performed in a tunnel under construction in Stockholm. Betongindustri AB, Stockholm, who were later contracted to deliver ready-mix concrete for shotcrete use during construction, supplied the concrete. The pre-construction trials started in 1997 and were completed in 1998.

2.3 Pre-construction trial results

Vattenfall Utveckling AB, Älvkarleby, undertook laboratory testing of shotcrete properties. All the requirements were fulfilled after only two rounds of trials. It was especially satisfying that freeze-thaw tests showed acceptable results. The final mix included Rescon Superflow 2000 as superplasticizer and Rescon AF 2000 as accelerator. The results from laboratory-tests for this mix-design are shown in Table 2.

Table 2. Test results for trial-mix shotcrete.

Property	Method	Specified	Result
Compressive strength (MPa)	SS 13 72 20	40	60
Post-crack flexural strength $f_{5,10}$ (MPa)	ASTM C1018	4.0	4.5
Post-crack flexural strength $f_{10,30}$ (MPa)	ASTM C1018	3.0	4.0
Frost resistance (kg/m³)	SS 13 72 44	0.5	0.15

2.4 Quality control

A number of tests were required to be carried out on the in-place shotcrete for Quality Assurance purposes during construction. These were all required in the project specifications. The tests included:
1. Fibre content
2. Thickness, measured in 25 mm diameter drilled holes.
3. Compressive strength, based on cubes sawed from panels sprayed during construction.
4. Flexural strength of beams sawed from panels sprayed during construction.
5. Adhesion, based on cores drilled and pulled off in-situ.
6. Freeze-thaw resistance

Frequency of testing depended on risk estimations and geological conditions. The compressive strength-tests were normally carried out once per 1000 m² of in-place shotcrete, and flexural tests once per 2000 m². Adhesion tests were done once per 1000 m². Freeze-thaw tests were only necessary in zones where frost was expected.

2.5 Machine development

In parallel with the pre-construction trials to develop the shotcrete mix for this project, machines were developed to suit the conditions existent in this project. Aliva AG, Switzerland, was contracted to supply concrete pumps, robotic arms, and the additive pump for shotcreting. AB Besab, Sweden, completed the carrier, compressor, and electrical equipment.

The maximum capacity of the concrete pump was 20 m³ per hour. However, this was reduced to 10-15 m³/h during practical spraying. The total vertical reach of the robot arm was 15 metres, and the unit could move five metres along the tunnel during spraying before re-location of the equipment was necessary.

2.6 *Construction*

To date, approximately 60% of the contract has been completed, which is equivalent to about 10 000 m³ shotcrete. Some changes in the mix-design were necessary during construction, the most important involved changing the superplasticizer to Master Builders Glenium 51. This was done because of some unexpected variations in viscosity in the concrete that influenced pumpability. More than 100 strength tests, including both compressive and flexural strength, have been completed during construction to date, and all show satisfactory and uniform results.

3 VIBRATION TESTS ON APPLIED SHOTCRETE

3.1 *Background*

Technical specifications for this project required that the shotcrete should reach a compressive strength of 6 MPa before loading, and that the maximum allowable Peak Particle Velocity (PPV) arising from blasting-induced vibrations was 150 mm/s after 24 hours. It was therefore necessary that spraying not be allowed closer than 50 metres from the blasting face. These restrictions tended to interrupt the construction cycle during this project. Similar projects have been successfully completed without such restrictions, e.g. the Arlanda Express railway tunnel north of Stockholm, where shotcreting was carried out much closer to the blasting face.

It was therefore desirable to determine how blasting vibrations affected the shotcrete. Ansell (2000) has studied how blasting-induced vibrations affect young shotcrete, and in particular has determined the maximum allowable PPV's associated with stress waves close to the location of a blast. It was therefore necessary to determine how far from a blast the Peak Particle Velocity decreases to the critical level. Swedish Rock Engineering Research (SveBeFo), the Swedish National Road Administration, and the contractors Selmer/Besab have assessed the last issue in the Southern Link project.

3.2 *Blasting tests*

The blast-holes were charged with Site Sentisised Emulsion (SSE) explosives. The hole length was 5.2 metres, and each hole was charged with approximately seven kilograms of explosive. All tests were performed without shotcrete on the tunnel-walls.

Accelerometers were installed in boreholes of 0.30 metres depth in the tunnel walls between 5 and 50 metres from the blasting face. Each installation included two accelerometers, one parallel with the tunnel axis and one perpendicular to the tunnel wall. The accelerometers were grouted into the boreholes and were connected to a measuring unit with one channel dedicated to each accelerometer. Seven blasting rounds were recorded, each involving 14 devices.

3.3 *Data analysis*

Peak Particle Velocity (PPV) data from four rounds were used for the final analysis (Reidarman & Nyberg 2000). The measured vibration directions at each station were both parallel and perpendicular to the tunnel axis. All together 69 values across and 98 values along the tunnel axis were used from the four rounds. For construction work it is necessary to know the maximum vibration level and the most interesting may be the level that no vibrations exceed.

3.4 *Results*

The tests showed that the first arrival Peak Particle Velocities (PPV's) were usually moderate in magnitude, even as close as five metres from the blasting location. This was particularly true of the registrations from those accelerometers located perpendicular to the tunnel wall. Recorded peak particle veloci-

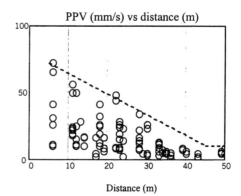

Figure 5. Measured Peak Particle Velocity (PPV) parallel to tunnel axis, as a function of distance from blasting face.

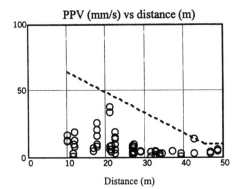

PPV (mm/s) vs distance (m)

Distance (m)

Figure 6. Measured Peak Particle Velocity, PPV, perpendicular to the tunnel axis.

ties (PPV's) for the tests are shown in Figures 5 and 6. It is apparent that peak vibration levels parallel to the walls are higher than the values perpendicular to the surfaces. For simplicity, a line representing the maximum vibration intensities (PPV=80-1.5 R) was used for both directions.

Based on the report by Nyberg & Reidarmann (2000) it was concluded that the measured vibration levels were far below the maximum permitted 150 mm/s required by the client. Comparing this study with Ansell (2000) also indicates that shotcreting might be performed close to the blasting face without causing damage to the concrete. However, more studies are needed to confirm this. Such a study can provide important input to future tunneling projects.

REFERENCES

Ansell, A. 2000. Dynamically loaded rock reinforcement. Doctoral Thesis, Bulletin 52, Dept. of Structural Engineering, Royal Institute of Technology, Stockholm.

Reidarman L. & Nyberg U., 1999. Vibrationer bakom front vid tunneldrivning, *SveBeFo Report 51* (in print), Stockholm, Sweden.

Nyberg U. & Reidarman L. 2000. Vibrationsmätning för nyutförd sprutbetong nära front i Södra Länken, Swedish Rock Engineering Research SveBeFo, Stockholm.

Shotcrete: Engineering Developments, Bernard (ed.) © 2001 Swets & Zeitlinger, Lisse, ISBN 90 5809 176 7

Reduction of rebound and dust for the dry-mix shotcrete process

W.Kusterle
Institute for Building Materials and Building Physics, University of Innsbruck, Austria

M.Pfeuffer
Heidelberger Zement Group Technology Center, Leimen, Germany

ABSTRACT: Results of investigations carried out at the University of Innsbruck demonstrate that dust formation in the nozzle region is mainly influenced by the wetting level, resulting from the ratio of water content and solid matter particles. Due to the development of a new pre-wetting nozzle which injects compressed air and water into the locally narrowed material stream, dust formation can be lowered by over 60 % in the dry-mix shotcrete process.

The amount of rebound in the dry-mix shotcrete process strongly correlates with the consistency of the sprayed concrete. In more than 120 laboratory tests the rheological properties of the mortars were correlated to the amount of rebound in the spraying process. Based upon these results a new powder admixture has been developed leading to a reduction of rebound of up to 45 % for the dry-mix shotcrete process combined with moist aggregates without altering water-cement ratio and without influencing strength development.

1 INTRODUCTION

Dry-mix shotcrete is typically used in tunnelling when very high-early strength is required under conditions of poor rock and high water ingress. The drawbacks of this process are low application rate, quite high rebound and dust formations. The amount of rebound and dust generated during application of dry-mix sprayed concrete are mainly influenced by factors arising from process and concrete technology (Morgan 1995).

More than 200 measurements in the laboratory and in large-scale tests as well as under construction conditions were used to study correlations between and causes of rebound and dust development, and to develop suggestions of how to minimize rebound and dust development. Although there exists a correlation between rebound and dust development, and the influencing factors are similar, the two topics will be discussed separately in a systematic survey.

2 DUST

Dust develops during many stages in tunnel construction, but dust concentration is highest when shotcrete is applied. The intensity of dust formation is influenced by various parameters.

For the dry-mix method, it depends on the delivery equipment, spraying rate, operating air pressure, distance from receiving surface, skill of the nozzleman, pre-wetting water quantity, type of nozzle, mix-

ture design, strength development of the binding agent, temperature of the mix and also on the ventilation in the tunnel (Pfeuffer & Kusterle 2000). One of the most important technical measures designed to reduce dust is pre-wetting. The dry-mix can already be sprayed with a cloud of water in a predampener after it has been drawn off from the storage silo. Wetting can also be undertaken by means of a pre-wetting nozzle (chamber), which can be installed in accordance with the setting behavior of the binding agent upstream of the normal nozzle located at the end of the concrete delivery hose.

The addition of water to the mix is carried out in the field at pressures of up to 10 MPa or on the basis of a mixture of compressed air and water. These wetting procedures have been investigated using different nozzle systems without resulting in any lasting success. In summary, these investigations revealed that none of these nozzle systems was capable of reducing the development of fine dust at the nozzle to a satisfactory degree without encrustation. The Institute for Building Materials and Building Physics at the University of Innsbruck thus decided to develop a novel nozzle system based on the addition of compressed air/water mixture, which guarantees high dust reduction while at the same time minimizing the problem of encrustation.

2.1 Design of the new nozzle system

Individual observations based on models of the pre-wetting nozzle system were used to analyze the fluid

Figure 1. Dust formation of the dry-mix sprayed concrete process.

processes within this system with regard to their geometrical form. Several modifications subsequently undertaken (Fig. 2), were tested in the laboratory and on site applications, and resulted in a considerable reduction of dust development and encrustation as compared with conventional nozzles.

The basic form of the modification is a narrowing of the wetting section (2) in order to alter the flow behaviour of the mix (Fig.2). The mix is accelerated through the narrowed section; at the same time, a so-called under-pressure suction effect is created. On the one hand, this prevents encrustation after the introductory zone and on the other, the reduction of pressure through the narrowed section enables the compressed air/water mixture to enter in a more straightforward manner. In other words, the air/water mixture can be fed in at the same pressure that is prevailing in the concrete supply hose. In this way, the flow behavior of the mix remains unchanged in its form and no turbulence occurs as is the case with conventional nozzles with different pressure conditions. In addition, the suction effect prevents the added

water being laterally displaced against the pipe wall. The supply pipes as well as the pre-wetting nozzle body therefore remain free of encrustation.

The compressed air/water mixture is added at an angle (<45°) via an arrangement of 2 supply lines (3) from the supply hose (4), so that a large pre-wetted area is attained in the flow cross-section. In this way, the compressed air/water mixture is prevented from de-mixing, settling on the pipe walls in the form of a film, or entering the flow as pure water and leading to encrustation.

2.2 Results for dust formation

In the lab and site tests, the problem of dust emission and encrustation was examined and the results obtained were compared with results achieved by conventional nozzles. The dry-mix method with oven-dry aggregates and spray cement SBM-T (cement with working time under one minute) was applied. Measurements were also carried out under the same test conditions. The pre-wetting nozzle (chamber) was installed 2.2 m upstream of the main nozzle in the concrete supply hose. By means of an installed flowmeter, the amount of water added (up to 30 % of the total amount of water) could be kept constant at the pre-wetting nozzle. The residual amount of water was regulated by the nozzle-man at the discharge nozzle at the end of the delivery hose. Figure 3 shows the results of the lab tests. Examples of a series of tests on site are presented in Figure 4.

At the moment, the newly developed nozzle system is being applied at selected construction sites (construction of subway tunnel in Dortmund and suburban railway in Stuttgart, Germany). The results of the observations show encrustation has be eliminated from the nozzle and a reduction of dust development by 70 % at a pre-wetting content of 15 % of the total water quantity. Practical experience in the use of the dry-mix method has shown, that the pre-wetting nozzle can either be set up roughly 2.2 m

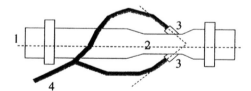

1 concrete delivery hose
2 narrowing of the wetting section
3 supply lines
4 supply line for air/water mixture

Figure 2. Schematic description of the pre-wetting nozzle (chamber): Pfeuffer-nozzle, situated upstream of and used in combination with a normal nozzle located at the end of the material delivery hose.

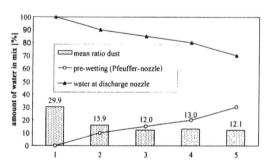

Figure 3. Variation of water used for pre-wetting at the newly developed nozzle (Pfeuffer-nozzle) at constant total water quantity (test in the laboratory: spray cement SBM-T and oven-dry aggregate).

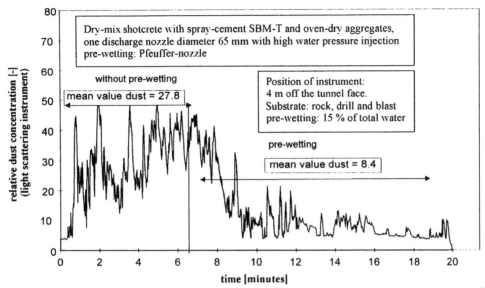

Figure 4. Dust formation in site tests: Dry-mix method with dry aggregates and SBM-T, pre-wetting approx. 15 % of the overall water quantity. Measurement of respirable dust being performed by means of a light-scattering instrument hund DM data. Conversion of relative dust concentrations into fine-dust concentrations obtained by gravity-based samplers can be done by conversion factors amounting from 1.1 to 1.5.

upstream of the discharge nozzle or directly downstream of the delivery equipment, depending on the setting behavior of the binding agent. In the process, up to 15 % of the required total amount of water - depending on the type of binding agent used - is added via compressed air at the pre-wetting stage. On account of the narrowed section, the pressure of the air/water mixture can be the same as the pressure existing in the concrete delivery hose without any constricted section. So far, no encrustation has occurred in the zone where the compressed air/water mixture is introduced. In this way, tedious cleaning jobs as are required in the case of conventional nozzles can be dispensed with. The robust design complies with requirements on site and has shown itself to be very user-friendly in day-to-day use. The pre-wetting stage, apart from minimizing dust, also reduces the wear inside the supply hoses during the utilization of oven-dry, especially sharp-edged material.

The new wetting system has been developed not only for tunneling applications. With the pre-wetting nozzle cross-sections correspondingly adjusted to the delivery hose diameter, the application of the dry-mix method using oven-dry material for constructing new buildings as well as for concrete repair purposes can be improved by using this pre-wetting nozzle.

3 REBOUND

Rebound is defined as the portion of the sprayed material that does not adhere to the substrate

(Armelin et al. 1997). It consists mainly of coarse aggregates and to a minor extent of cement and mixing water. The amount of rebound is influenced by the mixture composition (cements, aggregates, additives, admixtures, water-cement ratio), the adhesion capability of the mix, the process engineering (delivery equipment, material velocity at the nozzle, hose diameter) as well as on-site parameters (experience of the nozzleman, type of substrate, thickness of application, temperature, etc.) (Kusterle et al. 1997 & Lukas et al. 1995). The consistency of the sprayed concrete is influenced by the grain size distribution of the fine and coarse aggregates as well as the quantity and the fluidity of the cement paste. The concrete must be as plastic as possible so that the sprayed aggregates can be embedded in the fresh concrete. On the other hand the cohesion and the adhesion strength of the sprayed concrete to substrate interface must be greater than the weight of the applied shotcrete. The following theoretical claims can be derived for the fresh, sprayed concrete mix, where the cement paste is considered a viscous fluid (Bingham model), in which the aggregates are embedded (Fig. 5) (Banfill 1992 & Hattori et al. 1991):

1: The viscosity should be as low as possible in the proximity of the penetrating aggregates, in order to ensure that the aggregates will penetrate deeply and with dense particle packing.

2: The system fluid and granules should show an adequate yield stress so that the application is stable on the vertical wall.

Both requirements can only be combined if the

fluid has pseudo-plastic flow behavior with a corresponding yield stress.

3.1 *Laboratory tests*

A viscometer (Viskomat NT of the company Schleibinger Geräte) with a mortar paddle was used to test the flow properties of the cement paste (500 g CEM I 42.5 R, 150 g quartz sand of 0.5 mm maximum grain size, admixtures in powder form, 250 g of the total water quantity including the water content of the liquid admixtures). This is a rheometer of the Couette-system, e.g. a fixed measuring cell immerses into a rotating sample container that electronically measures the torque against rotational speed (Figure 5).

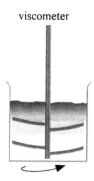

viscometer

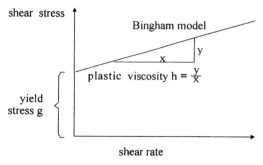

Figure 5. Flow curve of a Bingham model, showing yield stress g and plastic viscosity h.

Examination of the flow properties in more than 120 tests with various wetting techniques and admixtures confirmed that the thixotropic properties of the binder paste of the sprayed dry-mix before setting are the determining factor of rebound behavior in the dry-mix process. On this basis, an admixture in powder form was developed, which, when applied in the mixture, shows intrinsically viscose flow behavior (decrease in apparent viscosity when shear stress is increased) with corresponding relative yield stress before initial setting of the cement (Figure 6).

The results of the rheological test show a decreasing relative viscosity h in addition to the relatively marked yield stress g when the doses of the rebound minimizing agent in powder form is increased (Figure 6).

An adjustment of the water-cement ratio from 0.50 to 0.55 intensifies this behavior. This probably is due to the higher water-ratio that allows the powdered admixture to develop the intrinsic viscose behavior.

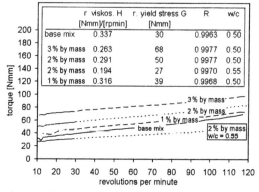

Figure 6. Effect of the newly developed agent (% by weight/weight of cement) to minimize rebound on the relative viscosity h and the relative yield stress g (expressed as G and H values).

3.2 *Results of spraying tests*

The spraying experiments took place at the shotcrete testing facilities at the University of Innsbruck, Institute for Building Materials and Building Physics, and at a construction site. The rebound-minimizing agent tested was added to the cement (spray cement Chronolith-ST). The mixing ratio of binder (cement and admixtures) to aggregate was set to 1:4.83 for each test. The maximum size of the aggregate used was 4 mm in the laboratory experiments and 8 mm at the construction site. The intrinsic moisture content ranged from 3.0 % to 3.5 %.

The use of the rebound-minimizing agent reduced the rebound in laboratory spray experiments depending on rheological characteristics (low relative viscosity with corresponding relative yield stress). The rebound was reduced by 51 % in comparison to the reference specimen at the same consistency when applying a dosage of the additive of 1.6 % by weight/weight of cement. The rebound is reduced by 40 % compared to the base mix at the same water-cement ratio and an approximately equal strength development (Figures 7 and 8).

The reduction in rebound obtained in laboratory experiments when applying the dry-mix method has also been proven in practical work. The rebound

170

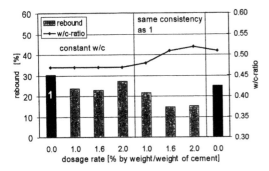

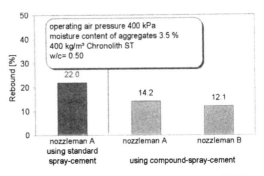

Figure 7. Effect of the newly developed rebound-minimizing agent on rebound behavior in the dry-mix process (380 kg/m³ Chronolith-ST; 3 % moisture in mix).

Figure 9. Effect of the newly developed rebound minimizing agent in compound-spray-cement for the dry-mix process (experiment at construction site).

behavior with damp aggregates could be improved for the dry-mix method even when different persons sprayed the concrete (person A, person B), with other conditions left unchanged (tunnel construction site, gun, substrate, measuring range, etc.). The rebound could be reduced by up to 45 % (absolute ratio = 12.1 %, at the construction site; Figure 9) when using a dosage of 1.6 % by mass in the compound-spray-cement at an approximately stable water-cement ratio.

4 CONCLUSIONS

Prolonged inhalation of fine dust particles when using the dry-mix shotcrete process may lead to health problems. A considerable contribution to industrial hygiene has been made by the development of a new pre-wetting nozzle (dust could be reduced by more than 60 %). The application of a newly developed powder admixture with intrinsically viscose flow properties leads to a pseudo-plastic surface on the sprayed concrete. By app-

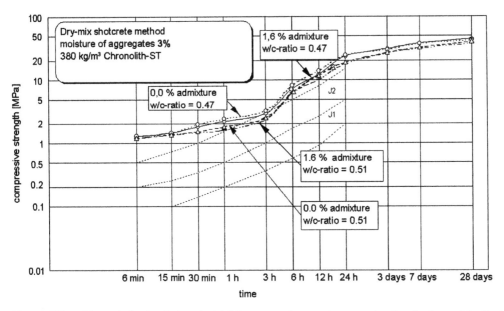

Figure 8. Effect of the newly developed rebound minimizing agent on the concrete compressive strength when applying the dry-mix process (experiments in the laboratory). Dosage calculated in % by weight/weight of cement. Determination of compressive strength by penetration test, stud driving method and on drilled cores. J1, J2, J3 showing early strength classes according to Austrian guideline on sprayed concrete.

lying this measure the rebound proportion could be reduced by about 45 % at an approximately stable constant water-cement ratio and concrete strength.

REFERENCES

Armelin, H. S., Banthia, N., Morgan, D. R. & Steeves, C. 1997. Rebound in dry-mix shotcrete. *ACI Concrete International* 19 (9): 54-60.

Banfill, P.F.G. 1992. Structural breakdown and the rheology of cement mortar. Theoretical and applied rheology: *Proc. XIth Int. Congr.*: 790-792.

Hattori, K. & Izumi, K. 1991. A new viscosity equation for non-Newtonian suspensions and its application. *Rheology of Fresh Cement and Concrete:* 83-92. London: Spon.

Kusterle, W.& Eichler, K. 1997. Tests with rebound behaviour of dry-sprayed concrete. *Tunnel* (5) 1997.

Lukas, W., Kusterle, W. & Pichler, W. 1995. Innovations in shotcrete technology. *Shotcrete for Underground Support VII:* Engineering Foundation.

Morgan, D. R. 1995. *Sprayed concrete: properties, design and application.* Edited by S.A. Austin and P.J. Robins. New York: Mc Graw-Hill.

Pfeuffer, M. & Kusterle, W. 2000. Improving pre-wetting technology for the dry-mix sprayed concrete method. *Tunnel* (2) 2000.

Pfeuffer, M. & Kusterle, W. 2000. Rheology and Rebound Behaviour of Dry-Mix Sprayed Concrete. *Magazine of Concrete Research* (1) 2001.

Shotcrete: Engineering Developments, Bernard (ed.) © 2001 Swets & Zeitlinger, Lisse, ISBN 90 5809 176 7

Phase transitions in shotcrete: from material modelling to structural safety assessment

Ch.Hellmich, J.Macht, R.Lackner, H.A.Mang
Vienna University of Technology, Vienna, Austria

F.-J.Ulm
Massachusetts Institute of Technology, Cambridge, MA, USA

ABSTRACT: This paper contains a contribution to material modeling of shotcrete in the framework of thermodynamics of chemically reactive porous media. The focus is on the practically highly relevant calcium aluminate hydration reaction of low sulfate cements, taking place immediately after the spraying of shotcrete. What is of practical interest, is the immediate strength. Afterwards, the calcium silicates dominate the hydration process and, hence, the thermal, elastic, plastic, creep and shrinkage behavior of the material. The scope of the paper ranges from basic material modelling over respective experiments – focussing on adiabatic testing – to structural finite element (FE) analyses of tunnels driven according to the New Austrian Tunneling Method (NATM).

1 INTRODUCTION

'Classical' shotcretes based on Portland cement and admixed accelerators with a high alkali content were used until the early 1990's. They caused severe health problems for the workers on site (harm of lungs and eyes). As a remedy, two major developments took place:

- alkali-free accelerators, e.g., (Pichler 1995),

- so-called shotcrete cements without accelerators, e.g., (Eichler 1999).

In this contribution, the focus is on proper thermochemomechanical material modelling of the second group of shotcretes, with emphasis on structural application of such models to tunneling.

As for ordinary cements, calcium sulfate (gypsum) is added in the standard manner in order to delay the setting of concrete, i.e., to prevent the uncontrolled reaction of calcium aluminate. Thus, only fine, thin, and short ettringite crystals are formed. They have no significant influence on the strength of the material, which is almost exclusively controlled by the hydration of calcium silicates. For shotcrete, however, an immediate, yet controlled setting is highly desirable. This was recently achieved by low sulfate cements with well-balanced components. As for such shotcrete cements, after the addition of water, calcium alumi-

nate hydrates (CAH) with a plate-like microstructure are formed, establishing a first skeleton and providing the required immediate strength of shotcrete, see, e.g., (Eichler 1999) and Figure 1(a). Later on, these CAH are integrated into the matrix formed by the calcium silicate hydrates (CSH), see Figure 1(b).

For the study of the kinetics of hydration of all kinds of concrete, adiabatic experiments play an important role; for general aspects, see, e.g., (Bye 1999); from a material modelling viewpoint, see, e.g., (Ulm and Coussy 1996). For shotcrete, such tests are *not* contained in standards, e.g., (Guideline 1997). Therefore, these tests are normally not performed. This has led to alternative strategies for determination of the characteristics of the ki-

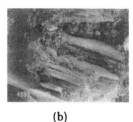

(a)　　　　　(b)

Figure 1. Microstructure of shotcrete: (a) initial formation of calcium aluminate hydrates, after (Eichler 1999); (b) subsequent formation of calcium silicate hydrates, after (Baroughel-Bouny 1994).

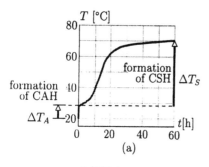

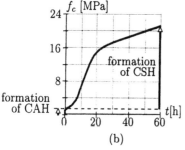

Figure 2. Typical results of experiments on shotcrete based on low sulfate cements: for (a) adiabatic and (b) strength evolution.

netics of shotcrete, based on strength evolutions, (Hellmich et al. 1999d). However, on behalf of the Institute for Strength of Materials of the Vienna University of Technology, adiabatic tests have recently been performed on shotcrete based on low sulfate cements at the laboratory of the Schretter and Cie cement plant in Vils, Tyrol, Austria. A typical result of such an adiabatic test is illustrated in Figure 2(a). On the other hand, a typical strength evolution with time of such a shotcrete is shown in Figure 2(b). Since there are well-known *chemo-thermal cross effects* (e.g., a linear one between the latent heat production and the specific mass of formed hydrates, see, e.g., (Bye 1999)) as well as *chemo-mechanical couplings* (e.g., a quasi-linear one between the strength and the specific mass of formed hydrates, see, e.g., (Byfors 1980)), the aforementioned tests can be used for (macroscopically) studying the kinetics of the underlying chemical reactions. In more detail, regarding Figure 2, one may divide the numerous chemical reactions between water and cement, influencing both (adiabatic) temperature evolution and strength growth, into two main groups as far as the kinetics is concerned:

1. The first group is dominated by the very fast reaction between calcium aluminate (C_3A) and water, lasting *several minutes*. The re-

spective reaction products are called calcium aluminate hydrates (CAH).

2. The second group is predominantly influenced by the formation of calcium silicate hydrates (CSH), including the integration of the aforementioned CAH. The underlying rate-determining process is roughly the diffusion of free water between the already formed hydrates in order to reach the unhydrated cement. Once it is reached, new hydrates are formed quasi-instantaneously. This diffusion process is obviously lasting longer than the reaction described under item 1. , namely, *several days*.

In the following, these two groups will simply be called the CA and the CS hydration reaction, respectively. Remarkably, the ratio between the temperature rises because of the CAH-formation and the CSH-formation, respectively, and the ratio between the respective strength growths are *not* the same. Therefore, if the CAH-formation is to be accounted for properly in a macroscopic material model, *one* internal variable is not sufficient for the description of the state of hydration. This internal state variable, called hydration degree, was employed in the models of Ulm and Coussy (1996) and Hellmich et al. (1999d). In contrast, a material model with *two* such state variables, one related to the specific mass of water bound in CAH and one referring to the respective mass bound in CSH, will be presented in the following.

2 MATERIAL MODELLING

2.1 Thermodynamic basis

As already mentioned, the kinetics of low sulfate cements can be studied preferably using adiabatic tests. This is the motivation to start with the simplest thermodynamic framework for the description of such chemothermal problems, which is that of closed, undeformable, chemically reactive porous media, see also (Hellmich et al. 1999c).

In the absence of external volume heat sources, for such a medium, the first law of thermodynamics reads as follows, e.g. (Coussy 1995),

$$\frac{de}{dt} = \dot{e} = -\text{div}\mathbf{q}, \qquad (2.1)$$

where e is the internal energy, and $\mathbf{q}$ is the heat flux vector, which is positive for an efflux. This law states that the change of the internal energy equals the external (surface) heat sources.

Under the aforementioned conditions, the second law of thermodynamics is given as, e.g., (Coussy 1995),

$$\dot{S} \geq -\mathrm{div}\frac{\mathbf{q}}{T}, \qquad (2.2)$$

with S denoting the entropy and T standing for the absolute temperature. This law states that the change of S is greater than or equal to the external (surface) entropy sources.

Inserting (2.1) into (2.2) and making use of the definition of the free Helmholtz energy, ψ, i.e.,

$$e = \psi + TS, \qquad (2.3)$$

results in the Clausius-Duhem inequality, reading, e.g. (Coussy 1995),

$$\varphi = -S\dot{T} - \dot{\psi} - \frac{\mathbf{q}}{T} \cdot \mathrm{grad}\, T \geq 0, \qquad (2.4)$$

with the dissipation φ.

In the present case, ψ depends on three state variables. One of them, T, is an external state variable. The extent of the CA hydration, ξ_A, and of the CS hydration, ξ_S, are internal variables. They are defined as $\xi_i = m_i/m_{i,\infty}$, $i = A, S$, with m_A (m_S) [kg m^{-3}] standing for the specific mass of water bound in calcium aluminate (silicate) hydrates and $m_{A,\infty}$ ($m_{S,\infty}$) standing for the respective mass at complete calcium aluminate (silicate) hydration. $m_{A,\infty}$ ($m_{S,\infty}$) is related to the CA (CS) content of the considered shotcrete, c_A (c_S) [kg m^{-3}]. The respective relations result from stoichiometric considerations (see (Byfors 1980), p.29). They are $m_{A,\infty} = 0.40 c_A$ and $m_{C,\infty} = 0.21 c_S$. Accounting for the dependencies of ψ on T, ξ_A, and ξ_S in (2.4) yields

$$(-S - \frac{\partial\psi}{\partial T})\dot{T} - \frac{\partial\psi}{\partial \xi_A}\dot{m}_A - \frac{\partial\psi}{\partial \xi_S}\dot{m}_S - \frac{\mathbf{q}}{T} \cdot \mathrm{grad}\, T \geq 0. \qquad (2.5)$$

The normality hypothesis states that the evolution of each state variable is independent of the evolution of all other state variables, e.g. (Coussy 1995). In this context, (2.5) must hold for chemically inert conditions ($\dot{\xi}_A = 0$; $\dot{\xi}_S = 0$) as well as for uniform temperature distributions (grad $T = \mathbf{0}$). This leads to the state equation for the entropy,

$$S = -\frac{\partial\psi}{\partial T} = S(\xi_A, \xi_S, T). \qquad (2.6)$$

Moreover, Equation (2.5) allows identification of

$$A_A = -\frac{\partial\psi}{\partial \xi_A} = A_A(\xi_A, \xi_S, T) \qquad (2.7)$$

and

$$A_S = -\frac{\partial\psi}{\partial \xi_S} = A_S(\xi_A, \xi_S, T) \qquad (2.8)$$

as the driving forces of the CA and CS hydration

reactions, correlated to the rate of the respective hydration extent, $\dot{\xi}_A$ and $\dot{\xi}_S$. A_A and A_S are referred to as chemical affinities (Coussy 1995).

Furthermore, from (2.5), $-(\mathrm{grad}\, T)$ can be identified as the driving force of the entropy flux $\mathbf{q}/T$. A linear isotropic relation between these quantities (Fourier's law) is reasonable for shotcrete applications,

$$\frac{\mathbf{q}}{T} = -\frac{1}{T}k\,\mathrm{grad}\, T, \qquad (2.9)$$

with the thermal conductivity coefficient k. A_A, A_S, and S are related via the Maxwell symmetries (see (2.6), (2.7), and (2.8))

$$\frac{\partial S}{\partial \xi_A} = \frac{\partial A_A}{\partial T} = -\frac{\partial^2\psi}{\partial T\partial \xi_A} = -\frac{L_A}{T}, \qquad (2.10)$$

$$\frac{\partial S}{\partial \xi_S} = \frac{\partial A_S}{\partial T} = -\frac{\partial^2\psi}{\partial T\partial \xi_S} = -\frac{L_S}{T}. \qquad (2.11)$$

L_A and L_S are the latent heats of the respective chemical reactions. As in physical chemistry (Atkins 1994), they are linked to the change of entropy in consequence of the respective chemical reaction, $\partial S/\partial \xi_A$ and $\partial S/\partial \xi_S$. They reflect the exothermal nature of the considered chemical reaction because $L_A > 0$ and $L_S > 0$.

Using (2.6), (2.7), (2.8), (2.10), and (2.11) in the expressions (2.1) to (2.4) leads to the following entropy balance law:

$$T\frac{\partial S}{\partial T}\dot{T} - L_A\dot{m}_A - L_S\dot{m}_S - A_A\dot{m}_A - A_S\dot{m}_S = -\mathrm{div}\mathbf{q}. \qquad (2.12)$$

The first term in (2.12) refers to a change of the internal energy resulting from a temperature variation. Consequently, $T\partial S/\partial T = \partial e/\partial T = C_V$ refers to the heat capacity per unit volume. The chemical dissipations, $A_A\dot{\xi}_A$ and $A_S\dot{\xi}_S$, which are related to the diffusion of free water as described under item 2. in the Introduction, are negligible in comparison with the latent heat productions, L_A and L_S, see (Ulm and Coussy 1995). Therefore, the balance law

$$C_V\dot{T} - L_A\dot{\xi}_A - L_S\dot{\xi}_S = -\mathrm{div}\mathbf{q} \qquad (2.13)$$

is used in the following.

2.2 Strong and weak couplings concerning the chemical affinities

Differentiation of (2.7) and (2.8) and consideration of (2.10) and (2.11) results in the (infinitesimally) incremental state equations:

$$dA_A = -\frac{\partial^2 \psi}{\partial \xi_A \partial T}dT - \frac{\partial^2 \psi}{\partial \xi_A^2}d\xi_A - \frac{\partial^2 \psi}{\partial \xi_A \partial \xi_S}d\xi_S$$
$$= -L_A dT - \kappa_A d\xi_A - \kappa_{AS}d\xi_S, \qquad (2.14)$$
$$dA_S = -\frac{\partial^2 \psi}{\partial \xi_S \partial T}dT - \frac{\partial^2 \psi}{\partial \xi_S^2}d\xi_S - \frac{\partial^2 \psi}{\partial \xi_S \partial \xi_A}d\xi_A$$
$$= -L_S dT - \kappa_S d\xi_S - \kappa_{AS}d\xi_A, \qquad (2.15)$$

with the equilibrium constants κ_A, κ_S, and κ_{AS}. Obviously, (2.14) and (2.15) satisfy the Maxwell symmetry $-\partial^2 \psi / \partial \xi_S \partial \xi_A = \partial A_S / \partial \xi_A = \partial A_A / \partial \xi_S = \kappa_{AS}$. The non-vanishing term κ_{AS} indicates the mutual dependencies of A_A and A_S on the state of both chemical reactions. This follows directly from the fact the free water is the reactant of *both* (macroscopical) chemical reactions.

For temperature ranges in standard concrete (shotcrete) applications (283 K < T < 373 K), changes dT are rather negligible as far as the thermodynamic imbalances dA_A and dA_S are concerned (*partial decoupling hypothesis*). Thus, only the forms of (2.14) and (2.15) specialized for isothermal conditions ($dT = 0$) will be considered in the following. Integration of these forms results in

$$A_A = A_A(\xi_A, \xi_S) \quad \text{and} \quad A_S = A_S(\xi_A, \xi_S). \quad (2.16)$$

2.3 Hydration kinetics

The kinetics laws for the two encountered chemical reactions read as:

$$\tau_A \frac{d\xi_A}{dt} = \mathcal{F}(\xi_A, \xi_S) \qquad (2.17)$$

and

$$\tau_S \frac{d\xi_S}{dt} = \mathcal{G}(\xi_A, \xi_S), \qquad (2.18)$$

where the functions $\mathcal{F}(\xi_A, \xi_S)$ and $\mathcal{G}(\xi_A, \xi_S)$ describe the reaction orders, see, e.g., (Atkins 1994) and (Ulm and Coussy 2000). τ_A and τ_S are the characteristic times of the reactions.

Considering a linear transform of time $t = \mathcal{T}s$, with a reference time (time unit) $\mathcal{T}$ (see, e.g., (Barenblatt 1996)) yields kinetics laws in dimensionless form:

$$\frac{\tau_A}{\mathcal{T}}\frac{\partial \xi_A}{\partial s} = \mathcal{F}(\xi_A, \xi_S) \qquad (2.19)$$

and

$$\frac{\tau_S}{\mathcal{T}}\frac{\partial \xi_S}{\partial s} = \mathcal{G}(\xi_A, \xi_S). \qquad (2.20)$$

Inspecting the first kinetics law at the time scale of τ_S. i.e.. $\mathcal{T} = \tau_S$, yields:

$$\frac{\tau_A}{\tau_S}\frac{\partial \xi_A}{\partial s} = \mathcal{F}(\xi_A, \xi_S). \qquad (2.21)$$

This indicates that for $\tau_A \ll \tau_S$ the first reaction can be considered at chemical equilibrium:

$$\tau_A \ll \tau_S \leftrightarrow \mathcal{F}(\xi_A, \xi_S) = 0. \qquad (2.22)$$

This is in fact the case since τ_A lies in the range of 5 to 15 minutes whereas the magnitude of τ_S lies within the range of decades of hours.

Accounting for (2.22), leads to

$$\mathcal{F}(\xi_A, \xi_S) = 0 \leftrightarrow \xi_A = \xi_A(\xi_S). \qquad (2.23)$$

Thus, there exists a unique CA hydration function $\xi_A = \xi_A(\xi_S)$ obtained from thermodynamic equilibrium of the CA hydration reaction relative to the time scale of the CS hydration reaction.

Since the CA reaction is always in equilibrium with respect to the time scale of the calcium silicate reaction, it follows that

$$\mathcal{G} = \mathcal{G}(\xi_A(\xi_S), \xi_S) = \mathcal{G}(\xi_S). \qquad (2.24)$$

The CS hydration is a thermo-activated process. This is standardly accounted for by an Arrhenius activation term, yielding τ_S in the form

$$\tau_S = \tau_{S,0}\exp\left[\frac{E_{a,S}}{R}\left(\frac{1}{T} - \frac{1}{T_0}\right)\right], \qquad (2.25)$$

where $\tau_{S,0}$ refers to a reference temperature T_0, e.g., $T_0 = 293$ K. Accounting for (2.24) and (2.25) and defining the chemical affinity (see (2.8)) as

$$A_S = \frac{\mathcal{G}}{\tau_{S,0}}\exp\left(\frac{E_{a,S}}{RT_0}\right), \qquad (2.26)$$

yields (2.18) in the form

$$\dot{\xi}_S = A_S(\xi_S)\exp\left(-\frac{E_{a,S}}{RT}\right), \qquad (2.27)$$

with the activation term $E_{a,S}/R \approx 4200$ K (see, e.g., (Bye 1999), p.110).

3 EXPERIMENTAL DETERMINATION OF MATERIAL PROPERTIES

3.1 Experimental program

For the determination of material properties of low sulfate cements for shotcrete, two tests in an adiabatic calorimeter were performed at the laboratory of the Schretter and Cie cement plant in Vils, Tyrol, Austria. One of them was a typical shotcrete mixture. The respective components are listed under "shotcrete I" in Table 1.

The respective test result is illustrated by the thick curve in Figure 3. The experiment had to be

Table 1. Shotcrete mixtures for adiabatic tests.

	shotc. I	shotc. II
cement content [kg/m^3]	380	280
aggregate content [kg/m^3]	1804	1907
water content [kg/m^3]	228	196
water/cement ratio [–]	0.6	0.7
specific mass ρ [kg/m^3]	2412	2383

terminated after 25 h since the large temperature rise caused problems with the experimental set-up. Therefore, a second shotcrete mixture with a lower cement content was composed, see "shotcrete II" in Table 1. The respective test result is illustrated by the thin curve in Figure 3.

The second test provides information permitting determination of T_∞ of the first test, as will be shown in the following: For an adiabatic experiment, div$\mathbf{q} = 0$, (2.13) reads as

$$C_V \dot{T}_{ad} - L_A \dot{\xi}_A - L_S \dot{\xi}_S = 0. \tag{3.1}$$

Integration over the entire experimental history results in

$$\int_{t_0}^{\infty} (C_V \dot{T}_{ad} - L_A \dot{\xi}_A - L_S \dot{\xi}_S) dt =$$
$$C_V(T_{ad,\infty} - T_0) - L_A - L_S = 0$$
$$\leftrightarrow \quad T_{ad,\infty} - T_0 = \frac{L_A + L_S}{C_V}. \tag{3.2}$$

T_0 denotes the temperature at which the shotcrete is placed in the calorimeter. It is given as $T_0 = 20$ °C. For the second experiment, $(T_\infty - T_0)_{II} = 72.0 - 20.0 = 52.0$ °C is known. The heat capacity of concrete per unit mass is constant $C_m = C_V/\rho = 1$ kJ/°C/(kg concrete) (see, e.g., (Byfors 1980), p.268). Hence, see Table 1, $C_V \approx 2400$ kJ/°C/(m^3 shotcrete). Inserting this value into (3.2) renders the total latent heat as $(L_A+L_S)_{II} = 2400 \cdot 52.0 = 125000$ kJ/(m^3 shotcrete II). The latent heats L_A and L_S stem without exception from

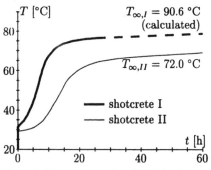

Figure 3. Temperature rises in adiabatic experiments.

the *cement qualities*, see, e.g., (Byfors 1980). Consequently, their sum can also be related to the mass of cement, see Table 1, $l_A + l_S = 125000/280 = 446$ kJ/(kg cement), independent from the considered shotcrete mixture. In turn, this enables calculation of the total latent heat of shotcrete I, see Table 1 for the respective cement content, $(L_A + L_S)_I = 446 \cdot 380 = 169000$ kJ/(m^3 shotcrete I); and, finally, of the desired adiabatic temperature change for the first experiment, $(T_\infty - T_0)_I = 169000/2400 = 70.6$ °C $\leftrightarrow T_{\infty,I} = 90.6$ °C.

3.2 Latent heats of CA and CS hydration reaction

As for the determination of L_A, (2.17) and (2.18) are inserted into (3.1), yielding

$$C_V \dot{T}_{ad} - L_A \frac{1}{\tau_A}\mathcal{F} - L_S \frac{1}{\tau_S}\mathcal{G} = 0. \tag{3.3}$$

Considering again a linear transform $t = \mathcal{T}s$, as in Subsection 2.3, yields

$$C_V \frac{\partial T_{ad}}{\partial s} - L_A \frac{\mathcal{T}}{\tau_A}\mathcal{F} - L_S \frac{\mathcal{T}}{\tau_S}\mathcal{G} = 0. \tag{3.4}$$

Considering (3.3) at the time scale of the CA hydration reaction, $\mathcal{T} = \tau_A$, yields for $\tau_A \ll \tau_S$,

$$C_V \frac{\partial T_{ad}}{\partial s} - L_A \mathcal{F} = 0 \leftrightarrow C_V \dot{T} - L_A \dot{\xi}_A = 0. \tag{3.5}$$

Integrating (3.5) from t_0 over a time interval lying within the considered time scale yields, see also Figure 2(a),

$$L_A = \frac{\Delta T_A}{C_V}. \tag{3.6}$$

Applying (3.6) to the two shotcrete mixtures yields

$$L_{A,I} = 2400 \cdot 10.5 = 25200 \text{ kJ/(m}^3 \text{ shotc. I)} \tag{3.7}$$

and

$$L_{A,II} = 2400 \cdot 7.9 = 18960 \text{ kJ/(m}^3 \text{ shotc. II)}. \tag{3.8}$$

Relating L_A to the cement mass reveals that, as expected, the latent heat because of CA hydration is an intrinsic *cement quality*, see also Subsection 3.1, $l_A = 25200/380 \approx 18960/280 \approx 67$ kJ/(kg cement). Using the numbers from Subsection 3.1, l_S can be calculated as $l_S = 446 - 67 = 379$ kJ/(kg cement). Hence, for shotcrete I, which is a typical mixture used for tunneling, $L_S = 379 \cdot 380 = 144000$ kJ/(m^3 shotcrete I), see also Table 2 concerning the input for the thermochemomechanical analyses decribed in Section 4.

3.3 Determination of kinetics characteristics

In order to quantify the kinetics characteristics

of low sulfate cements for shotcrete, the material functions $A_S(\xi_S)$ and $\xi_A(\xi_S)$ remain to be determined. This will be done here on the basis of adiabatic tests.

The very fast temperature rise under adiabatic conditions at the beginning of the hydration between cement and water results from the CA reaction, see Figure 2(a). For low sulfate cements, this is the only stage of the overall hydration reaction where calcium aluminate hydrates (Figure 1(a)) are formed. Therefore, the material function $\xi_A = \xi_A(\xi_S)$ is characterized by a large gradient for very small values of ξ_S and approaches very soon the (asymptotic) value $\xi_A = 1$;

$$\xi_A(\xi_S) = 1 - \exp\left(\frac{-\xi_S}{\hat{\xi}_S}\right), \qquad (3.9)$$

with $\hat{\xi}_S \approx 0.1\%$ (see Figure 4).

As for the determination of $A_S(\xi_S)$, i.e., of the kinetics characteristics of the CS hydration reaction, it should be recalled that the CA hydration reaction is in equilibrium with respect to the time scale of the CS hydration reaction. From (2.22) and (2.17), it follows:

$$\mathcal{F} = 0 \leftrightarrow \dot{\xi}_A = 0 \qquad (3.10)$$

Introducing (3.10) into (3.1) yields

$$C_V \dot{T}_{ad} - L_S \dot{\xi}_S = 0 \qquad (3.11)$$

Table 2. Material properties of shotcrete based on low sulfate cement.

chemical affinity $A_S(\xi_S)$ for shotcrete I and II; regression parameters	a_A	5.980
	b_A	18.02
	c_A	85.89
	d_A	7.377
latent heats of cement [kJ/(kg cement)]	l_A	67
	l_S	370
latent heats of shotcrete I [kJ/(m³ shotcrete I)]	L_A	25500
	L_S	144000

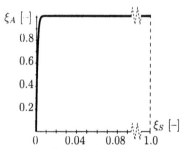

Figure 4. Material function for the reaction extent of calcium aluminate $\xi_A(\xi_S)$.

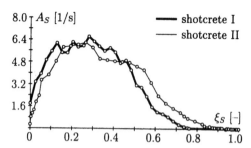

Figure 5. Chemical affinity for calcium silicate hydration for the two considered shotcrete mixtures.

for time scales considerably larger than τ_A. Regarding Figure 2(a), temporal integration of (3.11) results in

$$C_V(T_{ad}(t) - T_0 - \Delta T_A) - L_S \xi_S(t) = 0. \qquad (3.12)$$

From (3.12), (3.6), and (3.2) it follows that

$$\xi_S(t) = \frac{T_{ad}(t) - T_0 - \Delta T_A}{T_{ad,\infty} - T_0 - \Delta T_A}. \qquad (3.13)$$

Inserting (2.27) into (3.3) and accounting for (3.13) delivers A_S in parameter form with t as the parameter,

$$A_S(t) = \frac{1}{T_\infty - T_0 - \Delta T_A} \frac{dT_{ad}}{dt}(t) \exp\left(\frac{E_a}{RT(t)}\right). \qquad (3.14)$$

Combining (3.14) and (3.13) gives access to the desired material function $A_S(\xi_S)$. Evaluation of the $T(t)$-curves in Figure 3 yields the material functions $A_S(\xi_S)$ depicted in Figure 5. The chemical affinities for both shotcretes, exhibiting different cement contents and different water/cement ratios, are approximately the same. This allows to conclude that

- there is no influence of the cement content on the chemical affinity A_S;

- there is no practically relevant influence of the water/cement ratio on the chemical affinity as far as water/cement ratios ranging from 0.6 to 0.7 are concerned.

In order to facilitate the definition of the chemical affinity A_S in thermochemomechanical Finite Element analyses, the application of a nonlinear regression function is suitable, see also (Hellmich et al. 1999c),

$$A_S(\xi_S) = a_A \frac{1 - \exp(-b_A \xi_S)}{1 + c_A \xi_S^{d_A}} \quad \text{for } \xi_S > \bar{\xi}_S \approx 1\% \qquad (3.15)$$

and

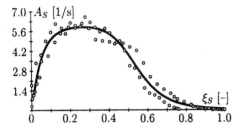

Figure 6. Nonlinear regression function for $A_S(\xi_S)$.

$$A_S(\xi_S) = A_S(\bar{\xi}_S) \text{ for } \xi_S \leq \bar{\xi}_S \approx 1\%. \quad (3.16)$$

The data of both shotcretes are used for determination of the regression parameters a_A, b_A, c_A, and d_A, see Figure 6 and Table 2. A correlation coefficient of $r = 98.0$ % indicates the suitability of the chosen regression function.

4 STRUCTURAL ANALYSES OF TUNNELS

4.1 Thermochemical analysis

4.1.1 Problem definition

The underlying thermochemical problem is given by (2.13) and (2.23), i.e.,

$$C_V \dot{T} - \tilde{L}(\xi_S)\dot{\xi}_S = -\text{div}\mathbf{q}, \quad (4.1)$$

with

$$\tilde{L}(\xi_S) = L_S + L_A \frac{\partial \xi_A}{\partial \xi_S}(\xi_S), \quad (4.2)$$

together with the Fourier heat conduction law, (2.9), and the Arrhenius law, (2.27). On the boundary $S = S_T \cup S_q$ of the domain, either the temperature can be prescribed, see, e.g., (Zienkiewicz and Taylor 1994),

$$T = \bar{T} \text{ on } S_T \quad (4.3)$$

or the heat efflux $q_n = \mathbf{q} \cdot \mathbf{n}$, i.e.,

$$q_n = \alpha(T - T^{env}) \text{ on } S_q. \quad (4.4)$$

$\mathbf{n}$ is the outward normal on a surface element dS, α is the radiation coefficient, and T^{env} is the environmental temperature. The initial and final conditions for each material point are given by

$$\begin{aligned} t = t_0 &: \quad \xi_S = 0; \quad T = T_0; \\ t = \infty &: \quad \xi_S = 1; \quad T = T_\infty. \end{aligned} \quad (4.5)$$

The nonlinearity of $\tilde{L}$ in (4.1) with respect to ξ_S renders the thermochemical problem (4.1) to (4.5) slightly more complicated than the one dealt with in (Hellmich et al. 1999c) in the context of the

nonlinear finite element method (FEM). The term in the global consistent tangent operator which is to be modified, reads as

$$\mathop{\mathbf{A}}_{e=1}^{n^e} \int_{V^e} \mathbf{N}^e \frac{1}{\Delta t_{n+1}} \left(L_S + L_A \frac{d\xi_A}{d\xi_S} \Big|_{n+1}^{(k)} \right.$$
$$\left. + \Delta\xi_{n+1}^{(k)} L_A \frac{d^2\xi_A}{d\xi_S^2} \Big|_{n+1}^{(k)} \right) \frac{d\xi_S}{dT} \Big|_{n+1}^{(k)} \mathbf{N}^e dV. \quad (4.6)$$

with the assembly operator $\mathop{\mathbf{A}}\limits_{e=1}^{n^e}$, the shape function $\mathbf{N}^e$, the time-step number n, the FE number e, the time increment Δt_{n+1}, and the number of the step in the global Newton-Raphson iteration, k.

4.1.2 Analyzed tunnel structure

Concentrating on tunnels with shotcrete linings and anchor bolts, which are constructed according to the New Austrian Tunneling Method, a thermochemical analysis of the Sieberg tunnel, situated in the Western part of Lower Austria, is described first. This tunnel was recently constructed as part of the high capacity railway line connecting Vienna and Salzburg. Crown, bench, and invert were excavated subsequently. As is typical for large parts of the tunnel, the installation of the bench took place about one month after shotcreting of the crown. At this time, the process of heating and cooling of the crown in consequence of hydration was almost completed. Therefore, all construction steps can be analyzed separately. Since, in addition, the problem is axisymmetric (Hellmich et al. 1999b), the FE mesh depicted in Figure 7 is sufficient for the thermochemical analysis. Missing material properties for shotcrete and soil are summarized in Table 3. In the considered geographical region, the mean air temperature in winter is about 0 °C; the soil temperature is constant at roughly 10 °C. The initial shotcrete temperature T_0 was 21 °C.

In order to investigate the influence of the explicit consideration of the heat shock because of CA hydration on the structural level, the respective results are compared with results obtained from a former material model with only one state variable for the overall hydration process, see (Hellmich 1999). For this former model, material properties of a shotcrete based on a low-sulfate cement produced by Lafarge were considered. Figure 8 shows that although the total latent heat of shotcrete I is smaller than the latent heat of the

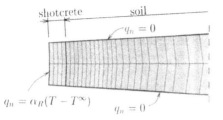

Figure 7. Thermochemical analysis: FE mesh and boundary conditions.

Table 3. Thermal and chemical material parameters for shotcrete and soil (clayey silt), after (Hellmich and Mang 1999).

SHOTCRETE I	
heat capacity C_V [kJ/m³K]	2400
thermal conductivity k [kJ/(m h K)]	12.6
thermal dilatation coefficient α_T [K⁻¹]	$1 \cdot 10^{-5}$
radiation coefficient shotcrete - air α_R [kJ/(m² h K)]	14.4
SOIL	
heat capacity C_V [kJ/m³K]	2300
thermal conductivity k [kJ/(m h K)]	7.2
thermal dilatation coefficient [K⁻¹]	$1 \cdot 10^{-5}$

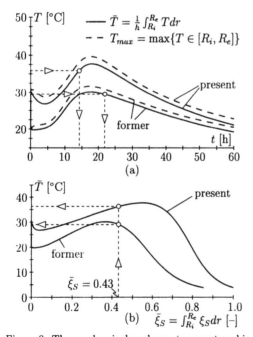

Figure 8. Thermochemical analyses: temperature histories in shotcrete tunnel shell obtained from former and present material model (R_i, R_e: radius of the interior and exterior shell surface).

Lafarge shotcrete ($l_A + l_S = 500$ kJ/(kg cement)), the heat *shock* caused by the CA hydration leads to a temperature in the tunnel shell, which is 8 °C *higher* than the one obtained with the former model.

Figure 8(b) shows the evolution of the mean temperature $\bar{T}$ (average value of T over the shell thickness) as a function of the analogous average value $\bar{\xi}_S$. Combination of Figures 8(a) and 8(b) allows determination of the time instant of the onset of chemical shrinkage, characterized by $\xi_S = 0.43$ (see Figure 9 showing the material function relating ξ_S to the shrinkage strain ε^s). For the analysis on the basis of the former material model, this time instant is obtained as $t = 22$ h. On the basis of the present material model, higher temperatures cause larger reaction rates, resulting in an onset of chemical shrinkage already 14 hours after shotcreting. This affects the axial forces in the tunnel shell as detailed in the next subsection.

Figure 10 shows the evolution of the temperature moment $\Theta = \frac{1}{h} \int_h T(r - r_0)dr$. It indicates the location of the temperature maximum within the shell section. This evolution is governed by the latent heat production in the shotcrete shell, the heat flow into the cavity as well as the conduction of heat into the ground material, for details see (Hellmich and Mang 1999). Θ is related to the temperature-induced curvature changes resulting

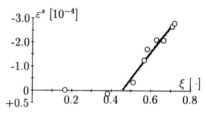

Figure 9. Material function for chemical shrinkage (Hellmich et al. 1999d).

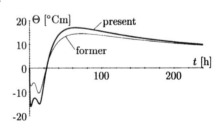

Figure 10. Thermochemical analyses: evolution of temperature moment Θ obtained from former and present material models.

in bending moments as will be described in the next subsection. Interestingly, the present formulation leads to larger temperature moments.

4.2 Chemomechanical analysis

The distributions and evolutions of the two reaction extents ξ_A and ξ_S and of the temperature T obtained from thermochemical analysis serve as input for the following chemomechanical analysis. On the basis of the reaction extents, the mechanical properties of shotcrete are known. They are related to the reaction extents by means of intrinsic material functions. Such intrinsic functions are determined from extended laboratory tests. With the exception of the intrinsic function for the compressive strength, all employed functions were taken from (Hellmich et al. 2000), whereby the variable ξ used in this reference is to be replaced by ξ_S. As regards the evolution of the compressive strength, the respective contribution of the CAH (see Figure 2(b)) requires an additional material function, see Figure 11.

The employed material model (see (Hellmich et al. 2000) and references therein) allows consideration of

- chemical shrinkage strains, ageing elasticity, and strength growth in consequence of hydration;

- *microcracking* of the hydrates resulting in permanent or plastic strains (Microcracking is controlled by means of a multi-surface material model involving the Drucker-Prager and the Rankine criterion (Lackner et al. 2000).);

- stress-induced *dislocation-like processes* within the hydrates resulting in flow (or long-term) creep strains;

- stress-induced *microdiffusion of water* in the capillary pores between the hydrates resulting in viscous (or short-term) creep strains.

Some typical mechanical properties of shotcrete used in the following analysis are given in Table 4.

4.2.1 Hybrid structural analyses of tunnels

In (Hellmich et al. 1999a) a hybrid method was presented, combining 3D *in-situ* displacement measurements with advanced material modelling of shotcrete in the framework of 3D chemomechanical FE analyses:

A ring with a width of 1 m is fictitiously cut out of the tunnel shell and modelled by 3D finite elements, Figure 12(a). Displacement fields are prescribed as boundary conditions at all surfaces at which forces are introduced into the investigated

Table 4. Mechanical material parameters for shotcrete.

final Young's modulus [MPa]	40800
Poisson's ratio [–]	0.2
final compressive strength $f_{c,\infty}$ [MPa]	39.6
final tensile strenth $f_{t,\infty}$	$f_{c,\infty}/10$

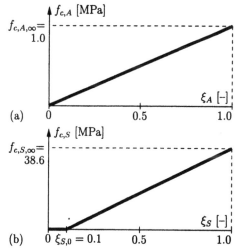

Figure 11. Intrinsic material functions for the compressive strength $f_c = f_{c,A} + f_{c,S}$: (a) contribution of CAH, (b) contribution of CSH.

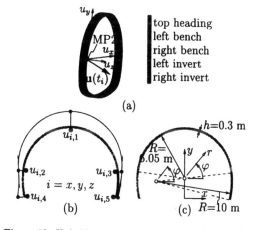

Figure 12. Hybrid method: (a) 3D FE model, (b) interpolated displacements as boundary conditions, (c) cross-section of Sieberg tunnel, Lower Austria.

part of the tunnel. These displacement fields are approximated by means of temporal and spatial interpolation of measurements, performed at discrete points (MP) on the interior surface of the tunnel shell, Figure 12(a) and (b).

This method delivers spatial fields of stresses as the main result. From these stresses, axial forces and bending moments can be computed:

$$n_\varphi = \int_h \sigma_\varphi dr, \quad n_z = \int_h \sigma_z dr, \quad (4.7)$$

and

$$m_\varphi = \int_h (r - r_0)\sigma_\varphi dr, \quad m_z = \int_h (r - r_0)\sigma_z dr, \quad (4.8)$$

with r_0 indicating the radius of the middle surface of the shell. Unlike as in the structural model used in (Hellmich et al. 1999a), the circumferential displacements are prescribed along the (longitudinal) shell sections at the ends of the top heading and of the benches, respectively, during respective construction states. Without this additional constraint, the analysis shows shear failure close to the shotcrete-soil interface. In reality, this kind of failure is prevented by rock bolts.

Herein, results for the measurement cross-section 1452 (km 156.990) of the Sieberg tunnel are presented, see Figure 12(c). The deformational state in this cross-section is illustrated in Figure 13 by means of the circumferential stretch $\epsilon_\varphi = 1/h \int_h \epsilon_\varphi\, dr$ and the longitudinal stretch $\epsilon_z = 1/h \int_h \epsilon_z dr$ at the top of the cross-section ($\varphi = 90°$). Compressive circumferential stretches prevail. Remarkably, they are *not* monotonically increasing. The longitudinal stretch ϵ_z lies within the compressive regime. It is significantly smaller than ϵ_φ.

Herein, the focus is on the influence of the description of the hydration process by means of two state variables ("present" material model) in comparison to the description employing one state variable only (Hellmich et al. 2000) ("former" material model).

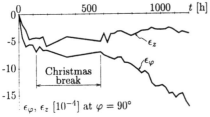

Figure 13. Stretches at km 156.990 of Sieberg tunnel (at $t = 0$ the top heading of the shell is installed).

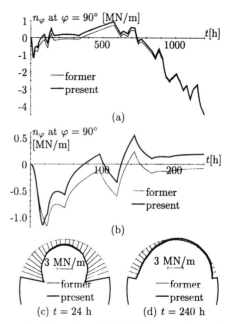

Figure 14. Circumferential axial forces at km 156.990 of Sieberg tunnel obtained from former and present material model: (a)(b) evolutions at $\varphi = 90°$; spatial distributions at (c) $t = 24$ h and (d) $t = 240$ h.

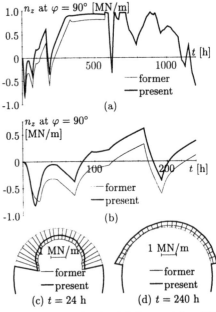

Figure 15. Longitudinal axial forces at km 156.990 of Sieberg tunnel obtained from former and present material model: (a)(b) evolutions at $\varphi = 90°$; spatial distributions at (c) $t = 24$ h and (d) $t = 240$ h.

Figure 15. It is particularly noteworthy that the maximal bearable load in the tensile regime is well described by the employed Rankine failure criterion, see Figure 15(a).

At early stages of the hydration, the higher temperatures in the shell obtained from the present material model (see Figure 8(a)) resulted in higher compressive loading in the circumferential direction of the shotcrete shell, see Figure 14. At $t \approx 14$ h, chemical shrinkage is starting. From $t \approx 18$ h on, chemical shrinkage and the deformation related to the temperature decrease during the cooling phase resulted in unloading of the shell. In some time intervals, even tensile axial forces in the circumferential direction are observed. The same is true for the longitudinal axial forces, see

The evolutions of the bending moment m_φ are shown in Figures 16(a) and (b). The larger (negative) temperature moments obtained in the present analysis as compared to the former analysis (see Figure 10) have resulted in larger (positive) bending moments at the beginning of the loading history. Conversely, larger (positive) temperature moments occurring later on lead to larger (negative) bending moments. Chemical shrinkage and cooling of the tunnel shell lead to a reduction of bending.

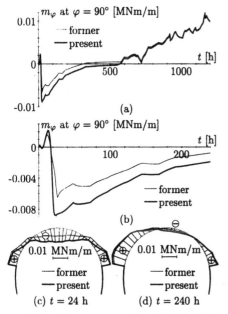

Figure 16. Bending moment at km 156.990 of Sieberg tunnel obtained from former and present material model: (a)(b) evolutions at $\varphi = 90°$; spatial distributions at (c) $t = 24$ h and (d) $t = 240$ h.

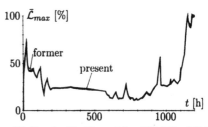

Figure 17. Level of loading at km 156.990 of Sieberg tunnel obtained from former and present material model.

In addition to understand the load carrying behavior of the shotcrete tunnel shell, determination of the level of loading (degree of utilization) for the purpose of a safety assessment is of crucial interest. For this purpose, the level of loading $\mathcal{L}$, amounting to 0% for the structure without loads and to 100% when the (compressive) strength is reached, is evaluated. Hereby, the actual stress state (represented by the stress tensor $\boldsymbol{\sigma}$) is extrapolated to the employed failure surface of Drucker-Prager f_{DP}, satisfying the condition

$$f_{DP}(\gamma\boldsymbol{\sigma}, f_c) = 0 \quad \rightarrow \quad \mathcal{L} = 1/\gamma . \qquad (4.9)$$

In Equation (4.9), γ represents the extrapolation coefficient and f_c is the actual compressive strength of shotcrete. The evolutions of the maximum value of the average of $\mathcal{L}$ over the shell thickness, $\bar{\mathcal{L}}_{max}$, obtained from the former and the present material model, are given in Figure 17. For both models, relatively large values of the level of loading, namely, up to 75%, are observed right after the installation of the top heading. The interruption of excavation during the Christmas break resulted in a decrease of the level of loading to 22%. The continuation of work after the Christmas break and the installation of the benches gave values of $\bar{\mathcal{L}}_{max}$ correlated to large stress redistributions in the soil. The extremely high value of $\bar{\mathcal{L}}_{max} = 98\%$ occurring at the end of the left bench is related to a local force transfer into the tunnel shell. Hence, this value does not affect the overall stability of the shell.

For all described loading states, the difference between the two employed material models is almost negligible. Future research work will show whether this holds only for the investigated case or whether this is generally true.

REFERENCES

Atkins, P. W. 1994. *Physical Chemistry*. Oxford University Press, Oxford, England, 5 edition.

Barenblatt, G. 1996. *Scaling, Self-Similarity, and Intermediate Asymptotics*. Cambridge University Press, Cambridge, England, 1 edition.

Baroughel-Bouny, V. 1994. Caractérisation des pâtes de ciment et des bétons - méthodes, analyse, interprétation [Characterization of cement pastes and concretes - methods, analysis, interpretations]. Technical report, Laboratoire Central des Ponts et Chaussées, Paris, France. In French.

Bye, G. 1999. *Portland Cement*. Thomas Telford Publishing, London, England, 2 edition.

Byfors, J. 1980. Plain concrete at early ages. Technical report, Swedish Cement and Concrete Research Institute, Stockholm, Sweden.

Coussy, O. 1995. *Mechanics of porous continua*. Wiley, Chichester, England.

Eichler, K. 1999. Bindemittel- und Verfahrenstechnologie von modernem Spritzbeton. In Kusterle, W., editor, *Spritzbeton-Technologie 99*, pages 195–202, Innsbruck, Austria. Institut für Baustofflehre und Materialprüfung, Universität Innsbruck. In German.

Guideline 1997. *Richtlinie für Spritzbeton [Guideline for shotcrete]*. Österreichischer Betonverein, Vienna, Austria. In German.

Hellmich, C. 1999. *Shotcrete as part of the New Austrian Tunneling Method: from thermochemomechanical material modeling to structural analysis and safety assessment of tunnels*. PhD thesis, Vienna University of Technology, Vienna, Austria.

Hellmich, C., Lechner, M., Lackner, R., Macht, J., and Mang, H. 2000. Creep in shotcrete tunnel shells. In Murakami, S. and Ohno, N., editors, *Creep in Structures 2000 - Proceedings of the fifth IUTAM Symposium on Creep in Structures*, Nagoya, Japan. In print.

Hellmich, C., Mang, H., and Ulm, F.-J. 1999a. Hybrid method for quantification of stress states in shotcrete tunnel shells: combination of 3D *in-situ* displacement measurements and thermochemoplastic material law. In Wunderlich, W., editor, *CD-ROM Proceedings of the European Conference of Computational Mechanics*, Munich, Germany.

Hellmich, C. and Mang, H. A. 1999. Influence of the dilatation of soil and shotcrete on the load bearing behavior of NATM-tunnels. *Felsbau, Rock and Soil Engineering*, 17(1):35–43.

Hellmich, C., Ulm, F.-J., and Mang, H. 1999b. Thermochemomechanical couplings for shotcrete: application to NATM-tunneling. In Jones, N. and Ghanem, R., editors, *CD-ROM Proceedings of the 13th ASCE Engineering Mechanics Division Conference*, Baltimore, MD, USA.

Hellmich, C., Ulm, F.-J., and Mang, H. A. 1999c. Consistent linearization in finite element analysis of coupled chemo-thermal problems with exo- or endothermal reactions. *Computational Mechanics*, 24(4):238–244.

Hellmich, C., Ulm, F.-J., and Mang, H. A. 1999d. Multisurface chemoplasticity I: Material model for shotcrete. *Journal of Engineering Mechanics (ASCE)*, 125(6):692–701.

Lackner, R., Hellmich, C., and Mang, H. 2000. Numerical treatment of multisurface chemoplasticity with special emphasis on brittle material failure and creep. *International Journal for Numerical Methods in Engineering*. Submitted for publication.

Pichler, W. 1995. *Umweltneutraler Spritzbeton im Trockenspritzverfahren für die Neue Österreichische Tunnelbaumethode; alkalifreie Erstarrungsbeschleunigung, Entwicklung und Technologie [Nonpolluting shotcrete in the dry-mix shotcrete method for the New Austrian Tunneling Method; alkali-free accelerators, development and technology]*. PhD thesis, University of Innsbruck, Innsbruck, Austria. In German.

Ulm, F. and Coussy, O. 2000. Environmental chemomechanics of concrete. In *Lecture Notes in Physics*. Springer, New York. In print.

Ulm, F.-J. and Coussy, O. 1995. Modeling of thermomechanical couplings of concrete at early ages. *Journal of Engineering Mechanics (ASCE)*, 121(7):785–794.

Ulm, F.-J. and Coussy, O. 1996. Strength growth as chemo-plastic hardening in early age concrete. *Journal of Engineering Mechanics (ASCE)*, 122(12):1123–1132.

Zienkiewicz, O. and Taylor, R. 1994. *The Finite Element Method*, volume 1. McGraw-Hill, London, England, 4. edition.

Shotcrete: Engineering Developments, Bernard (ed.) © 2001 Swets & Zeitlinger, Lisse, ISBN 90 5809 176 7

Evaluation of shrinkage-reducing admixtures in wet and dry-mix shotcretes

D.R.Morgan, R.Heere & C.Chan
AMEC Earth & Environmental Limited, Burnaby, British Columbia, Canada

J.K.Buffenbarger
Master Builders Inc., Cleveland, Ohio, USA

R.Tomita
Taiheiyo Cement Corporation, Japan

ABSTRACT: A systematic study was conducted to evaluate the effects of incorporation of shrinkage-reducing admixtures in wet and dry-mix shotcretes. It was demonstrated that such admixtures have no adverse effects on the properties of the plastic and hardened shotcretes and are very effective in reducing drying shrinkage and preventing restrained drying shrinkage induced cracking.

1 INTRODUCTION

Shotcrete, by virtue of its need to have good adhesion and cohesion, so that it does not sag, slough or fall out when applied to vertical or overhead surfaces, typically has a high cementing materials content (portland cement and supplementary cementing materials such as fly ash and silica fume). As such, it typically has a higher autogenous and drying shrinkage capacity than conventional cast-in-place structural concretes. This makes shotcrete more vulnerable to shrinkage-induced cracking and de-lamination, unless special shotcrete mixture design, installation and curing procedures are adopted.

A systematic study was conducted to evaluate the effects of incorporating shrinkage-reducing admixtures in wet and dry-mix shotcretes. The influence of such admixtures on the properties of the freshly applied shotcrete, such as: slump, air content, setting time, thickness of buildup and rebound was evaluated. In addition the effect of shrinkage-reducing admixtures on the basic properties of the hardened shotcrete was evaluated, i.e. compressive strength, boiled absorption and volume of permeable voids, and tensile bond strength.

The main purpose of the study, however, was to assess the benefits of incorporation of shrinkage-reducing admixtures in wet and dry-mix shotcretes with respect to reducing shrinkage and mitigating or preventing restrained drying shrinkage induced cracking.

2 SHRINKAGE-REDUCING ADMIXTURES

Shrinkage-reducing admixtures (SRA's) were first developed in Japan in 1982 (Sato et al., 1983, Tomita et al., 1983). On October 15, 1985, U.S. Patent number 4,547,223 was granted to Goto et al. for the invention, the main component being a poly-oxyalkylene alkyl ether, a lower alcohol alkylene oxide adduct. Since this invention, interest in this technology has grown and on September 17, 1996, U.S. Patent Number 5,556,460 was granted to Berke et al. for an SRA with a similar base com-position. Several low viscosity, water soluble SRA's have now been developed by Taiheiyo Cement and Sanyo Chemical Industries in Japan.

These admixtures function by reducing capillary tension and the tensile forces that develop within the concrete pores as it dries. As pores become less than fully saturated, a meniscus forms at the air water interface due to surface tension. The surface tension of the pore solution meniscus exerts an inward compressive pulling force on the side of the pore wall. The compressive forces exerted on all pores ranging from 2.5 to 50 nanometers in the concrete matrix are the primary cause of drying shrinkage.

SRA's are primarily used as integral, liquid admixtures, but some can be applied topically to concrete surfaces or impregnated upon organic fillers for applications requiring powdered admixtures (Sato et al. 1983, Tomita et al. 1983, Berke et al. 1997, Nmai et al. 1998, Nmai et al. 1999, Nmai & Seow 1999). The shrinkage-reducing admixture Tetraguard AS20, supplied by Master

Builders Inc., was used in the wet-mix shotcrete evaluated in this study. The shrinkage reducing admixture Tetraguard PW, supplied by Taiheiyo Cement, was used in the dry-mix shotcrete evaluated in this study. Hereafter, these two shrinkage-reducing admixtures will be referred to as SRA-W and SRA-D, respectively. Collectively, they will be referred to as SRA's.

3 SHOTCRETE MIXTURE DESIGN AND BATCHING

3.1 Wet-mix shotcrete

The wet-mix shotcrete mixture design selected is similar to mixtures commonly used in North America for permanent shotcrete linings in tunnels and mines, slope stabilization and infrastructure rehabilitation projects, except that the water-reducing and air entraining admixtures normally used in such shotcretes were not added in this study. This was done in order that the influences of the SRA's, independent of any other chemical admixtures, could be evaluated.

The SRA-W was added to the plain control wet-mix shotcrete mixture at nominal addition rates of 2.0% and 4.0% by mass of cement, respectively. The actual as-batched proportions, in kg/m^3 for the plain control shotcrete mix (WPC) and the two mixes with SRA's (WA2 and WA4), are given in Table 1.

Table 1: As batched wet-mix shotcrete mix proportions, based on saturated surface dry (SSD) aggregates.

Material Mix Designation	Mix Proportions kg/m^3		
	WPC	WA2	WA4
Cement Type 10	403	404	406
Silica Fume	45	45	46
Fly Ash, class F	30	30	30
Coarse Aggregate (10-2.5 mm)	504	505	508
Sand	1138	1141	1147
Water	190	173	162
Tetraguard AS20	0.00	7.98	15.98
Air Content (as shot)	2.8%	3.5%	3.4%
TOTAL	2311	2307	2315
Tetraguard AS20 by mass of cement, %	0	2.0	3.9
Water*/Cementing materials ratio	0.40	0.38	0.37

* The *water* in the water/cementing materials ratio calculation includes the liquid SRA.

In order to maintain tight control over the mixture proportions, all shotcrete mixtures were precision dry-batched in a commercial dry-bagging plant. Bone-dry materials were used and all dry ingredients were mass-batched. The materials were premixed in a rotary pan mixer with counter rotating paddles, before being discharged into 30 kg paper bags.

The wet-mix shotcrete was mixed in a paddle-type mixer unit attached to a 75 mm internal diameter swing valve shotcrete piston pump. Typically fourteen 30 kg bags of shotcrete were batched at a time. The required water content to provide a slump in the range of 30 to 40 mm was added to the mixer unit together with the SRA-W and the shotcrete mixed for approximately 5 minutes before being discharged into the pump hopper. Note that the addition of the SRA's resulted in water reductions of 4.7% and 6.3% for the mixes WA2 (2% SRA) and WA4 (4% SRA) respectively, relative to the plain control mix WPC.

3.2 Dry-Mix Shotcrete

The dry-mix shotcretes evaluated in this study also had the same nominal mix proportions as the plain control wet-mix shotcrete, except that, being produced by the dry-mix shotcrete process, they had a lower water demand as-shot. Actual as-batched mixture proportions (based on aggregates in a SSD condition) for the plain control dry-mix shotcrete (DPC) and dry-mix shotcretes with SRA-D is given in Table 2. The SRA-D was added at 2% and 4% by mass of cement to the shotcretes for mixes DA2 and DA4, respectively.

In the case of the dry-mix shotcretes, the dry powdered SRA-D was pre-blended in with the bone dry materials at the dry batch plant. The 30 kg bags of dry bagged shotcrete were then continuously fed into a pre-moisturizing auger, which discharged the shotcrete into a rotary barrel feed shotcrete gun, attached to a 38-mm internal diameter hose. The remaining water was added at the water ring at the nozzle with control of water addition exercised by the nozzleman in the usual way.

Table 2. As-batched dry-mix shotcrete mix proportions, based on saturated surface dry (SSD) aggregates.

Material Mix Designation	Mix Proportions kg/m^3		
	DPC	DA2	DA4
Cement Type 10	400	400	400
Silica Fume	45	45	45
Fly Ash, class F	30	30	30
Coarse Aggregate (10-2.5 mm)	450	450	450
Sand	1210	1210	1210
Water (estimate)*	180	180	180
Tetraguard PW	0	8.0	16.0
TOTAL	2315	2323	2331

* Water includes all water added in a pre-moisturizing auger and at the water ring at the nozzle.

4 SHOTCRETE PROPERTIES

The as-batched slump and air content of the wet-mix shotcretes were recorded. In addition, the air content of the shotcrete, as-shot into an ASTM C231 air pressure meter base, was determined. The results of these tests are given in Table 3. The slumps and air contents of all mixtures, as batched, were similar.

The mixtures with SRA, however, tended to lose less air on shooting than the plain control mix. Also shown in Table 3 are the results of rebound and setting time tests. It can be seen that the addition of the SRA's to either the wet or dry-mix shotcretes had little influence on rebound. Rebound was, as expect-ed, considerably lower in all the wet mix shotcretes (9 to 11%) compared to the dry-mix shotcretes (25 to 27%).

With respect to setting time, the addition of 2% by mass of cement of the SRA-W had little effect on either initial or final setting time of wet-mix shotcrete (WA2). At 4% by mass of cement there was however, an increase in both initial set (by 40 minutes) and final set (by 65 minutes) in the wet mix shotcrete (WA4).

Table 3. Properties of Plastic Wet-Mix Shotcrete

Property	Wet-Mix Shotcrete Designation		
	WPC	WA2	WA4
SRA Addition Rate by			
Mass of Cement (%)	0.0	2.0	3.9
Slump As-batched (mm)	40	30	30
Air Content			
As-Batched (%)	4.4	4.0	3.2
As-Shot (%)	2.8	3.5	3.4
Rebound (%)	10	9	11
Setting Time to			
ASTM C1117			
Initial Set (hr:min)	3:55	3:50	4:35
Final Set (hr:min)	6:35	6:05	7:40

Table 4. Properties of Plastic Dry-Mix Shotcrete

Property	Dry-Mix Shotcrete Designation		
	DPC	DA2	DA4
SRA Addition Rate by			
Mass of Cement (%)	0.0	2.0	3.9
Rebound (%)	25	27	25
Setting Time to ASTM			
C1117			
Initial Set (hr:min)	4:00	1:25	1:10
Final Set (hr:min)	6:10	6:15	10:00

Table 5. Properties of Hardened Wet-Mix Shotcrete

Property	Wet-Mix Shotcrete Designation		
	WPC	WA2	WA4
SRA Addition Rate by			
Mass of Cement (%)	0.0	2.0	3.9
Compressive Strength to			
ASTM C39 at:			
7 days (MPa)	33.5	38.7	31.4
28 days (MPa)	56.6	52.0	52.5
Absorption after			
Immersion and Boiling			
(%)	6.4	4.8	4.6
Volume of Permeable			
Voids (%)	14.0	10.6	10.2
Tensile Bond Strength to			
CSA A23.2-6B at:			
8 days (MPa)	-	-	-
29 days (MPa)	-	-	-
33 days (MPa)	1.9	1.8	2.1

In the dry-mix shotcretes, the mixtures with SRA-D (mixes DA2 and DA4) appeared to have a greater stiffness as shot and this was reflected in the substantially earlier initial setting times of these mixes compared to the plain control mix (DPC) as shown in Table 4. The mix DA2 with 2% by mass of cement addition of SRA-D had about the same final setting time as the plain control mix (DPC). By contrast there was a nearly 4 hour delay in final setting time in the mix DA4 with 4% by mass of cement of SRA-D, compared to the plain control dry-mix shotcrete (DPC).

5 HARDENED SHOTCRETE PROPERTIES

5.1 Compressive Strength

Compressive strength tests were conducted to ASTM C39 on 77-mm diameter x 100-mm long cores drilled from standard ACI 506.2 test panels with 600 x 600 x 125 mm dimensions. Tests were conducted on pairs of cores at ages 7 and 28 days. The test panels had been field cured for 3 days, prior to being de-moulded and cured in a moist room at $23\pm1°C$ and $98\pm2\%$ relative humidity until the time of testing in accordance with ASTM C1140. Test results are given in Tables 5 and 6 for the wet and dry-mix shotcretes respectively.

The addition of 2% SRA-W to the wet-mix shotcrete (WA2) resulted in a small increase in compressive strength at 7 days and small decrease in compressive strength at 28 days relative to the plain control mix (WPC). There was a slight reduction in compressive strength at both 7 and 28 days for the mix WA4 with 4% SRA-W addition. In the dry-mix shotcretes the addition of 2% SRA-D (DA2) had no significant effect on compressive strength at 7 or 28 days. By contrast, at 4% SRA-D addition rate, mix

Table 6: Properties of Hardened Dry-Mix Shotcrete

Property	Dry-Mix Shotcrete Designation		
	DPC	DA2	DA4
SRA Addition Rate by			
Mass of Cement (%)	0.0	2.0	4.0
Compressive Strength to			
ASTM C39 at:			
7 days (MPa)	42.6	42.2	35.0
28 days (MPa)	55.5	54.7	46.7
Absorption after			
Immersion and Boiling			
(%)	7.0	6.4	6.7
Volume of Permeable			
Voids (%)	15.2	13.8	14.5
Tensile Bond Strength to			
CSA A23.2-6B at:			
8 days (MPa)	1.8	2.0	1.2
29 days (MPa)	2.5	2.7	2.8
33 days (MPa)	-	-	-

DA4 had a reduced compressive strength at both 7 and 28 days, compared to the plain control dry-mix shotcrete (DPC). It should, however, be noted that all mixtures tested readily met typical performance specifications for structural quality shotcretes of 30 MPa at 7 days and 40 MPa at 28 days, commonly specified in North America.

5.2 Boiled Absorption and Permeable Voids

The ASTM C642 test is commonly used in specifications in North America to quantify the quality of wet and dry-mix shotcretes. Maximum allowable values of 8% absorption after immersion and boiling and 17% volume of permeable voids are commonly specified for structural quality shotcretes. All the wet and dry-mix shotcrete mixtures tested readily satisfied these requirements, as shown in Tables 5 and 6. It is, however, of interest to note that all the shotcretes with SRA's, for both the wet and dry-mix shotcretes, produced lower values of absorption after immersion and boiling and volume of permeable voids, compared to their respective plain control shotcretes.

5.3 Tensile Bond Strength Tests

Tensile bond strength tests were conducted to CSA A23.2-6B on pairs of cores extracted from the restrained shrinkage test chamber. The tested shotcretes had been applied to a mature (3 months old), dry substrate shotcrete, with an as-shot surface finish. A minimum tensile bond strength of 1.0 MPa at 28 days is commonly specified for structural quality shotcretes applied to properly prepared concrete or shotcrete substrates. All of the shotcretes tested readily met this requirement, with most of the results being close to, or in excess of 2.0 MPa, as shown in Tables 5 and 6. Clearly, incorporation of the SRA's in the shotcretes did not have any detrimental effect on shotcrete bond.

6 SHRINKAGE TESTING

6.1 Unrestrained Shrinkage Testing

Sets of three 75×75×275-mm prisms were diamond saw cut from the standard ACI test panels at age 3 days. Steel studs were epoxied into the ends of the prisms which were then placed in a drying chamber, with a temperature of 23 ± 1°C, 50 ± 4% R.H. and controlled rate of evaporation. Unrestrained free shrinkage length change tests were conducted in accordance with ASTM C341. The average length change for the sets of three prisms, for each of the mixtures evaluated, is shown in Figure 1 for the wet-mix shotcretes and Figure 2 for the dry-mix shotcretes.

In the wet-mix shotcrete series, the plain control mix (WPC) exhibited an average shrinkage of 880

microstrain at 40 days. This is typical of wet-mix shotcretes made without water reducing admixtures. By contrast, wet-mix shotcretes with 2% and 4% SRA-W exhibited 40-day shrinkage of only 550 and 500 microstrain, respectively. There is a small but consistent difference between the shrinkage of these two mixes, with the 4% SRA-W mix exhibiting a consistent 40 to 70 micro-strain lower shrinkage at all ages after 7 days compared to the 2% SRA-W mix. Compared to the plain control wet-mix shotcrete, the mixes with 2% and 4% SRA-W displayed 50 to 70% less shrinkage at about 16 days. At later ages, the relative difference was not quite as pronounced, but is still substantial.

In the dry-mix shotcrete series, the plain control mix (DPC) exhibited average shrinkage of 690 microstrain at 40 days. This is typical of conventional dry-mix shotcretes. By contrast, dry-mix shotcretes with 2% and 4% SRA-D exhibited 40-day drying shrinkage values of only 430 and 460 microstrain respectively. The 4% SRA-D mix exhibited an almost constant 15 microstrain lower shrinkage than the 2% SRA-D mix at all ages past 7 days to 30 days. Compared to the plain control dry-mix shotcrete, the mixes with 2% and 4% SRA-D displayed shrinkage values in the range of 40 to 50% less than the plain control mix at about 10 days. At

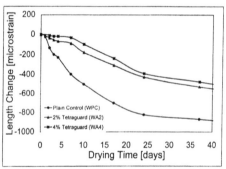

Figure 1. Unrestrained free shrinkage in wet-mix shotcretes

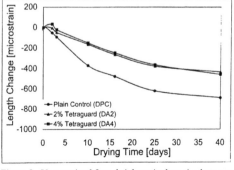

Figure 2. Unrestrained free shrinkage in dry-mix shotcretes

later ages, the relative effect is not quite as pronounced, but is still substantial.

6.2 Restrained Shrinkage Testing

Restrained shrinkage testing was conducted on shotcrete sprayed onto a substrate consisting of a heavy mesh reinforced shotcrete, approximately 75 to 100-mm thick, which was produced 3 months prior to the application of shotcrete test samples. A final flash layer of shotcrete was applied during the shooting to produce a rough as-shot surface, which provided the restraint to the freshly applied shotcrete test samples in the restrained shrinkage tests.

Shotcrete test strips approximately 5 m long, 300-mm wide and 50-mm thick were shot onto the substrate described above. For each of the wet and dry-mix shotcretes evaluated, two shotcrete strips were produced. One strip was shot inside the conditioning chamber, for exposure to a hot, windy, drying environment, and the other strip was sprayed onto the outside of the conditioning chamber for exposure to a relatively cool and humid ambient environment. After each strip was shot, the surface was struck off and finished to a smooth finish with a steel trowel. Plastic sheets were temporarily used to cover finished surfaces during the shooting of new strips, in order to prevent contamination of the test sample during the shotcreting of adjacent test strips. The strips were otherwise left uncovered and exposed in their respective environments.

The conditioning chamber consisted of an enclosed, insulated chamber (shotcrete lined steel shipping container) with internal heaters, a dehumidifier and a fan to create air convection. Internal temperatures in the range of 35 to 39°C and relative humidities in the range of 25 to 35% were maintained for the duration of the test. The water evaporation rate was determined by placing beakers of water in various positions in the chamber and measuring the loss of water mass with time. An evaporation rate of 295 $g/m^2/h$ was determined for the interior of the chamber and 20 $g/m^2/h$ for the exterior of the chamber. This corresponded very well with the theoretically calculated rates of evaporation determined using the ACI 305R charts. Exterior temperatures ranged from 3 to 15°C and relative humidities ranged from 45 to 100%, but were mostly above 70%. The shotcrete test strips in both exposure environments were monitored for any crack development immediately after the strips were sprayed and for a period of one month.

In the wet-mix shotcrete series, the plain control mix subjected to the aggressive drying environment in the conditioning chamber developed substantial cracking over the surface of the strip. Most of the cracks formed in a predominantly longitudinal orientation along the length of the strip. The majority of the cracks formed at an early age, within 4 to

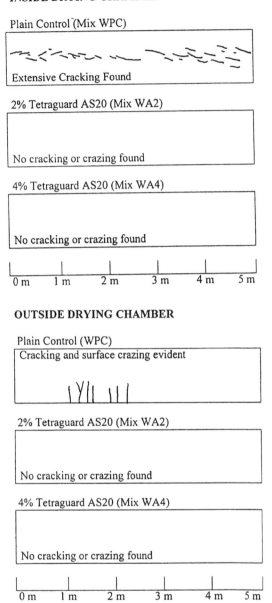

Figure 3. Crack mapping in restrained drying shrinkage test for wet-mix shotcretes.

5 hours after shooting and most of the remaining cracks formed over the next 24 hours, either as new cracks, or as extensions to existing cracks. Crack widths varied from approximately 0.0 to 0.7 mm. By contrast neither the 2% SRA-W nor the 4% SRA-W wet-mix shotcretes developed any visible cracks throughout the entire monitoring period.

In the outdoors environment, the plain control

INSIDE DRYING CHAMBER

Plain Control (Mix DPC)

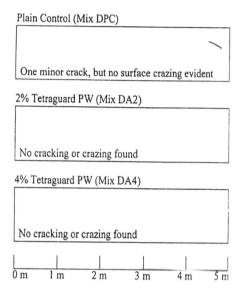

OUTSIDE DRYING CHAMBER

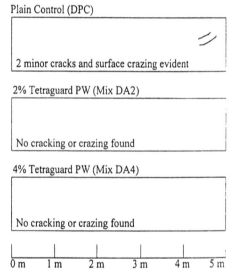

Figure 4. Crack mapping in restrained drying shrinkage test
For dry-mix shotcretes.

wet-mix shotcrete strip developed some predominantly vertical hairline cracks (0.0 to 0.2 mm wide) several weeks after spraying. The cooler and more humid outdoors environment delayed crack formation on this outdoor strip. The short transverse cracks primarily formed at the edge of the strip, near mid-length of the strip. Crazing also developed over the surface of the plain control

shotcrete strip. Again, neither the 2% SRA-D mix nor the 4% SRA-D mix developed any visible cracks or crazing throughout the entire monitoring period.

In the dry-mix shotcrete series, the plain control mix subjected to the adverse interior environment of the conditioning chamber developed a minor crack at the corner of the strip. This crack formed approximately 7 hours after shooting. Neither the 2% SRA-D mix nor the 4% SRA-D mix have developed any visible cracks to date.

In the outdoors environment, the plain control dry-mix shotcrete strip developed some hairline cracks at the corner of the strip approximately 8 hours after shooting. Crazing also developed over the surface of the plain control shotcrete strip. Again, neither the 2% SRA-D nor the 4% SRA-D dry-mix shotcretes developed any visible cracks or crazing.

The differences in crack development between the plain control wet and dry-mix shotcretes are likely due to differences in mix proportions (particularly water demand) of the in-situ shotcrete. The addition of SRA-D has clearly been very effective in inhibiting restrained drying shrinkage crack formation for both the internal and external exposure conditions, for both wet and dry-mix shotcretes.

7 CONCLUSIONS

The following conclusions are drawn from the test program:
- The shrinkage-reducing admixture Tetraguard AS20 (SRA-W) can be conveniently added to shotcrete as a liquid admixture in wet-mix shotcrete. The shrinkage reducing admixture Tetraguard PW (SRA-D) can be conveniently added as a dry powdered ingredient, batched together with pre-bagged materials in dry-mix shotcrete.
- Wet or dry-mix shotcretes containing these SRA's can be batched, mixed, shot and finished in the same way as ordinary wet or dry-mix shotcretes.
- Rebound and build-up characteristics of wet and dry-mix shotcretes do not appear to be adversely affected by the addition of SRA's and appear similar in behavior during shooting to ordinary shotcretes.
- SRA-W, when added at 2% by mass of cement to shotcrete, does not appear to influence the final setting time of the mix. However, when added at 4% by mass of cement to shotcrete, the SRA-W appears to cause a retardation of the mix at final set.
- The addition of SRA's at 2% by mass of cement appears to slightly increase the early age compressive strength (to 7 days) of both wet and

dry-mix shotcretes. At 28 days however, there is a slight reduction in the compressive strength compared to the plain control shotcrete. The addition of SRA's at 4% by mass of cement appears to slightly reduce the compressive strength of both wet and dry-mix shotcretes at all ages compared to the plain control shotcrete. Strengths of both wet and dry-mix shotcretes with SRA's however, remain well above commonly specified values for structural shotcretes of 30 MPa at 7 days and 40 MPa at 28 days.

- The addition of SRA's at 2 or 4% by mass of cement to wet or dry-mix shotcrete appears to modestly improve the boiled absorption and volume of permeable voids of the shotcrete. Good to excellent quality shotcrete can be produced with this admixture added at the given dosages.
- The bond strengths of the plain control shotcrete and shotcretes with 2% and 4% by mass of cement of SRA's addition are similar. They are considered excellent for structural quality shotcrete.
- Unrestrained shrinkage testing of the shotcretes evaluated indicates that SRA additions at 2% and 4% by mass of cement to shotcrete can reduce drying shrinkage considerably (between 20 to 70% reduction at early ages). The absolute amount of shrinkage reduction provided by SRA additions depends on the shrinkage potential of the base mix.
- The difference in shrinkage reduction between mixes batched with 2% and 4% by mass of cement addition of SRA's appears to be small in both wet-mix and dry-mix shotcretes.
- The addition of SRA's at 2% or 4% by mass of cement to shotcrete appears to be effective in reducing the restrained shrinkage cracking of wet and dry-mix shotcretes, even in an adverse simulated desert environment, such as that used in this test program.

In summary, the SRA's evaluated in this study were demonstrated to be effective in reducing drying shrinkage and drying shrinkage induced cracking in both wet-mix and dry-mix shotcrete applications. Addition rates of 2% by mass of cement of these SRA's appeared to be suitable for markedly reducing shrinkage and eliminating restrained drying shrinkage cracking.

Finally, it should be noted that the SRA's were the only chemical admixtures added to the shotcrete in this study. If other chemical admixtures such as superplasticizers, air entraining admixtures, or shotcrete accelerators are also added to the shotcrete, then tests would need to be conducted to assess performance.

ACKNOWLEDGEMENTS

The work detailed in this paper was sponsored by Master Builders Inc. and ASANO Co. Ltd., and was conducted at the Burnaby, British Columbia laboratories of AMEC Earth & Environmental Limited.

REFERENCES

ACI 506.2, Specification for Shotcrete. American Concrete Institute, Farmington Hills, MI.

ASTM C39, Standard Test Method for Compressive Strength of Cylindrical Concrete Specimens. ASTM, West Conshohocken, PA.

ASTM C231, Standard Test Method for Air Content of Freshly Mixed Concrete by the Pressure Method. ASTM, West Conshohocken, PA.

ASTM C341, Standard Test Method for Length Change of Drilled or Sawed Specimens of Hydraulic-Cement Mortar and Concrete. ASTM, West Conshohocken, PA.

ASTM C642, Standard Test Method for Density, Absorption, and Voids in Hardened Concrete. ASTM, West Conshohocken, PA.

ASTM C1117, Standard Test Method for Time of Setting of Shotcrete Mixtures by Penetration Resistance. ASTM, West Conshohocken, PA.

ASTM C1140, Standard Practice for Preparing and Testing Specimens from Shotcrete Test Panels. ASTM, West Conshohocken, PA.

CSA A23.2-6B, Method of Test to Determine Adhesion by Tensile Load.

Berke, N.S., Dallaire, M.P., Hicks, M.C., and Kerkar, A., 1997. New Developments in Shrinkage-Reducing Admixtures, *Superplasticizers and Other Chemical Admixtures in Concrete*, SP-173, American Concrete Institute, Farmington Hills, MI: 971-998.

Nmai, C., Tomita, R., Hondo, F., and Buffenbarger, J.1998. Shrinkage-Reducing Admixtures for Concrete, *ACI Concrete International*, 20 (4): 31-37.

Nmai, C., Mullen, B., Fletcher, K., 1999. Comparative Evaluation of Shrinkage-Reducing Admixtures for Concrete, In *Concrete99, Concrete Institute of Australia 19th Biennial Conference Proceedings*, Baweja et al. (eds.), May: 84-90.

Nmai, C. and Seow, K.H., 1999. Shrinkage-Reducing and Other Durability Enhancing Admixtures for Concreting in the 21st Century. In *24th Conference Proceedings on OUR WORLD IN CONCRETE AND STRUCTURES*, Singapore, August 24-26.

Sato, T., Goto, T., and Sakai, K., 1983. Mechanism for Reducing Drying Shrinkage of Hardened Cement by Organic Additives, *Cement Association of Japan (CAJ) Review*: 52-54.

Tomita, R., Takeda, K., and Kidokoro, T. 1983. Drying Shrinkage of Concrete Using Cement Shrinkage Reducing Agent, *Cement Association of Japan (CAJ) Review*: 198-99.

Shotcrete: Engineering Developments, Bernard (ed.) © 2001 Swets & Zeitlinger, Lisse, ISBN 90 5809 176 7

A new chemical approach to obtain high early strength in shotcrete

R.Myrdal, R.Hansen & G.Tjugum
Rescon Mapei AS, Sagstua, Norway

ABSTRACT: To increase early compressive strength in wet mixed shotcrete a new two component chemical system has been investigated. This involves the combination of a 'dormant' accelerator, blended with a set consistency control admixture, and a commercial shotcrete accelerator. The effect of the 'dormant' accelerator starts once it is mixed with the shotcrete accelerator. An approximate 100% increase in one hour compressive strength values have been measured in small scale laboratory experiments. Two to three times higher compressive strengths have been measured after three hours, compared to concrete activated only by the conventional shotcrete accelerator. The effect on one day strength was insignificant. The new chemical admixture is a proprietary blend of phosphonates, carbonates, alkanol-amines and hydroxy carboxylates.

1 INTRODUCTION

In modern wet mixed shotcrete technology, rigorous requirements for high early compressive strength are often difficult to comply with. The focus on early compressive strength has become a dominant issue, and compressive strength values in the order of ~1MPa after one hour have been required for some tunnel projects. With the new generation of alkali-free shotcrete accelerators (Myrdal 1999, Erlien 1999, Garshol & Melbye 1996), this can be achieved provided all factors affecting early compressive strength development are optimised: factors such as type of cement, concrete mix design, and temperature. Unless these factors are taken into account during planning and application of shotcrete, poor results are often achieved. Consequently, more efficient shotcrete accelerators are required.

However, another chemical approach is possible. This paper presents the ideas and first experimental attempts to develop a new chemical system to speed up early compressive strength development in shotcrete. So far, only small scale laboratory test results are available and are presented.

2 A NEW CHEMICAL APPROACH

The idea is to combine two separate chemicals: (1) an alkali-free shotcrete accelerator added at the nozzle of the spraying equipment, and (2) a liquid chemical admixture that is added in the mixer at the concrete plant. Both chemicals contain accelerators, but the type added in the mixer should not react with the cement until it is activated by the shotcrete accelerator. The new chemical cannot be pre-blended with the shotcrete accelerator due to chemical reactions that lead to precipitation of solid products. It can, however, be blended with the superplasticizer or the retarder (the consistency control admixture often used in modern shotcrete). In this way the number of admixtures handled at the concrete plant is not increased.

The main problem with this approach has been to make a system with practically no risk of uncontrolled setting and hardening of the concrete in the period between mixing and shotcreting. It is imperative that the new accelerating admixture is 'kept asleep' until it interacts with the shotcrete accelerator. This could be accomplished by blending the admixture with retarding agents. In this way the accelerator can be kept inactive for several hours in typical concrete mixes used for shotcreting.

Based on these ideas, a R&D project was initiated with the goal of developing a new consistency control admixture containing 'dormant' accelerators. A proprietary blend of phosphonates was used as the main retarder in combination with carbonates, alkanolamines and hydroxy carboxylates. The three latter chemicals may have a retarding or accelerating effect, depending on chemical types, dosage and mutual interactions.

It is believed that this approach is different from the Meyco®TCC system, which is based on incompatible chemistries to produce an instant slump loss of the shotcrete (Coverdale 1996), rather than increased early compressive strength.

3 LABORATORY EXPERIMENTS

3.1 Concrete mix design

All experiments were carried out with an ordinary Portland cement concrete containing silica fume, with a water/binder ratio of 0.41. An initial slump of 210±10 mm was obtained by using a polycarboxylate-based superplasticizer. 10 Litre concrete samples were mixed by hand using an electrical drill. The mix design (based on 1 m^3 of concrete) is shown in Table 1. All tests were carried out at 23±1°C.

3.2 Chemical additives

Approximately five minutes after mixing, aqueous solutions of the formulated chemical blends were added to the fresh concrete samples. Four different blends were tested, denoted A1 to A4. A1 was a pure phosphonate-based chemical, while A2 to A4 contained proprietary blends of carbonates, alkanolamines and hydroxy carboxylates. Table 2 gives a general overview of the formulations.

All blends contained phosphonates to maintain fluidity and workability of fresh concrete. The other chemical components were considered potential cement hydration accelerators, especially in combination with alkali-free shotcrete accelerators. The admixture dosage was 1.5 % by cement weight in all tests.

The slump of the concrete mixes containing the additives was measured as a function of time, first at 15 minutes after addition, then after some hours, in

Table 1. Concrete mix design.

Material	kg/m^3
Portland cement, Norcem CEM I-42.5-R	485
Silica fume, Elkem	25
Aggregate, 0-8 mm natural sand (FM=3.1)	1530
Total water, w/(c+s)=0.41*	210
Superplasticizer, *Mapefluid X404	3.2

* Binder weight = cement + silica fume (c + s)

Table 2. Proprietary formulations of additives*.

Chemical components	Chemical blend notation			
	A1	A2	A3	A4
Phosphonates	X	X	X	X
Carbonates	--	X	X	X
Alkanolamines	--	X	X	X
Hydroxy carboxylates	--	X	X	X

* The relative amounts of chemicals in formulations A2 to A4 differ.

some cases up to 22 hours. The concrete was remixed for about 10 seconds prior to every slump measurement, but was otherwise kept undisturbed in closed buckets.

3.3 Shotcrete accelerator

After addition of the chemical blends (A1 to A4), a shotcrete accelerator was added to the concrete and mixed in by hand using a high speed electrical drill. The accelerator dosage was 8 % by cement weight in all tests. The mixing time was approximately 20 seconds. During this period of time the concrete began to set.

For compressive strength measurements the set concrete was placed in cube moulds (100×100×100 mm) and vigorously vibrated on at vibrating table for another 20 seconds. The filled moulds were thereafter kept open to air at 23±1°C. All tests were carried out with a commercial liquid alkali-free shotcrete accelerator (®Mapequick AF 2000 - an aqueous solution based on aluminium sulphate). After demoulding (1 to 24 hours after addition of the shotcrete accelerator) the cubes were weighed and the compressive strength was measured. All measured compressive strength values presented below are the average of three samples.

Note that the compressive strength values obtained by this experimental procedure are generally lower than normally achieved in real shotcreting using spraying equipment. The reason for this is probably more efficient and immediate mixing of shotcrete accelerator in sprayed concrete, and better compaction, compared to manual mixing and placement of the concrete in moulds. Long experience with this technique, however, has shown that reliable results relative to sprayed samples are obtained when comparing shotcrete accelerators.

4 RESULTS AND DISCUSSIONS

4.1 Concrete slump

The effect of chemical additives A1 to A4 on concrete slump is given in Table 3. The results show that the pure phosphonate-based retarder (A1) prolonged the concrete workability, as expected, compared to the reference concrete mix. The incorporation of potential accelerators A2 to A4 produced almost immediately slump loss, especially A3 and A4. From a consistency control point of view, A2 seemed the most promising admixture containing 'dormant' accelerators.

Table 3 shows that by addition of a small amount of A2 (1.5% by cement weight) the workability time of the concrete could be prolonged several hours. It appeared that the potential accelerating effects of the chemicals in blend A2 were hindered by the phosphonate retarders. Additional tests with A2, A3 and

Table 3. Concrete slump as a function of time after addition of different chemical additives - dosage 1.5 % by cement weight.

Time	Concrete slump / mm*				
	Reference**	A1	A2	A3	A4
15 minutes	210	220	190	170	170
2 hours	100	220	180	110	130
6 hours	0	210	170	--	--
15 hours	--	--	150	--	--
22 hours	--	--	110	--	--

* Not measured: '- -'
** Concrete mix shown in Table 1.

Table 5. Compressive strength as a function of age of fresh concrete and curing time*.

Age of fresh concrete**	Compressive strength / MPa		
	1 hour	3 hours	24 hours
Reference, 2 hours	0.75	1.13	22.5
A2, 2 hours	1.42	3.10	24.3
A2, 6 hours	1.09	1.71	23.6
A2, 15 hours	0.79	1.87	24.2

* A2 dosage: 1.5 % by cement weight. Shotcrete accelerator dosage: 8 % by cement weight.
** Age at the time of shotcrete accelerator addition.

A4 without phosphonates showed concrete setting only a few minutes after addition.

4.2 Early compressive strength development

Table 4 shows the compressive strength values measured on the hardened cubes after 1 hour, 3 hours and 24 hours. The strength values given in Table 4 were obtained in cases where the conventional alkali-free shotcrete accelerator was added 2 hours after the concrete was mixed (i.e. the fresh concrete was two hours old when the shotcrete accelerator was mixed in).

The compressive strength values obtained with the reference concrete are typical values measured in laboratory experiments with this type of accelerator. However, the low slump of the reference concrete at the time the shotcrete accelerator was added (see Table 3) could indicate problems of efficient mixing of the shotcrete accelerator. Therefore, a fresh (30 minutes old) reference concrete with a slump of 210 mm was tested to see if this affected early strength. The strength values measured were 0.80 and 1.22 MPa after one and three hours respectively, indicating a small and insignificant effect.

By addition of A1 (pure retarder), the early strength values decreased (see Table 4). After 24 hours the retarding effect on strength development was no longer noticeable.

Table 4 shows that all three accelerator blends (A2 to A4) gave higher early strength compared to the reference and retarded concretes. A2 appeared the most efficient with an almost 100 % increase in one hour strength compared to the reference con-

crete. After three hours A2 gave two to three times higher strength than the reference. The effect on one day strength was insignificant.

As shown in Table 3, A2 also gave the best slump retention among the blends containing accelerators. Therefore, further tests were carried out only with A2. The next step was to see if the age of the fresh concrete (containing A2) affected early strength development. Table 5 shows the effect of time of shotcrete accelerator addition.

The results given in Table 5 show that the effect of the accelerators in A2 decreased with the age of the concrete upon addition. The reason for this behaviour was not investigated, but the slump loss with time indicates some hydration and agglomeration of cement particles in this period. This phenomenon is intriguing and it will be studied further in another project.

5 CONCLUSIONS AND FUTURE WORK

The results presented demonstrate a promising new way of increasing early compressive strength in shotcrete. It is evident that synergistic effects caused by proper chemical combinations in concrete are capable of speeding up the cement hydration process. A doubling of early compressive strength values has been achieved by introducing a new chemical admixture in the fresh concrete.

Since all experiments were carried out manually with only small laboratory concrete batches, the results obtained should be interpreted with care.

The following investigations will be carried out in the next stage of this project: (1) fundamental laboratory study of the interactions between proprietary chemical blends and cement, and (2) full scale shotcreting in the field to evaluate the effect of these chemicals on compressive strength development.

Table 4. Compressive strength as a function of time after addition of alkali-free shotcrete accelerator (dosage 8%).

Concrete*	Compressive strength / MPa		
	1 hour	3 hours	24 hours
Reference	0.75	1.13	22.5
A1	0.57	0.78	23.7
A2	1.42	3.10	24.3
A3	0.99	2.98	25.0
A4	0.82	1.92	23.4

* Concrete without chemical additives (reference) and with the chemical additives described in Table 2.

REFERENCES

Coverdale, T. et al. 1996. Total Consistency Control for Wet Shotcrete – A Slump Killing System for Improved Performance. In N. Barton et al. (eds), *Proc. Second International Symposium on Sprayed Concrete. Gol, Norway, 23–*

26 September 1996: 182-192. Oslo: Norwegian Concrete Association.

Erlien, O. 1999. On Site Experience with Alkalifree Shotcrete Accelerators in Norway. In N. Barton et al. (eds), *Proc. Third International Symposium on Sprayed Concrete. Gol, Norway, 26–29 September 1999*: 227-236. Oslo: Norwegian Concrete Association.

Garshol, K.F. & Melbye, T.A. 1996. Practical Experience with Alkali-Free, Non-Caustic Liquid Accelerator for Sprayed Concrete. In N. Barton et al (eds), *Proc. Second International Symposium on Sprayed Concrete. Gol, Norway, 23–26 September 1996*: 257-268. Oslo: Norwegian Concrete Association.

Myrdal, R. 1999. Modern Chemical Admixtures for Shotcrete. In N. Barton et al (eds), *Proc. Third International Symposium on Sprayed Concrete. Gol, Norway, 26–29 September 1999*: 373-382. Oslo: Norwegian Concrete Association.

Shotcrete: Engineering Developments, Bernard (ed.) © *2001 Swets & Zeitlinger, Lisse, ISBN 90 5809 176 7*

Pull-out mechanisms of twisted steel fibers embedded in concrete

A.E. Naaman & C.Sujivorakul
Department of Civil and Environmental Engineering, University of Michigan, USA

ABSTRACT: This paper describes some of the parameters leading to the unique bond behavior of newly developed steel fibers, called Torex fibers, which are polygonal in cross section and twisted along their length. Pull-out tests, whereby a single fiber is pulled out from a cementitious matrix, are used to study the bond behavior of these fibers. The results of several series of pull-out tests are described and analyzed based on the following parameters: 1) the cross-sectional shape of the fiber, that is, a triangular or a square cross section, 2) the number of ribs (induced by twisting) per unit length of fiber, 3) the compressive strength of the matrix (from 10 to 50 MPa) and the tensile strength of the fiber, and 4) the embedded length of the fiber. Some comparison with smooth and hooked steel fibers is also provided. The high efficiency of the new fibers implies that a lesser volume of them is needed to achieve a certain level of composite performance. It is concluded that the new fibers are also a key to advancing the development of both strong and ductile high performance fiber reinforced cement composites

1 INTRODUCTION AND BACKGROUND

Cementitious materials such as mortar and concrete have an inherent weakness in resisting tension. The addition of discontinuous fibers to such matrices not only enhances their cracking resistance but also leads to numerous improvements in their mechanical properties. The success of composite action depends on the transmission of forces between the fiber and the matrix. Fiber-matrix bond is recognized as a fundamental parameter in the mechanics of fiber reinforced cement composites. Enhancing bond between the fiber and the matrix is also a key to the development of high performance fiber reinforced cement composites (HPFRCCs). The components of bond are generally classified as follows: (1) physical and/or chemical adhesion between a fiber and a matrix; (2) frictional resistance between a fiber and a matrix; (3) mechanical anchorage which is due to the particular geometry or deformation of the fiber such as in twisted, crimped or hooked fibers; and (4) fiber-to-fiber interlock, or entanglement which exists only at very high fiber contents such as in the case of SIFCON (slurry-infiltrated fiber concrete), or SIMCON (slurry infiltrated mat concrete).

Numerous experimental investigations have focused on understanding and evaluating bond (Banthia & Trottier 1994, Naaman & Najm 1991, Guerrero 1999, Naaman 1999, Guerrero and Naaman 2000). Pull-out tests are generally used as an indirect method to measure bond. They lead to a pull-out load versus end slip relationship from which various bond parameters can be extracted. Also, analytical models have been developed to investigate the mechanisms of bond during fiber pullout (Alwan et al. 1999, Chanvillard & Aitcin 1996, Nammur & Naaman 1989, Naaman et al. 1991a, Sujivorakul et al. 2000).

1.1 Twisted polygonal fibers

Recent studies at the University of Michigan have led to the development of steel fibers with optimized geometry (low order polygonal or substantially polygonal in section) that offer a ratio of lateral surface area to cross sectional area larger than that of round fibers. This ratio has been defined as the fiber intrinsic efficiency ratio (*FIER*). An increase in the *FIER* of a fiber leads to a direct increase in the contribution of the adhesive and frictional components of bond. In order to increase the *FIER* of a fiber a non-round polygonal shape is needed. Moreover, unlike a fiber with a round section, a fiber of polygonal section can be twisted, developing ribs, and thus creating a very effective mechanical bond component. One of the key features of the new fibers is that they show, under direct pull-out, a unique bond stress versus slip response not observed to date by any other fiber on the market. This response is characterized by a slip hardening bond behavior that translates into an equivalent bond strength increasing with in increase in slip, up to about 80% of the embedded length (Naaman 1998 and 1999, and Guerrero 1999).

1.2 *Advantages of twisted fibers in shotcrete*

Wet shotcreting of a concrete mixture containing steel fibers requires special considerations involving both the matrix and the fiber. The matrix must maintain an acceptable consistency and flowability for the shotcrete operation, and the fiber parameters, such as fiber volume fraction and aspect ratio, are limited on the upper side by practical mixing considerations (ACI 506). Generally the higher the volume fraction and the longer the aspect ratio of the fibers the better is the performance of the composite. However, mixtures with longer fibers are more difficult to "shotcrete;" and, everything else being equal, fibers with hooked ends are more difficult to flow in the pressurized shotcrete tubing than straight fibers, because they tend to interlock. Since twisted fibers were found to be more efficient than any other steel fiber on the market, including hooked-ends fibers, a smaller volume fraction or a shorter aspect ratio of them is needed to achieve the same level of performance for a specified composite, particularly in improving its mechanical properties such as post-cracking strength and toughness. Thus, for shotcrete operations, triangular or square fibers twisted along their longitudinal axis offer both the benefits of being easier to mix and flow in the shotcrete tubing, and have a reduced volume fraction for same level of performance. Equivalently, for same volume fraction as other steel fibers, they lead to a composite with a significantly higher post-cracking strength and toughness.

2 OBJECTIVE

The main objective of this research is to study the components of bond in twisted polygonal steel fibers embedded in cement based matrices. Parameters investigated are the geometry of the fiber (cross-sectional shape, number of twists), the strength of the fiber and the matrix, and the embedded length of the fiber.

3 RESEARCH SIGNIFICANCE

The main significance of this research is to develop high performance fiber reinforced cement composites (HPFRCCs) which exhibit multiple cracking and strain hardening behavior at lower cost

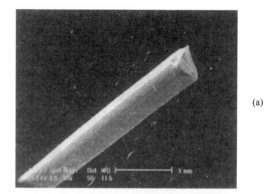

(a)

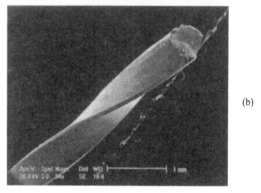

(b)

Figure 1 Microscopic view of a triangular twisted fiber; (a) before twisting, (b) after twisting.

Table 1 Type and properties of fibers

Fiber ID	Equiv. Diameter, d_e (mm.)	Cross-Sectional Area (mm^2)	Cross-Sectional Shape	Tensile Strength (MPa)	*FIER* per unit length
Twisted Fibers					
T7S-B	0.79	0.49	Square	2000	5.71
T3T-B	0.327	0.084	Triangular	2000	13.93
T5T-A	0.5	0.196	Triangular	2500	10.29
T3T-A	0.3	0.071	Triangular	2500	17.15
Hooked Fibers					
H5C-C	0.5	0.196	Circular	1300	8.0
Smooth Fibers					
S5C-A	0.5	0.196	Circular	2500	8.0

when the twisted polygonal steel fibers are used as reinforcements. These advanced fibers are expected to increase the limits of structural and non-structural applications of HPFRCCs. Figure 1 illustrates the configuration of a steel fiber having triangular cross-sectional shape before and after twisting.

Table 2 Composition of matrix mixtures by weight ratio and their properties

Mortar ID	Cement	Fly Ash	Sand	Melment	water	$f'c$, MPa
Mix-1	0.3	0.7	3.5^3	None	0.575	10.5
Mix-2	0.8	0.2	1.0^1	None	0.45	38
Mix-3	0.8	0.2	1.0^2	0.03	0.45	48.5

[1] Sand ASTM-50-70
[2] Sand ASTM-270
[3] Flint Sand

4 EXPERIMENTAL PROGRAM

4.1 *Test parameters*

The experimental program described here focused on evaluating the frictional and mechanical components of bond of twisted steel fibers subjected to pull-out from a cementitious matrix. Four different types of twisted steel fibers were used; they are identified by their tensile strength, size and cross sectional shape as shown in Table 1. The first letter of the identification denotes the type of fibers (T = twisted fibers, H = hooked fibers, S = smooth fibers); the second letter denotes approximate value of equivalent diameter of the fibers in 10^{-1} mm; the third letter denotes the cross-sectional shape of the fibers (T = triangular, S = square, C = circular); and the last letter denotes the tensile strength of the fibers (A = 2500 MPa, B = 2000 MPa, C = 1300 MPa). Moreover, for comparison, hooked and smooth steel fibers were also used in the tests. The

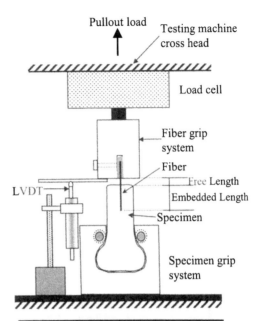

Figure 3 Fiber pullout setup

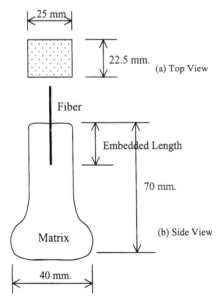

Figure 2 Layout of typical pullout specimens

hooked fibers (trade name Dramix, by Bekaert) were round in cross section with a diameter of 0.5 mm and a length of 30 mm. Their properties are also given in Table 1.

Three different mortar matrices (Mix-1, Mix-2, and Mix-3) were used as shown in Table 2. The component materials were Portland Cement Type III, silica sand (type 50-70, i.e. passing ASTM sieve #50 and retained on sieve # 70, and type 270, i.e. passing ASTM sieve #270), flint shot blasting sand, and fly ash type C. The size of the sieves conforms to ASTM specification E11. The average

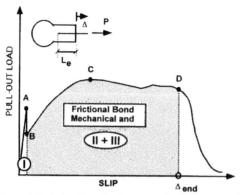

Figure 4 Typical pull-out load versus end slip response of twisted fibers

data acquisition system, and stored in a computer file. The pullout load versus end slip relationship was then plotted using this data file. All tests were performed under displacement control. The load was applied at a rate of 0.2 mm/minute (0.008 in./min.) at the beginning of the test up to an end slip of 2.5 mm (0.1 in.). After that the rate was changed to 1.27 mm/min. (0.05 in./min.) until the fibers were totally pulled out.

5 ANALYSIS OF RESULTS

5.1 *Typical pull-out behavior of twisted fibers*

Figure 4 shows a typical pull-out load versus end slip response of twisted steel fibers, which has been identified and described by Naaman (1999). Most of the results described here are identical to this response. The initial ascending part is almost linear up to point A (Fig. 4) mainly due to combined adhesive-frictional bond. The elongation of the fiber is too small compared to the slip of the fiber and can be neglected, since the free length (Fig. 3) is minimized during the test. The adhesive bond vanishes after point A. As a result, a sudden drop is observed from point A to point B. It is also believed that this drop is due to some untwisting of the fiber in the portion between the grip and the free surface of the specimen. Then the pull-out load recovers and keeps increasing up to a certain level (Point C) at which it tends to stabilize. Point D represents the load prior to failure. The most surprising behavior of twisted fibers is that a high level of pull-out load is maintained up to very large slips, about 70 to 90% of the embedded length. This behavior is due to successive untwisting and locking of the embedded portion of the fiber during slip, and can be described as a "pseudo-plastic" response. Experimental results that will be discussed next show that three parameters affect the pull-out load versus slip response of twisted fibers: (a) geometry of the fiber (cross-sectional shape, number of twists); (b) mechanical properties of the fiber; and (c) mechanical properties of the matrix. These parameters significantly influence the frictional and mechanical components of bond.

compressive strength of these mortars were obtained from 75 × 150 mm (3 × 6 in.) cylinders. Mix-1 is a low strength matrix, while Mix-2 and Mix-3 are medium strength and high strength matrices, respectively.

4.2 *Preparation for test specimens*

The test specimens were prepared by using plexiglass molds. The specimens used had a half dog-bone shape with dimensions 25.4 × 22.9 mm (1.0 in. × 0.9 in.) for the cross section and 71.1 mm (2.8 in.) in length where a fiber is embedded with a specific embedded length as shown in Figure 2. Five specimens were prepared for each series of test. The mortar matrices were prepared in a food type (Hobart) mixer. The matrices were poured into the mold mounted on a vibration table. After finishing the specimens, the molds were placed in a 100 percent relative humidity environment at room temperature, then they were removed and cured in a water tank for 14 days at room temperature. Before testing, the specimens were removed from the water tank and kept in laboratory air environment for at least 48 hours before testing.

4.3 *Test setup and testing procedure*

Figure 3 shows the test set-up used. The top end of the fiber was held by a specially designed grip attached to the load cell of a testing machine while the body of the half dog-bone specimen was restrained by a grip used to test standard ASTM tensile briquettes of mortar. A linear variable differential transducer (LVDT) was placed as shown in Figure 3 in order to measure the slip at the free end of the fiber, defined here as the differential movement between the fiber end and the fixed matrix specimen. The values of the pull-out force and the corresponding end slip were recorded by a

5.2 *Number of twists along the length of the fiber*

Typical pull-out load versus slip response of fibers with different twist ratios are shown in Figures 5 to 8. Five different number of twists of triangular fibers T3T-B (no twist, 6, 12, 18 and 24 ribs/in.) are compared for the low strength matrix (Mix-1, f'c = 10.5 MPa) and the medium strength matrix (Mix-2, f'c = 38 MPa) in Figures 5 and 6, respectively. It is observed that an increase in the number of twists increases the peak pull-out load. The pseudo-plastic behavior, where a maximum or plateau pull-out load

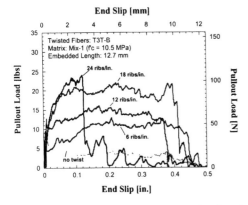

Figure 5 Influence of the numbers of twists in triangular fibers T3T-B embedded in Mix-1.

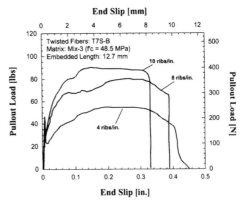

Figure 8 Influence of the numbers of twists in square fibers T7S-B embedded in Mix-3.

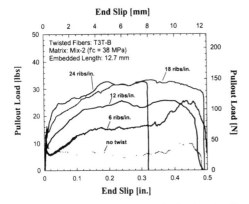

Figure 6 Influence of the numbers of twists in triangular fibers T3T-B embedded in Mix-2.

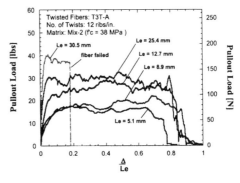

Figure 9 Influence of the embedded length in triangular fibers T3T-A embedded in Mix-2.

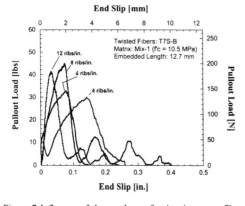

Figure 7 Influence of the numbers of twists in square fibers T7S-B embedded in Mix-1.

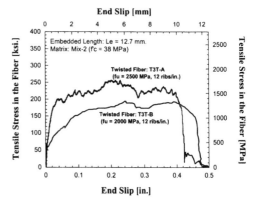

Figure 10 Influence of the fiber tensile strengths

201

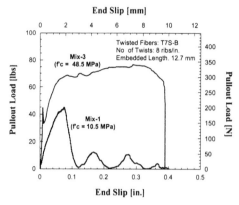

Figure 11 Influence of matrix strength on pull-out response of T7S-B fibers

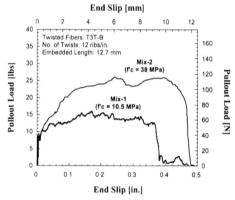

Figure 12 Comparison of T3T-B fibers embedded in different matrix strengths.

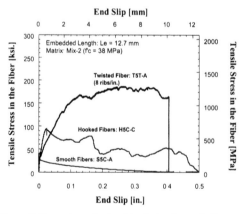

Figure 13 Comparison of twisted, hooked and smooth fibers

is maintained almost up to complete pull-out of the fiber, is observed for all number of twists, except for

the 24 ribs/in. In this last case, the load drops, because the shearing stress related to pull-out load, is higher than the shearing resistance of the matrix. Thus, a tunnel of the matrix material along the fiber fails. It is noted that the pseudo-plastic behavior up to large slips can be obtained only if the fiber tends to untwist during pullout.

The pseudo-plastic behavior is not observed for the square fibers, T7S-B, embedded in the low strength matrix (Mix-1, f'c = 10.5 MPa), even when the twist ratio is small, as shown in Figure 7. This can be explained in part by the low value of the fiber intrinsic efficiency ratio (FIER) of the square fibers used. An increase in the value of FIER increases the interfacial area between the fiber and the matrix; thus the frictional bond and the shearing resistance of the matrix are increased. The FIER of the square fiber T7S-B is about 40% that of the triangular fiber T3T-B as shown in Table 1. Their peak pull-out load drops suddenly as a result of the failure of the tunnel of matrix by shear along the length of the fiber, before the fiber undergoes untwisting and pull-out. In this case, the peak pull-out load is less sensitive to the number of twists. However, increasing the strength of the matrix will help obtain the desirable pseudo-plastic behavior for the square fibers, T7S-B, as shown in Figure 8.

5.3 Embedded length of the fiber

A comparison of the influence of embedded length on the pull-out response of triangular fibers, T3T-A, is shown in Figure 9. In this figure, the end slip (Δ) is normalized by the embedded length (Le) of the fiber. The reason is to illustrate that for all embedded lengths a pseudo-plastic behavior is observed up to about 80 % of the embedded lengths, except for Le = 30.5 mm. In this last case, the fiber fails in tension, at a stress of about 2500 MPa. It is observed that an increase in the embedded length increases the peak pull-out load and initial slope of the ascending branch. However, the peak pull-out load is not proportional to the embedded length.

5.4 Tensile strength of the fiber

Figure 10 provides a comparison of the pull-out load versus slip response of twisted fibers having different tensile strengths. The fibers compared are triangular T3TA (fu = 2500 MPa, 12 ribs/in.) and T3TB (fu = 2000 MPa, 12 ribs/in.). They have almost the same cross sectional shape, equivalent diameter and number of twists. It is observed that an increase in the tensile strength of the fiber increases the peak pull-out load, and the initial slope of the ascending branch. This is due to the effect on the mechanical bond of untwisting the fiber during pullout. Note that the torsional resistance of the fiber increases with its yield or tensile strength. This, in

202

turn, leads to an increase in the mechanical component of bond.

5.5 *Matrix compressive strength*

Pull-out curves of twisted fibers embedded in matrices of different strengths are shown in Figures 11 and 12. Figure 11 shows the response of twisted square fibers T7S-B with 8 ribs/in. embedded in Mix-1 and Mix-2, whereas figure 12 shows the response of twisted triangular fiber T3T-B with 12 ribs/in. embedded in Mix-1 and Mix-3. For the case of square fiber T7S-B which has a small value of *FIER*, an increase in the matrix strength increases the peak pull-out load, the initial slope of the ascending branch, and leads to pseudo-plastic behavior (Fig. 11). The pseudo plastic behavior does not materialize for the lower strength matrix as a result of the failure of the matrix by shear along the fiber. On the other hand, if the fiber has a larger value of *FIER* as in the case of triangular fiber T3T-B, the pseudo-plastic behavior can occur even at low values of matrix strengths (Fig. 12). Additional effects of matrix strengths can be found by analyzing Figs. 5 to 8.

5.6 *Comparison of twisted, hooked and smooth fibers*

Figure 13 shows the tensile stress developed during pull-out of the fiber for three different fibers (twisted, hooked and smooth fibers). These fibers have the same cross-sectional area. It is seen that the pull-out load versus slip response of twisted fibers is much better than that of hooked or smooth fibers, and only the twisted fiber develops a pseudo-plastic behavior. It is also observed that the pull-out energy (the area under the curve) for the twisted fiber is about 250% that for the hooked fiber and 1500% that for the smooth fiber, respectively.

6 CONCLUSIONS

This paper describes parameters influencing the particular bond characteristics of newly developed polygonal twisted steel fibers. The pull-out load versus end slip response obtained from pull-out tests represents a good predictor of the tensile response of the cracked composite. The new fibers show a pseudo-plastic behavior, whereby a maximum or plateau pull-out load is maintained almost up to complete pull-out of the fiber; the new fibers also lead to a higher peak pull-out load and significantly higher pull-out energy than prior fibers. This implies that a lesser volume of them is needed to achieve a certain level of composite performance. It is believed that the new fibers are a key to developing both strong and ductile high performance fiber reinforced cement composites.

ACKNOWLEDGMENTS

This research was supported in part by a grant from the National Science Foundation to the NSF Center for Advanced Cement-Based Materials. Any opinions, findings, and conclusions expressed in this study are those of the authors, and do not necessarily reflect the views of NSF or the ACBM Center.

REFERENCES

ACI Committee 506. Guide to Shortcrete and State-of-the-Art Report on Fiber Reinforced Shortcrete. *ACI Manual of Concrete Practice (Part 5)*: 506R1-506R41 & 506.1R1-506.1R13.

Alwan, J. M., Naaman, A. E. & Guerrero, P. 1999. Effect of Mechanical Clamping on the Pull-Out Response of Hooked Steel Fibers Embedded in Cementitious Matrices. *Concrete Science and Engineering* 1: 15-25.

Banthia, N. & Trottier, J. F. 1994. Concrete Reinforced with Deformed Steel Fibers, Part I: Bond-Slip Mechanisms. *ACI Materials Journal* 91(5): 435-446.

Chanvillard, G. & Aitcin, P. C. 1996. Pull-out Behavior of Corrugated Steel Fibers. *Advanced Cement Based Materials (ACBM)* (4): 28-41.

Guerrero, P. 1999. Bond Stress-Slip Mechanisms in High Performance Fiber Reinforced Cement Composites. *Ph.D. Thesis*. University of Michigan, Ann Arbor: 249 pages.

Guerrero, P. & Naaman, A. E. 2000. Effect of Motar Fineness and Adhesive Agents on Pullout Response of Steel Fibers. *ACI Materials Journal* 97(1): 12-20.

Naaman, A. E. 1998. New Fiber Technology: Cement, Ceramic, and Polymeric Composites. *Concrete International* 19(7): 57-62.

Naaman, A. E. 1999. Fibers with Slip Hardening Bond. *High Performance Fiber Reinforced Cement Composites - HPFRCC 3*. H. W. Reinhardt and A. E. Naaman, Editors, RILEM Pro 6, RILEM Publications S.A.R.L., Cachan, France, May 1999: 371-385.

Naaman, A. E., Namur, G. G., Alwan, J. M. & Najm, H. S. 1991. Fiber Pullout and Slip. I: Analytical Study. *Journal of Structural Engineering*, ASCE 117(9): 2769-2790.

Naaman, A. E. & Najm, H. 1991. Bond-Slip Mechanisms of Steel Fibers in Concrete. *ACI Materials Journal* 88(2): 135-145.

Nammur, G. Jr. & Naaman, A. E. 1989. Bond Stress Model for Fiber Reinforced Concrete Based on Stress-Slip Relationship. *ACI Materials Journal* 86(1): 45-57.

Sujivorakul, C., Waas, A. M. & Naaman, A. E. 2000. Pullout of a Smooth Fiber with an End Anchorage. *Journal of Engineering Mechanics*, ASCE 117(9): 986-993.

Shotcrete: Engineering Developments, Bernard (ed.) © 2001 Swets & Zeitlinger, Lisse, ISBN 90 5809 176 7

Load bearing capacity of steel fibre reinforced shotcrete linings

U.Nilsson & J.Holmgren
Concrete Structures, Dept. of Structural Engineering, KTH, Stockholm, Sweden

ABSTRACT: An experimental investigation of the flexural behaviour of Steel Fibre Reinforced Concrete (SFRC) has been carried out in order to improve design methods for SFRC tunnel linings and similar applications. A large number of suspended slabs and beams with fixed ends were tested, with the intention of gaining a clearer understanding of how these structural elements behave during failure. The calculations, which were based on yield line theory, were in poor agreement with the test results in that the estimated capacities were consistently lower than the actual load bearing capacity of the specimens. From the tests it was clear that compressive arch action enhanced the load bearing capacity of the specimens and that this phenomenon led to the error in the calculated values. The effect of compressive arch action is currently ignored in design methods for applications such as tunnel linings.

1 BACKGROUND

1.1 *Earlier research*

In Sweden research on shotcrete as a means of support for hard rock started in 1973. This research is still on-going. Some fundamental findings are described below.

The primary mode of failure of good quality shotcrete linings on hard rock is through adhesion failure. The primary failure load is independent of the lining thickness for normal shotcrete thicknesses. The critical adhesion stresses are concentrated in a narrow band along cracks where transverse load is transferred. After adhesion failure, collapse of the lining may be avoided if it is reinforced and anchored by rock bolts, or if it is acting as an arch that can transfer the load down to a fixed support (Hahn & Holmgren 1979, Holmgren 1979).

Load transfer from shotcrete to rock anchor is of utmost importance for the load carrying capacity of the lining and thus the total economy of the design. Steel fibre reinforcement provides better performance in this respect than bar reinforcement. If bar reinforcement is used it should be a mild steel in order to avoid a brittle failure in a collapse situation (Holmgren 1983, 1985a).

The primary mode of failure in the case of dynamic loading is also an adhesion failure after which the dynamic energy must be absorbed by the reinforced shotcrete and/or the rock anchors. The energy absorption capacity of the reinforced shotcrete is small compared to the capacity of a rock anchor, which is allowed to yield over a large part of its length. In general, steel fibre reinforced shotcrete performs better than bar reinforced shotcrete in a dynamic situation (Holmgren 1985b).

1.2 *Structural design of shotcrete linings*

Three basic mechanisms of load transfer can be distinguished in hard rock applications if rock burst problems are excluded:

1. The gravity load from loosening blocks is transferred to the surrounding rock by means of shotcrete, which adheres to the rock surface.
2. The gravity load from loosening blocks is transferred to the tunnel floor by means of shotcrete acting as an arch.
3. The gravity load from loosening blocks is transferred to the rock above the loosening part by means of shotcrete interacting with rock anchors. The shotcrete must be reinforced in order to carry the load between the anchors.

The latter mechanism is the subject of this article.

If it is assumed that a rock mass cannot support itself even if an adhering shotcrete lining is applied, a rock-anchored reinforced shotcrete lining must be adopted. Methods for the determination of the volume of the loosened rock mass will not be treated here. This should always be determined in co-operation with an engineering geologist.

The gravity load from the loosened rock has been treated as a uniform load acting on the reinforcement. A tunnel supported by shotcrete and rock bolts may then, from a design point of view, be considered as a slab supported by columns where each column represents a rock bolt. Under loading, this slab works as a number of edge restrained circular slabs with the load at the centre. Furthermore, the interaction between rock bolts and the shotcrete may also be considered as a beam with fixed ends according to Figure 1.

Statically indeterminate structures made of ordinary reinforced concrete are often designed using yield line theory because of its simplicity and the efficient use of materials compared to more conservative designs based on elastic theory. It is therefore interesting to determine if yield line theory is applicable for SFRC as well.

For the structure described above, the following applies. The bolt force is given by $F_b = qc_1c_2$, where q is the load from the rock and c_1 and c_2 are the distances between bolts in two perpendicular directions, as can be seen in Figure 1. For $0.8 < c_1/c_2 < 1.25$ the sum of positive and negative bending moment is obtained from $m'+m = 0.16qc_1c_2$ (local yield line pattern). For other values of this quotient a global yield line pattern is more likely and $m'+m=0.125qc^2$, where c is the maximum value of c_1 and c_2. For fibre reinforced shotcrete, $m' = m$, and so $m'+m$ can be re-placed by $2m$ in the above expressions (Holmgren 1992).

The value of m in the above expressions must not exceed the yield line moment capacity of the fibre reinforced cross-section. In addition, a rock bolt cover plate assembly must be placed outside the shotcrete layer to anchor the lining. The value of the nominal local shear stress around this plate must not exceed the design shear strength of the concrete.

1.3 *Finding a design value for the moment capacity of a steel fibre reinforced section*

As tables in handbooks have been prepared for a number of slab cases assuming conventional reinforcement, it is desirable to find a way of using these for fibre-reinforced concrete slabs as well. It is also desirable to base the calculation of the bending-moment capacity of the fibre-reinforced cross section on a recognised method of defining the mechanical properties of fibre-reinforced concrete. For example, performance could be based on flexural toughness parameters determined in accordance with ASTM C1018.

Consider a beam strip with fixed ends, as shown in Figure 2. If yield occurs at both supports and in the middle of the span, the widths of the cracks will be twice as large in the middle as at the supports. The load-bearing capacity may be expressed as a function of the bending moment at the support and at mid-span. What determines the load-bearing capacity of the beam is always the sum of the bending moment at a change in angle equal to θ and the bending moment at a change in angle equal to 2θ at mid-span. If the moment crack-rotation relation (m - θ) is known, a curve for the sum of the bending moments can easily be constructed.

In this case the following applies:

$$q = \frac{8}{L^2}[m(\theta) + m(2\theta)] \qquad (1)$$

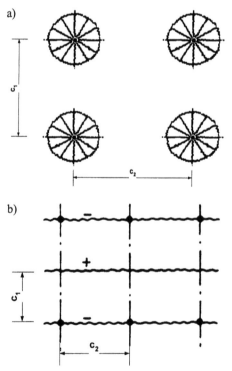

Figure 1. a) Local yield line pattern, b) global yield line pattern.

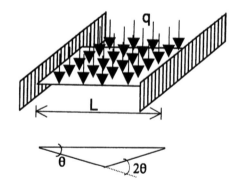

Figure 2. Built-in single span beam strip with uniformly distributed load.

206

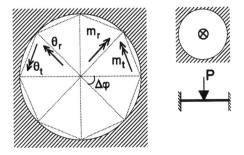

Figure 3. Built-in, circular slab with a point load acting at its centre.

Now consider a circular slab built-in around its periphery and subjected to a point load at its centre according to Figure 3. Here the load-bearing capacity can also be expressed as a function of the positive and negative bending moments, that is, bending moments along radial and tangential yield lines. The ratio between the widths of cracks in the two yield lines will be dependent on the number of radial yield lines that form. The curve for the sum of the bending moments can be constructed from the $(m - \theta)$ curve for various assumptions of the number of radial yield lines.

In the case shown in Figure 3, the following applies:

$$P = \frac{2\pi \tan \dfrac{\Delta \varphi}{2}}{\dfrac{\Delta \varphi}{2}} \left[m(\theta_r) + m(\theta_t) \right] \qquad (2)$$

where $\theta_t = \dfrac{\theta_r}{2 \sin \dfrac{\Delta \varphi}{2}}$ \qquad (3)

It is advisable to specify the requirements for the mechanical properties of fibre-reinforced concrete, as described below, including both the maximum strength and the residual strength. This provides a starting point for yield-line theory design of SFRC structures.

If a required value for the maximum strength is given, this can be replaced by

$$\frac{R_{5.10}}{100} f_{cr} \qquad (4)$$

determined in accordance with ASTM C1018. If a required value of the residual strength is given, this can be replaced by

$$\frac{R_{10.50}}{100} f_{cr} \qquad (5)$$

determined in accordance with ASTM C1018.

A rigid plastic model of material behaviour could be based on a simplified expression for the yield moment in the steel fibre reinforced section by averaging these two residual strengths (Alemo et al. 1997):

$$m = \frac{R_{5.10} + R_{10.50}}{200} f_{cr} \frac{h^2}{6} \qquad (6)$$

This yield moment capacity can be used for calculations using conventional yield line theory.

The use of the average of the residual strengths in accordance with Eqn. 6 provides a usable value of the bending moment capacity in the event of deflections that are up to several times the cracking deformation if the fibre-reinforced concrete shows strain hardening behaviour. For the case of a strain softening fibre-reinforced concrete, it is inappropriate to calculate the load-bearing capacity according to yield-line theory in the normal sense as the load steadily decreases when the deflection increases. But if the load case involving forced deformation is taken into account, this a relevant concept. The load-bearing capacity calculated in accordance with the equations above then refers to the residual load-bearing capacity after a forced deformation of a certain magnitude has taken place. This load case is very real in ground support. One of the great advantages of a reinforced lining is that movements in the rock do not immediately result in collapse of the strengthening medium.

The analysis above is founded on the assumption that the moment – angular rotation $(m - \theta)$ curve for a slab is similar to the curve that is obtained from flexural testing of simply supported beams. As in yield line theory for ordinarily reinforced concrete slabs, it is also assumed that compressive arch action exists only as a phenomenon that increases the factor of safety without being included in the calculation of the load carrying capacity of the slab.

In order to study the effects of the hypotheses above, the following experimental investigation of statically indeterminate beams and slabs was conducted.

2 EXPERIMENTAL INVESTIGATION

2.1 General

A series of SFRC beams and slabs were produced in order to conduct tests to investigate the theories described above (Nilsson 2000). The concrete used was a standard prescription for shotcrete used in tunnel linings in Sweden. Although the specimens produced in this investigation were intended to simulate a portion of a sprayed tunnel lining, all specimens were produced by casting instead of spraying. This was done because sprayed specimens are difficult to produce with good dimension tolerances in thickness and levelness. The concrete mix details are given in Table 1.

Table 1 Concrete mix proportions

Ingredient	Quantity (kg/m³)
Aggregate (0-8 mm)	1530
Cement	520
Silica Fume	19
Superplasticiser	2.7
Water	157
Water/cement ratio	0.44

Many specimens were made with prefabricated notches in order to initiate rupture in those notches. This was done to make it easier to compare the test results with an expression for the bending moment capacity in the beams and the slabs. The fibre content in the beam-tests was 60 kg/m³ and 100 kg/m³, respectively. In the slab tests, the fibre content was 40 kg/m³ and 60 kg/m³ respectively. These two latter amounts were chosen because they are very common in applications such as tunnel linings. The dosage rate of 100 kg/m³ in the beam tests was chosen in an attempt to obtain a strain hardening type of SFRC. The fibres used in this investigation were Dramix RC-65/35-BN. The length of these fibres is 35 mm and the diameter is 0.55 mm. The following experiments were conducted (Table 2).

Table 2. Summary of specimens.

Specimen Set	Number
Beams with fully clamped edges and prefabricated notches	12
Circular slabs with fully clamped edges	16
Circular slabs with fully clamped edges and prefabricated notches	24

The variables examined within each set of specimens included: dosage of fibres, specimen dimensions, and location of prefabricated notches. To determine the influence of compressive arch action in the specimens, eight slabs and four beams were manufactured. These were cast with the same type of concrete as the other specimens, except that there were no fibres in the concrete. Before testing, deep notches were sawn in the specimens to eliminate the tension capacity of the concrete during failure. This meant that the resulting load bearing capacity of the specimens could be attributed to compressive arch action alone.

2.2 Testing of beams

The beams were of uniform rectangular cross section having nominal dimensions of 100 mm height, 250 mm width and a length of 1860 mm. The beams were restrained against rotation and translation at both ends and had a span of 1500 mm (see Figure 4). They were tested under third-point loading. The experimental program included four types of beams, varying in fibre content and position of prefabricated

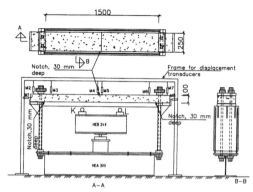

Figure 4. Test set-up for beam with a notch in the middle.

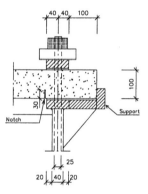

Figure 5. Support conditions for the beams.

notches. The notches were 50 mm deep in the unreinforced beams and 30 mm deep in the fibre reinforced beams.

The measuring equipment enabled recording of load and beam displacements. The data up to maximum load was recorded at load steps of 0.2 kN, then every 0.2 mm of displacement. In Figure 5 the support conditions for the beams are detailed.

The load was applied using a hydraulic actuator that was fixed in the test rig. A steel girder of type HEB 240 was mounted on top of the actuator to distribute the load at two equi-distant points along the beam through two half-round steel bars according to Figure 4.

2.3 Testing of slabs

A total of 40 slabs with fibre reinforcement were manufactured. The slabs were of circular form and were cast with two different spans and two different thicknesses, which resulted in the combinations according to Table 3 for each fibre content.

The thinner slabs of both sizes were tested without notches, while notches were sawn in the thicker slabs. In most of the slabs the depth of the notches was 10 mm, but some had 20 mm deep notches because a pre-mature punching failure occurred in the first tested slabs that had 10 mm deep notches. Of the type 2 slabs, two were sawn with an angle between the notches of 60 degrees, two were sawn with an angle of 45 degrees, and finally two were sawn with an angle of 20 degrees between the notches. The same procedure was carried out with the type 4 slabs. Finally, 8 slabs (4 of each span) were manufactured with plain concrete in order to study the effect of compressive arch action. The depth of these slabs was 40 mm and the notches were 20 mm deep.

Table 3. Tested slabs of each fibre content

Type	Span (mm)	Thickness (mm)	Number
1	682	30	4
2	682	40	6
3	936	30	4
4	936	40	6

The connection between the load-cell and the slab can be seen in Figure 6. To enhance the resistance to punching failure, the load-cell had a spherical form and a steel plate was glued to the bottom of the slab with a diameter of 200 mm and a thickness of 5 mm.

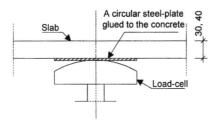

Figure 6. The connection between the slab and the load-cell.

The slabs were restrained against rotation and translation at the edges by eight very large clamps that were held together by bolts. The bolts were tightened with a torque of 140 Nm. The width of the support upon which the slabs were resting on was 125 mm. In addition, a steel ring was placed around the slab to resist the horizontal forces that arose during each test. The edge support conditions for the slabs can be seen in Figure 7.

A hydraulic actuator applied the load, which was also recorded during the tests. The displacements at different points of the slab and the support-ring were measured and recorded with linear variable displacement transducers. Data were recorded in steps of 0.2 kN up to maximum load, then every 0.2 mm of displacement of the load-cell. Six displacement transducers were placed on top of the support-ring in order to check the edge restraint.

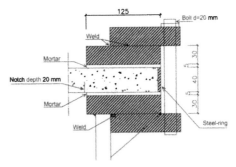

Figure 7. Support conditions for the slabs.

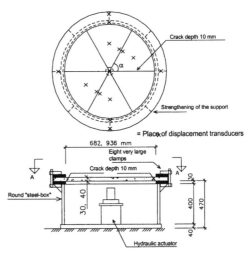

Figure 8. Test set-up for the slabs.

2.4 Results from the beam tests

Almost every beam failed as intended by a bending failure at the prefabricated notches. It was observed that the shape of the loading-displacement curves differed from the normal shape of steel fibre reinforced concrete, since the load-bearing capacity generally increased significantly after the cracking load was reached. The results also indicated that the beams made of plain concrete displayed almost the same peak load capacity as the reinforced beams. However, the capacity to carry load at large deformations was a great deal less for the unreinforced beams. The final condition of one of the beams with a notch in the middle and a fibre content of 60 kg/m^3 can be seen in Figure 9.

Examples of load/displacement curves obtained in the tests are shown below. Figures 10 and 11 indicate a comparison between fibre reinforced beams and beams made of plain concrete.

209

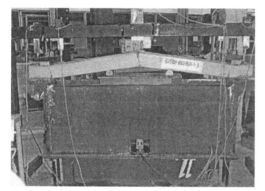

Figure 9. Final condition of a fibre reinforced beam with a notch in the middle.

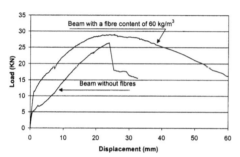

Figure 10. A comparison between a fibre reinforced beam (60 kg/m³) and a beam made of plain concrete.

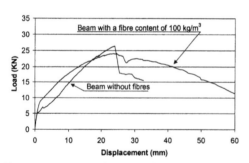

Figure 11. A comparison between a fibre reinforced beam (100 kg/m³) and a beam made of plain concrete.

2.5 Results from the slab tests

The typical failure mechanism for the slabs was a tough bending failure followed by punching at the centre, which generally occurred after a large deformation. The slabs with the smaller diameter had the greatest peak load capacity, while the bigger slabs generally displayed the toughest type of failure. As can be seen in Figure 12, a large number of radial cracks developed during each test in the slabs without pre-fabricated radial notches. The slabs with pre-fabricated notches displayed a failure mode in which radial yield lines mainly developed within the notches.

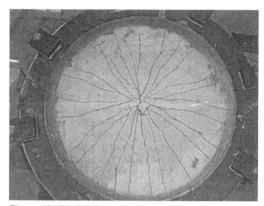

Figure 12. Final failure of a fibre reinforced slab without notches.

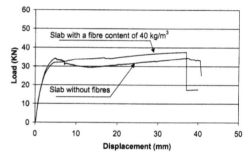

Figure 13. A comparison between a fibre reinforced slab (936 mm span and 40 kg/m³) and a slab made of plain concrete.

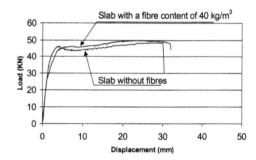

Figure 14. A comparison between a fibre reinforced slab (682 mm span and 40 kg/m³) and a slab made of plain concrete.

The peak load capacity and the toughness were more or less the same for the slabs whether they were reinforced with fibres or not. Examples of load/displacement curves obtained in the tests can be seen below. Figure 13 displays a comparison of slabs with a span of 936 mm, a thickness of 40 mm and prefabricated notches with an angle between them of 20 degrees. In Figure 14, a comparison is made between slabs with a span of 682 mm, a thickness of 40 mm and prefabricated notches with an angle between them of 20 degrees.

3 ANALYSIS

Calculations of structural load capacity in the present specimens have been performed according to yield line theory. According to this theory, the cracking pattern of the specimens is important with respect to energy absorption during failure. Therefore it was desirable to conduct tests that produced a consistent pattern of cracking. Yield lines in the slabs and in the statically indeterminate beams were treated like the cracked section of a statically determinate beam. The beams used for comparisons with the slabs and end-restrained beams were cast at the same time and were separated into two categories. One set was called 'reference beams' (which had the same depth as the slabs and the statically indeterminate beams), while the others were produced and tested in accordance with the recommendations of the standard-test method, ASTM C1018.

Examples of the comparison between test results and the calculated load capacities can be seen below. The comparison is presented in diagrams where the load-displacement curve from the tests describes the average value obtained from one type of test specimen. The theoretical load capacities are presented as numerical solutions based on the moment crack-rotation relationship in the cracked section of the reference and ASTM beams according to equations (2) and (7).

3.1 Slabs

The slab is loaded with a point load in the middle, as can be seen in Figure 3. A positive ultimate moment m_r per unit length causes yield on the upper surface of the slab. A yield line formed by a positive moment is referred to as a positive yield line. A negative ultimate moment m_t per unit length causes yield in the bottom of the slab along a negative yield line. The load bearing capacity of the slabs is calculated according to Equation (2).

Figure 15 shows a comparison between calculated post-crack load capacity and test results for a slab

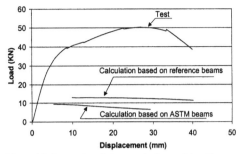

Figure 16. Comparison between a test on a slab with a span of 682 mm and numerical solutions according to yield line theory based on reference and ASTM beams.

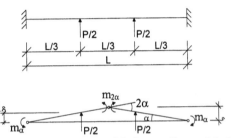

Figure 17. Yield line pattern of the beams with a notch in the middle.

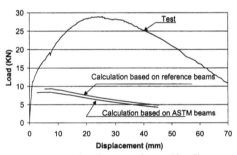

Figure 18. Comparison between a beam with a fibre content of 60 kg/m³ and numerical solutions according to yield line theory based on reference and ASTM beam tests.

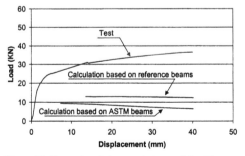

Figure 15. Comparison between a test on a slab with a span of 936 mm and numerical solutions according to yield line theory based on reference and ASTM beams.

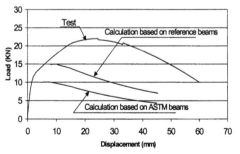

Figure 19. Comparison between a beam with a fibre content of 100 kg/m³ and numerical solutions according to yield line theory based on reference and ASTM beam tests

211

with a span of 936 mm and a fibre content of 60 kg/m³. In figure 16, results for a slab with a span of 682 mm and a fibre content of 60 kg/m³ can be seen.

3.2 Beams

The beam is loaded by two point loads according to Figure 17. A positive ultimate moment $m_{2\alpha}$ causes yield on the top surface of the beam. Furthermore, at the supports of the beam, a negative ultimate moment m_α causes yield in the bottom surface of the beam.

The failure load is described by:

$$P_\alpha = \frac{6}{L} \cdot (m_\alpha + m_{2\alpha}) \qquad (7)$$

In Figure 18, a comparison between calculated post-crack load capacity and test results for a beam with a fibre content of 60 kg/m³ can be seen. Figure 19 displays the results for a beam with a fibre content of 100 kg/m³.

4 DISCUSSION AND CONCLUSIONS

It was shown that yield line theory based on the rotation capacity at the yield lines was not able to predict the load bearing capacity of the test specimens. The calculated capacities were consistently lower than the actual capacities of the specimens. This can possibly be explained by the fact that compressive arch action is the main load carrying component of the specimens and that failure will occur when the concrete reaches its compressive strength limit in the compressed zones of the specimens.

It was also observed that the calculations based on ASTM beam tests consistently resulted in lower load capacities compared with calculations that were based on reference beams. The explanation of this is probably that a superior fibre effect was obtained in the reference beams, due to the fact that they are thinner than the ASTM beams.

It is assumed that the arch effect arises at a deformation level where the concrete stresses are reasonably small when the slab is fibre reinforced. When the slab is bar reinforced it is assumed that the compressive stresses in the concrete due to bending alone are already in the vicinity of the concrete strength when the arch action arises. The remaining capacity of the compression zone is thus rather limited in the later situation. This assumption requires a thorough theoretical investigation which remains to be performed.

In a structure that consists of several slab elements, such as a bolted tunnel lining, compressive arch action will most likely arise in the element that is overloaded if the surrounding elements can resist horizontal forces. The structure will then carry the

load by a resistance-moment between compressive forces in the element, instead of equilibrium between the tension zone and the compressive zone of the concrete section as in ordinary bending. This is very beneficial with respect to the peak load capacity of the structure, because the best load carrying characteristic of concrete is its compressive strength.

Additionally, it must be noted as has been indicated in these tests, that a large amount of fibres might result in a weaker structure. The reason for this is that the fibres used, which are glued together before they are mixed into the concrete, produce a large amount of air when the water dissolves the glue. A higher air content is then obtained and consequently a lower compressive strength of the concrete will result. This may explain the fact that the beams with a fibre content of 100 kg/m³ displayed a lower load bearing capacity compared with the beams that had a fibre content of 60 kg/m³.

Finally, it was noteworthy that compressive arch action appeared to be so significant in the tests, because the specimens were thin in relation to the span. However, the analyses indicated that compressive arch action must be the main load carrying component of the specimens, which might shed some light to the problem of safety against failure in designs based on yield line theory.

REFERENCES

Hahn, T, & Holmgren, J., 1979. Adhesion of shotcrete to various types of rock surfaces and its influence on the strengthening function of shotcrete when applied on hard jointed rock. Proceedings 4th International Congress on Rock Mechanics Montreux 1979, vol 1: 431-439, A. A. Balkema.

Holmgren, J., 1979. Punch loaded shotcrete linings on hard rock, Swedish Fortifications Administration, Report no 121:6, FortF/F dnr 2492F, (also Swedish Rock Mechanics Research Foundation, Report No 7:2/79), Stockholm, doctoral thesis.

Holmgren, J., 1983. Tunnel linings of steel fibre reinforced shotcrete, Proceedings 5th International Congress on Rock Mechanics Melbourne 1983, section D: D311-314, Australian Geomechanics Society.

Holmgren, J., 1985a. Bolt anchored steel fibre reinforced shotcrete linings, Swedish Rock Mechanics Research Foundation, Report No 73:1/85, Stockholm.

Holmgren, J. 1985b. Dynamiskt belastad bergförstärkning av sprutbetong, Swedish Fortifications Administration, Report A4:85, Stockholm.

Holmgren, J. 1992. Handbok i bergförstärkning med sprutbetong. Vattenfall Vattenkraft, 162 87 Vällingby.

Alemo, J, Holmgren, J, Skarendahl, Å. 1997. Stålfiberbetong för bergförstärkning. Provning och värdering., CBI Report 3:97, Stockholm.

Nilsson, U., 2000. Load bearing capacity of steel fibre reinforced shotcrete linings, Licentiate Thesis, Department of structural engineering, Royal Institute of Technology, Stockholm.

Shotcrete: Engineering Developments, Bernard (ed.) © 2001 Swets & Zeitlinger, Lisse, ISBN 90 5809 176 7

Durability of steel fibre reinforced shotcrete with regard to corrosion

E.Nordstrom
Vattenfall Utveckling AB, Alvkarleby, Sweden

ABSTRACT: The paper describes results from ongoing exposure tests examining steel fibre corrosion in cracked shotcrete. The purpose of the work is to shed light on questions about the long term durability of steel fibres in cracked concrete. In field exposure tests beams with different composition and crack widths are exposed to three different environments. After 2.5 years of exposure there is a continued increase in residual strength noted for small crack widths. For larger crack widths the increase has stopped. Measurements of the chloride contents show higher levels than in an earlier evaluation and the content increases closer to the crack opening. The following conclusions can be drawn from the second evaluation: 1) Samples exposed to freeze-thaw salts show greater evidence of fibre corrosion, 2) An increased crack width leads to increased corrosion, 3) Longer fibres corrode more than short ones at the same crack width, 4) The amount of corrosion decreased with an increased depth from the crack opening.

1 INTRODUCTION

Durability against corrosion of steel fibre reinforcement has been proved to be good, especially in concrete without cracks (Shroff 1966). However, results from research on cracked concrete is limited. It has been stated that steel fibres are more resistant under conditions in which conventional reinforcement exhibits extensive corrosion (Ohama 1987). No full explanation of the active mechanisms exists.

Ongoing research at Vattenfall Utveckling AB has attempted to define the active mechanisms for initiation and propagation of steel fibre corrosion in cracked concrete. Another objective is to determine the effect on loadbearing capacity, assuming corrosion is ongoing. This knowledge will make it possible to estimate the service life of SFRC.

In the following, results from ongoing field exposure tests after 2.5 years exposure are presented. Previous results were presented by Nordstrom (1999).

2 FIELD EXPOSURE TESTS

To be able to examine the effect of different parameters such as crack width, fibre dimension (fibre length), exposure conditions, and addition of accelerators on the rate of corrosion, field exposure tests are ongoing. The purpose of the tests is to determine the effect of corrosion in cracks on residual load-bearing capacity. The field exposure tests where started in September 1997. The test program and the second examination after 2.5 years of exposure is presented below. The field exposure tests are part of a larger study that will also consist of accelerated exposure tests. The results of the field exposure tests will also be correlated to the performance of a structure exposed to actual conditions.

2.1 *Test program*

The test program and the parameters examined are shown in Figure 1. There are three different exposure sites in addition to the control site in the laboratory:

1 National Road 40 (Rv 40) close to Borås, outdoor along a motor highway.
2 River Dal at Älvkarleby, outdoor with specimen partly immersed in river water.
3 Eugenia tunnel, a road tunnel in Stockholm.

2.2 *Specimens*

Four different concrete mix types were used, and these are listed in Table 1. The accelerator used was based on sodium silicate. The dosage of fibres added was 65-70 kg/m^3. Standard Portland cement (trade name Degerhamn Std P in Sweden) was used.

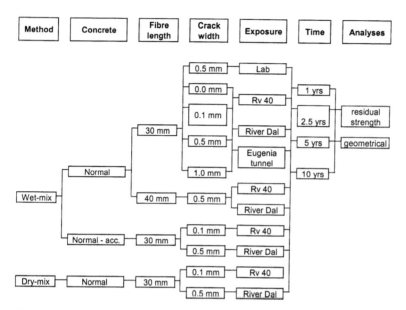

Method	Concrete	Fibre length	Crack width	Exposure	Time	Analyses

Figure 1 Field exposure test program

Table 1 Mix types

Mix	Method	Accelerator	Fibres
WA30	wet-mix	yes	Dramix 30/0.5
WA40	wet-mix	yes	Dramix 40/0.5
W30	wet-mix	-	Dramix 30/0.5
D30	dry-mix	-	Dramix 30/0.5

Table 2 Mix design used in field exposure tests

		WA30	WA40	W30	D30
w/c		0.42	0.42	0.42	-
cement	(kg/m³)	510	510	510	500
gravel, 0-8 mm	(kg/m³)	1202	1202	1202	815
gravel, 4-8 mm	(kg/m³)	-	-	-	286
sand, 2-5 mm	(kg/m³)	-	-	-	260
sand, 0-1 mm	(kg/m³)	298	298	298	138
plasticiser	(%/kg C)	1.4	1.4	1.4	
accelerator	(%/kg C)	3.5	3.5		
fibre dosage	(kg/m³)	70	70	70	65

To produce the specimens, large slabs where shot (2×1.25 m) after which beams where sawn from the slabs. Cracking of the beams (see Figure 2) was performed by applying a flexural load. In general, the flexural test was carried out in accordance with the ASTM C1018 method except that the beam dimensions differed and rate of deflection was 0.25 mm/min.

2.3 Exposure environment

The conditions at the three different exposure sites can be described as follows. Conditions in the laboratory (Lab) consisted of 65 % RH, +20 °C. Conditions at the motorway (Rv 40) were typical of out-door exposure along a highway with deicing salts in the winter and no shelter from rain. At the river (River Dal) the specimen were immersed to half the beam depth in fresh water and were not sheltered from rain. In the road tunnel (Eugenia tunnel) some deicing salts were present in the winter, the beams were sheltered from direct rain, and there was an increased concentration of acidifying gases.

The relative humidity (RH) and temperature were recorded at all sites. In additional, the water temperature in the River Dal and the amount of precipitation along Rv40 was measured.

2.4 Evaluation after exposure

Two beams for every combination of mix design and exposure condition were brought back to the lab after 2.5 years exposure. One of the two beams was subjected to flexural loading up to 5 mm deflection. The other beam was cut into small plates at different distances from the crack mouth, as shown in Figure 2.

2.4.1 Residual strength
During cracking prior to exposure, the deformation and load were recorded. After exposure, the beams

Figure 2. Dis-membering of beams (measurements in mm).

214

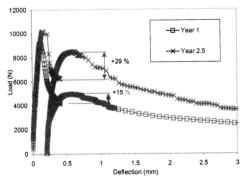

Figure 3. Residual strength before and after 1 and 2.5 years of exposure (mix WN30, w = 0.1 mm, RV40).

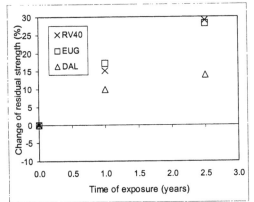

Figure 4. Development of residual strength with time of exposure (w = 0.1 mm)

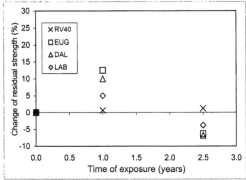

Figure 5. Development of residual strength with time of exposure (w = 0.5 mm)

where again subjected to flexural load. The purpose of this was to determine the effect of exposure on the residual strength. A typical test result is shown in Figure 3 where it can be seen that the residual strength increased (for a crack width of $w = 0.1$ mm) after exposure. In Figure 4, 5 and 6 the trend after 1

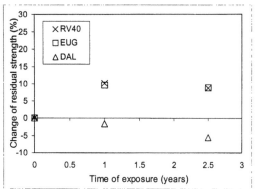

Figure 6. Development of residual strength with time of exposure (for crack width w = 1.0 mm)

Table 3 Change of residual strength (in percent) after exposure

Crack width	Motorway Rv40		Road tunnel Eugenia		River Dal		Lab	
(mm)	1998	2000	1998	2000	1998	2000	1998	2000
0.1	15	29	17	28	10	?	-	-
0.5	1	1	12	6	10	?	5	?
1.0	10	9	10	9	-2	?	-	-

and 2.5 years of exposure is shown. Beams with a crack width of 0.1 mm show a larger increase in residual strength compared to specimens with larger crack widths.

2.4.2 Carbonation
The carbonation depth was determined by spraying a phenolphthalein solution onto the newly cracked surface. It was found that all the specimens were carbonated to a depth of only a few millimetres. The outdoor exposure conditions and the relatively short time of exposure are the possible reasons for this.

2.4.3 Chloride content
The chloride content was determined in the beams subjected to repeated flexural load. Drilling debris was collected from each beam at different levels both from an exterior surface and at the crack surface. The chloride concentration was measured with the Rapid Chloride Test method (Petersen 1991). Generally the chloride content was higher after 2.5 years exposure than after 1 year. From Figure 7 it can be seen that the trend for chloride content after 2.5 years is higher closer to the crack opening.

2.4.4 Corrosion of fibres crossing cracks
Due to the presence of potentially corroded fibres in the plates extracted from the beams, there was a risk of tensile failure if the concrete was broken down by crushing. Break down by freezing was therefore used.

The plates where dried out fully at 200°C before being subjected to a 98% vacuum in a vessel. Before atmospheric pressure was restored, the vessel was filled with tap water. When letting air into the vessel the pressure gradient in the plates made them saturated with water by suction. Freezing of the saturated plates immersed in water was performed with a temperature cycle from +20°C to –25°C and back to +20°C in 24 hours. The freeze-thaw cycles was ongoing for approximately 3 weeks. After this the concrete was fully degraded and the fibres crossing the crack could easily be taken out for examination.

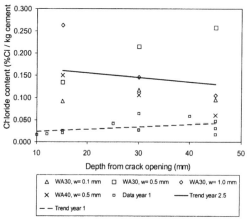

Figure 7. Chloride content at crack surface for WA30 and WA40 samples exposed at motorway Rv40 after 1 and 2.5 years.

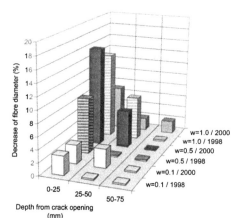

Figure 8. Decrease of fibre diameter for mix-type WA30 after 1 and 2.5 years of exposure at Rv40.

Table 4 Depth of corroded area on samples WA30 at Rv 40 (distance in mm from crack opening)

Crack width (mm)	Motorway Rv40 Exposure	
	1998	2000
0.1	15	15
0.5	30	40
1.0	5	60

From examination of the crack surface, additional information about the amount of corrosion can be found. In Table 4 the depth from the crack opening where corroded fibres could be found is shown. The depth of corrosion on single fibres was also measured by using a micrometer. Some of the results can also be seen in Figure 8 and 9.

3 DISCUSSION

The expectations after 2.5 years of exposure was to measure the development of corrosion in mix-types showing corrosion in the evaluation at year 1. It was also to control if any other mix-types had initiated corrosion. The number of samples brought back to the lab was too small to make accurate statistical treatment of the results. Still it will give an indication of the behaviour.

3.1 Residual strength

The increase in residual strength with exposure was probably caused by continued hydration of the concrete. Continued hydration results in an increased anchorage strength between fibre and concrete which gives an increased residual strength. This result has also been found by other authors, for example, Hannant et al. (1975). The increase is still higher for smaller crack widths and the effect of autogenous healing could be one reason. Some of the thinner cracks were completely filled with deposits. There is a trend that larger crack widths show the same or lowered residual strength at 2.5 years than at year 1 and this is probably due to brittle tensile failure of the corroded fibres.

3.2 Chloride content

The chloride contents where higher at 2.5 years compared to the evaluation after 1 year. The reason is that samples where brought back to the lab in the winter time. The addition of deicing salts and no (or limited) wash-out from rain is typical for the winter

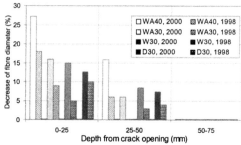

Figure 9. Decrease in fibre diameter for w = 0.5 mm after 1 and 2.5 years of exposure at Rv40.

time. The chloride profile also indicated a higher concentration of chlorides closer to the crack mouth.

3.3 *Extent of corrosion*

Surprisingly, corrosion was initiated after only one year of exposure at Rv40. A slight but not fully consistent trend is that the corroded area was more extensive at larger crack widths. The amount of corrosion was quite small and of a "surface" type. This is unlikely to have affected the residual strength.

It can be seen (in Figure 8) that the extent of fibre diameter reduction decreased with increased depth into the crack. It also appears that the 40 mm fibres show more severe damage from corrosion. This supports the theory that the galvanic effects related to the ratio of anode to cathode area is important. Samples made with the dry-mix method show more durable performance. This can probably be explained by a lower permeability causing a higher electrical resistance between anode and cathode.

The effect of crack width on the extent of corrosion of fibres is no longer clear after 2.5 years. Fibres in the beams with a crack width of 0.5 mm had corroded more than those with a crack width of 1 mm. This supports the theory that crack width mainly affects the time to initiation of corrosion (Schiessl & Raupach, 1997). Thsese authors suggest that during propagation of corrosion, other parameters (such as concrete quality, micro-climate etc.) determine the rate of corrosion and crack width is of minor importance.

3.4 *Practical consequences*

Since corrosion of steel fibres has been initiated after only 2.5 years, the expected performance in a 150 year scenario should be questioned. There are two possible practical consequences of this, but these also depend on the results of further research. Firstly, it is possible to achieve a sufficiently long service-life in steel fibre reinforced shotcrete if cracks occur. Repair of cracks, use of more corrosion-resistant fibres, or addition of an extra layer of fibre reinforced shotcrete will then be required. Secondly, the rate of corrosion propagation is so slow for narrow crack widths that cracks up to a certain critical width only decrease the service-life and residual strength to a limited extent.

4 CONCLUSIONS

After 2.5 years of field exposure, the following conclusions can be made about the performance of cracked steel fibre reinforced shotcrete:
1 The residual flexural strength continues to increase after exposure for small crack widths. This is probably due to increased anchorage strength for the steel fibres due to continued hydration of the concrete. For larger crack widths the trend is that corrosion leads to similar or reduced residual strength.
2 The chloride content is higher closer to the crack. The possible reason for this is that evaluation took place during the winter and therefore there was limited wash-out of chlorides by rain.
3 The amount of corrosion is limited at 2.5 years and is most severe for specimens subject to a high degree of exposure to deicing-salts at the motorway during the winter time.
4 The 40 mm long fibres corroded more than the 30 mm long fibres at the same crack width.
5 The amount of corrosion decreased with an increasing depth from the crack opening.
6 Samples made with the dry-mix method demonstrated more durable performance.

5 REFERENCES

Hannant, D.J. & Edgington, J., 1975. Durability of steel fiber concrete. In *Proc. RILEM symposium on fibre reinforced cement and concrete*, 159-169.

Nordstrom, E., 1999. Durability of steel fiber reinforced sprayed concrete with regard to corrosion. In *Proc. 3ʳᵈ Int Symposium on Sprayed Concrete, Gol, Norway, 26-29 September.*

Petersen, C.G., 1991. Rapid Chloride Test, The RCT-Method. *Concrete Repair Bulletin*, International Concrete Repair Institute, April.

Shroff, J.K., 1966. The effect of a corrosive environment on the properties of steel fiber reinforced Portland cement mortar. *M.S. Thesis, Clarkson College of Technology, Potsdam, NY.*, 130.

Schiessl, P. & Raupach, M., 1997. Laboratory studies and calculations on the influence of crack width on chloride-induced corrosion of steel in concrete. *ACI Material Journal*, 94 (1), 56-62.

Shotcrete: Engineering Developments, Bernard (ed.) © 2001 Swets & Zeitlinger, Lisse, ISBN 90 5809 176 7

Tele-operated shotcrete spraying with the Meyco Robojet Logica

N.Runciman & G.Newson
INCO Limited, Mines Research, Copper Cliff, Ontario, Canada

M.Rispin
Master Builders Inc., Cleveland, Ohio, USA

ABSTRACT: In 1999, INCO Limited Mines Research, in Sudbury, Ontario, Canada, took delivery of a Meyco Robojet Logica Spraymobile, a computerized concrete spraying system. The Spraymobile is a self-contained mobile machine capable of applying spray-on ground support in underground applications. The Logica component of the Robojet is designed to control the operation of the Robojet, as well as scan and map the tunnel geometry prior to the spraying operation. The Robojet can be operated in various modes, from manual to automatic operation. In the automatic mode the computer controls the standoff distance, angle, and motion of the spraying jet.

In 2000, INCO Limited, working with Meyco Equipment of Switzerland (a division of Master Builders Technologies), detailed a plan to create a tele-operation package for controlling the spraying process from a surface location at INCO's Mines Research Test mine. With the Robojet Logica operating in automatic mode, the operator at the machine becomes a process monitor. Since the Logica system controls the key spraying parameters of nozzle distance and nozzle angle to the sprayed surface, the operator could easily monitor and control the Robojet Logica from a remote location. The benefits of operating underground mobile mining equipment (drills and LHD's) from a surface location has previously been demonstrated at INCO Limited. However, the ability to control a ground support operation (shotcrete spraying) from the surface has not been attempted until now. This paper will describe the Spraymobile shotcreting system, the development of the tele-operation package for the Robojet Logica, as well as modifications to the original Spraymobile for implementing the tele-remote package on the Robojet Logica at the Mines Research Test mine.

1 INTRODUCTION

INCO Limited, Sudbury Operations, in Sudbury, Ontario, Canada, is one of the mining industry's largest producers of nickel. The increasing demand for lower cost mining applications has increased the demand for high technology solutions in all aspects of the mining cycle. To meet the needs of cost reduction and safer work environments, INCO has adapted by meeting these challenges through the development of underground mobile equipment automation. Some examples of this can be seen in the development and installation of an underground communication system capable of transmitting voice, data and video (Baiden & Scoble 1992), and the implementation of tele-operated LHDs and up-hole production drills at INCO's Stobie mine (Baiden 1996).

At INCO, prior to 1998, one of the underground mining activities that had not been the subject of the development of advanced equipment automation technologies was ground support. This changed when INCO identified sprayed ground support, more specifically the wet shotcrete application process, as the preferred method for automating, and as a result acquired the Meyco Robojet Logica (Runciman et al. 1999).

The conventional robotic shotcrete spraying process requires a skilled operator with the ability to apply a specified thickness and quality of concrete to the walls and back (roof) of the drifts. To do this, the operator uses a remote control typically containing two to four joysticks, which control all the boom movements of the spraying manipulator. The spraying process can be a very difficult, tiring and hazardous operation. If a conscious effort is not made to maintain quality, cost will increase, and efficiency and productivity will soon drop. With this in mind, INCO believed that if the quality and productivity of the spraying process could be enhanced, then costs associated with spraying shotcrete could be reduced.

In 1998, INCO Mines Research initiated a research project that would see further development of the Meyco Logica technology in collaboration with Meyco Equipment, a division of Master Builders Technologies, a Swiss based manufacturer of mechanized sprayed concrete machines for the underground construction and mining industries. The scope of this project was to evaluate, and jointly further develop the Logica, a prototype computer control system on the Robojet shotcrete spraying manipulator, with the ultimate goal being teleoperated spraying from surface. The Logica component of the Robojet is designed to control the operation of the Robojet, as well as scan and map the tunnel geometry prior to the spraying operation.

Since 1999, INCO has been testing the Robojet Logica and together with Meyco Equipment, is developing the automatic mode of operation of the Logica system. The Robojet can be operated in various modes, from manual to automatic operation. It is the automatic mode with the adaptation of a tele-operation control package that is discussed in this paper.

2 MEYCO ROBOJET LOGICA

The Meyco Robojet Logica is a prototype computer controlled concrete spraying robot manipulator. The Robojet Logica is based on the proven kinematic principle of the Meyco Robojet. The manipulator is equipped with motion control, which is software controlled and programmable. The Meyco Robojet has been mechanically proven with the sale of over 120 units worldwide (Tschumi 1998). The existing machine is equipped with an industry standard computer, a redesigned control console, as well as a modified hydraulic system. The Robojet manipulator, concrete pump, Logica system, and associated electrical, hydraulic and mechanical systems are mounted on a Normet NC-98 carrier (for this particular project) to create a mobile Spraymobile system capable of operating in most mining areas (Figure 1).

Although the Robojet spray manipulator was developed for the tunnelling industry, the size of the complete Spraymobile is within the dimensions of a typical underground drill jumbo. However, modifications were made to the Spraymobile to conform to the smaller working areas associated with underground mining. Some of these modifications include: a smaller accelerator tank (from 600 liters to 300 liters), re-routing the concrete hoses along the side of the carrier, and a replacement trailing cable.

The Robojet manipulator, with its eight degrees of freedom, is equipped with the necessary sensors at each joint to record the absolute position of each joint with respect to the next, and report this to the main computer. The Robojet also has the ability to

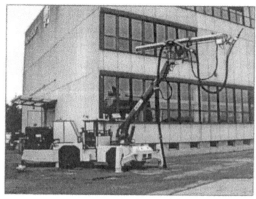

Figure 1. Meyco Robojet Logica Spraymobile

Figure 2. Operator console for the Logica.

scan and map the working area. To measure the tunnel dimensions and determine its work envelope, the lance head has been fitted with a scanner laser system.

The Robojet is capable of operating in three modes of operation. These include: manual, semi-automatic, and automatic. While spraying concrete during tele-remote operation, the Robojet will be operated in automatic mode.

Manual mode operation of the Robojet allows the operator to manipulate the spraying nozzle using a single "space" joystick. The space joystick allows the operator to point the nozzle in the correct location to spray the concrete (Figure 2). The operator does not need to be preoccupied with positioning the manipulator's joints while spraying. The result is simplified training and a shorter learning period for operating the Robojet.

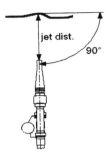

Figure 3. Stand-off distance and nozzle angle.

The semi-automatic mode allows the operator to control the movement of the nozzle along the walls of the drift. The computer (Logica) controls the stand-off distance and the angle of the spraying jet through the software (Figure 3). To be able to control these two spraying parameters, the laser scanner is used to create a three-dimensional work envelope in the computer (Figure 4). The computer uses this three-dimensional grid to calculate the perpendicular position of the nozzle and the spraying angle of the nozzle (Tschumi 1998).

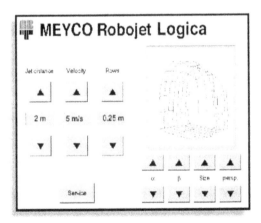

Figure 4. Logica PC control panel.

The Robojet Logica also has the capability of operating in automatic mode. The computer controls nozzle distance, nozzle angle and the velocity of the jet along the surface of the rock. The laser scanner is used the same way as for semi-automatic mode. The machine performs the spraying process automatically while the speed of the nozzle along the rock surface, and the row distance, remain constant. Row distance refers to the horizontal distance between the sprayed passes, as shown in Figure 5.

3 BENEFITS OF TELE-OPERATION CONTROL

The benefits described here are those that can be achieved through tele-remote spraying. Presently, shotcrete is still loaded into the hopper of the Spraymobile by an operator at the rear of the machine. Communication between the surface operator and the operator loading material at the rear of the machine is via two-way radios and the video system.

Operating the Robojet Logica system in automatic spraying mode allows the operator to monitor the spraying process, as compared to physically controlling the spraying manipulator with a set of joysticks.

Safety and productivity are the key benefits in mining applications when controlling the spraying process from a tele-remote location. Through tele-operation, the operator can control the spraying process without actually being at the face. This allows the operator to operate the machine from the clean and safe environment of a surface control room. In addition, since the operator is located in a surface control room, there is no need for personal protective equipment (filter mask, safety hat, glasses and boots). All of this equipment contributes to operator fatigue, tiredness and leads to a reduction in the quality of the sprayed surface.

Productivity gains can be realized when a single operator can monitor multiple machines in different areas of the mine, or at different mine sites. The machines would only need to be trammed to the workplace and setup. All spraying functions would be done automatically by computer control. The operator would be responsible for selecting the points which identify the scanned area end points. Operating in this mode also allows the opportunity to operate the machine during shift changes, without compromising the quality of the sprayed concrete layer, as the computer is maintaining all spraying parameters.

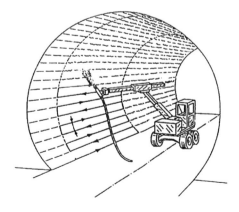

Figure 5. Row distance (Tschumi 1998).

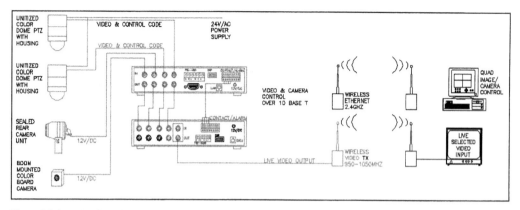

Figure 6. Schematic of the wireless video system for the tele-operated Spraymobile

4 EQUIPMENT MODIFICATIONS

The existing Meyco Robojet Logica Spraymobile operating at INCO's Research Mine required several modifications to allow the Spraymobile to be operated from a remote location.

A video system was installed on the Spraymobile to provide a set of operator eyes which are able to monitor all parts of the scanning and spraying operation. The video system is a combination of live video output, and ethernet video output and control. All live video is transmitted over a wireless broadband network. Video onboard the machine is provided by two pan-tilt-zoom cameras for monitoring the spraying operation and one rear facing camera to monitor the concrete hopper. A single camera mounted inside the laser box housing for visualization of the drift walls was provided to assist the operator in selecting the start and end points of the scanned area. Figure 6 shows the layout for the video system.

The existing Logica control package was also upgraded to receive new software for tele-operation control. The system hardware upgrade included a new CPU board with high speed serial ports and Direct Memory Access (DMA) to the CPU memory, which allowed faster data transfer between the laser and the computer. The scanning laser was also upgraded to provide faster scanning rates and a simpler laser housing design. The original Logica system utilized a point laser scanner which was integrated with the main translational joint (the lance) and a rotating joint of the Robojet. This system required that a rotational joint turn while the laser is scanning. After the scan, the lance would index the laser along the tunnel axis. The result was a scanned grid that required five to seven minutes to complete. The density of the grid yielded by this process was approximately 12 inches (30 cm) square. Although this system worked successfully, it was slow and complex,

relying on multiple joint movements to complete the entire scanning process.

To overcome these deficiencies, a new laser system was installed which allows scanning to take place in approximately 45 to 60 seconds. To accomplish this, the new laser (a Sick LMS 200) was fixed to the end of the lance on the Robojet (Figure 7). Since this laser incorporates a rotating mirror that deflects the laser through its complete scanning angle, rotating the joint of the Robojet to obtain a full scan is not required. Scanning is continuous and can be completed while the lance of the Robojet is moving at a constant rate along the drift axis.

Other modifications to the machine include a selector switch onboard the Spraymobile which can be switched to local or remote if the machine is to be operated using the tethered remote console, or from a surface location, respectively. All safety features (emergency stops, fire suppression, etc.) of the original system were maintained as an operator is currently present at the rear of the machine to load the shotcrete into the Spraymobile. A remote emergency stop and fire suppression switch were added to the surface console. As well, a self-detonating fire suppression system was installed to reduce damage should a fire break out during the time when the un-

Figure 7. New laser scanner assembly.

222

derground shotcrete mixer delivery truck operator is not at the Spraymobile, or in an area of the machine where the cameras are not pointed.

5 TELE-OPERATION CONTROL SYSTEM

The tele-operation control unit was designed to operate over the ethernet network at INCO's Research mine. The tele-operation package is a duplicate of the Robojet control console, visualization PC and the concrete pump controller, located in an operator station on surface. The mine network already has the infrastructure in place for machine control and video support, as well as the ability to support multiple machines over a single control ethernet network.

The Robojet Logica system is a combination of Interbus-S and serial control. A Motorola-based VME processor board, as well as the accompanying controller boards for the Interbus system, are the base components of the system. All of the original Robojet code was programmed using software specially designed for robot control.

To achieve the tele-operation solution for the spraying process, the software and hardware were modified to accept a new version of the control code for the robot. Additionally, since the Logica system was constantly evolving at Meyco Equipment in Switzerland, the newest software platform of the system in Switzerland had to be merged with the older software system of the Logica operating at INCO. After merging the two software versions, a successful test yielded the base software system for tele-operation platform. Figure 8 shows the test set-up using a hard-wired ethernet connection to the Robojet and Suprema pump.

The tele-operation control unit for the surface station is simply a replication of the operator pendant, visualization computer, and the Suprema concrete pump input menu panel. A schematic of the system is shown in Figure 9. The figure illustrates the control hardware onboard the Spraymobile on the right hand side of the page. Link "A" in the figure represents a connection between the Suprema pump and the VME computer. This connection was created to integrate the existing PLC and display of the Suprema with a touch screen panel on the operators console. This allows the operator to control concrete pump output and the dosing rate of the accelerator pump from the console. The original system only allowed the operator to turn the concrete pump and dosing pump on and off from the control-pendant. Any modification in the pumping rates needed to be changed at the panel, located at the back of the machine.

Figure 8. Testing the Robojet and Suprema system.

The tele-operation system will allow the surface operator to control all functions that are present on the original Robojet Logica system, as well as the ability to operate in a 'local' mode with the original remote control pendant plugged directly into the machine underground.

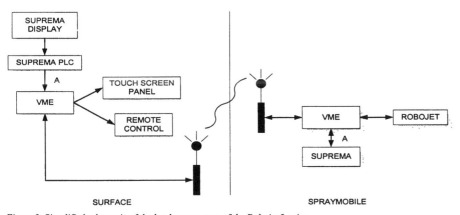

Figure 9. Simplified schematic of the hardware system of the Robojet Logica.

6 FUTURE WORK

At the time of writing of, INCO was preparing a list of specifications for the next generation computerized, tele-operated, sprayed ground support Spraymobile.

This second generation Spraymobile with the Logica technology is currently in the planning stage as one of INCO's producing mines, Stobie mine. This mine is a world leader in the application of steel fiber reinforced wet shotcrete (O'Hearn et al. 1998), and has expressed a desire to acquire the existing Logica Spraymobile from Mines Research to use in their production operations. Stobie currently uses tele-operation and other automation technology in their drilling and hauling aspects of the mining operation.

Stobie's desire to acquire the existing Spraymobile is driven by two reasons: to apply tele-operation technology to their ground control operation, and to reap the benefits of strict thickness control in their shotcrete application process, a benefit that is offered by the Logica technology.

Some key elements of the specification list for the second generation Logica Spraymobile include: diesel/hydraulic powerpack for all Robojet and concrete pumping, self contained (requiring no external connection to air or water services), integrated teleoperated tram guidance system (for remote tramming from workplace to workplace), integrated teleoperation system for spraying concrete or a polymeric thin ground support membrane, and hydraulically adjustable concrete hopper access.

The feasibility and attractiveness of two approaches are being evaluated. With one approach, the unit is more or less designed conventionally to spray material (concrete or polymer) that is fed to it at the workplace from a separate service vehicle. Alternatively, the unit is designed to carry all the material to be sprayed, negating the need for a separate service vehicle.

The goal of the second generation Logica Spraymobile is to automate the entire ground support process of the development cycle. The attractiveness of the second option above is that only one vehicle is required, reducing capital costs and the technical challenge of automatically indexing the service vehicle to the sprayer. Unfortunately, size constraints in terms of carrying sufficient quantities of concrete to spray a complete round are counter to the need to keep the unit as small as possible in view of the size of typical mine openings.

7 CONCLUSIONS

The tele-operation system was commissioned in August 2000, with system testing ongoing throughout 2000. The benefits of controlling the shotcrete spraying operation, and potentially the shotcrete loading function, from a remote location will demonstrate significant benefits in health, safety and productivity.

However, with any system that radically changes the way underground personnel control the process, their acceptance of this solution as a practical advance in the shotcreting process is critical to proceed with further developments in sprayed concrete equipment and process technology. A key part of this aspect has been INCO management's stalwart support of the introduction and integration process, and input on an ongoing basis from the Research Miners intimately involved with the project.

REFERENCES

Baiden, G.R. and M. Scoble 1992. Mine-Wide Information System Development at INCO Limited. *CIM Bulletin*, May, 85(160): 65-70.

Baiden, G. 1996. Future Robotic Mining at INCO Limited – The next 25 years. *CIM Bulletin*. January: 36-40.

O'Hearn, B., H. Buksa, and S. Walker 1998. Stobie Signals Shotcrete Success. *Engineering and Mining Journal*. August, 42-44.

Runciman, N., S. Espley and M. Rispin 1999. Testing and Evaluation of the Meyco Robojet Logica at INCO Limited. *Proceedings of the 5th International Symposium on Mine Mechanization and Automation (ISMMA)*. 14-16 June 1999: Sudbury, Ontario, Canada, 7 pages.

Tschumi, O. 1998. Automated Shotcrete Application. *Proceedings of the CIM Montreal 1998*. 3-7 May: Proceedings on CD media, paper number 095, 5 pages.

Shotcrete: Engineering Developments, Bernard (ed.) © 2001 Swets & Zeitlinger, Lisse, ISBN 90 5809 176 7

Classification of steel fibre reinforced shotcretes according to recent European standards

H.Schorn
University of Technology Dresden, Germany

ABSTRACT: Several years ago a European Standard organization (CEN) was established to coordinate the production of European Technical Standards. One of these standards deals with shotcrete, or in the terminology of the European standard – "sprayed concrete". The standard also covers steel fibre reinforcement, but not polymer or glass fibre reinforcement. Steel fibre reinforced sprayed concretes are classified not only according to compressive strength and environmental conditions as ordinary concretes are, but by residual strength classes combined with deformation classes determined from a particular bending test which does not appear to be useful for other type of fibre reinforcement.

1 EUROPEAN STANDARDIZATION

In the European community some years ago it was decided to harmonize important technical standards to secure a free trade between all countries of the European Union (EU). The aim was to create a uniform body of standards meeting modern needs applying throughout the single European market. This task is the responsibility of the Joint European Standards Institution, CEN/CENELEC and ETSI. CEN is an association comprising the national standard bodies of the European countries as named in Figure 1. For developing European Standards several Technical Committees (TC) are responsible for different technical items. In the case of shotcrete TC 104 "Concrete" established a Working Group (WG) 10, see Figure 2. In other cases sub-committees (SC) were established for coordinating several Working Groups. In WG 10 a group of experts and delegates from several European countries started the project. The German Standard Organization (DIN) served as a coordination centre. Working Group 10 produced a shotcrete standard in two parts and additional standards for particular testing methods:

Part 1 Definitions, Specifications, and Conformity
Part 2 Execution (Strengthing of ground, upgrading and repair, new structures, etc.)
Several sections for Testing Sprayed Concrete

The title of the standard is not "Shotcrete" as might be expected from commonly used terminology

Table 1. European standards organisations participating in CEN and their abbreviations.

Germany (DIN)	Sweden (SIS)
Greece (ELOT)	Switzerland (SNV)
Belgium (IBN/BIN)	Iceland (STRÍ)
Portugal (IPQ)	Italy (UNI)
Netherlands (NEN)	Spain (AENOR)
Ireland (NSAI)	France (AFNOR)
Norway (NSF)	United Kingdom (BSI)
Austria (ON)	Czech Republic (CSNI)
Luxembourg (SEE)	Denmark (DS)
Finland (SFS)	

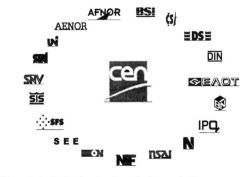

Figure 1. Insignia of national standards organisations.

in literature. The correct term, which is independent of present or former company names, is, in the opinion of European experts, "Sprayed Concrete". The standard regulates the wet-process as well as the dry-process. Present discussions in the committee have concluded, so the standard is ready to be sent to

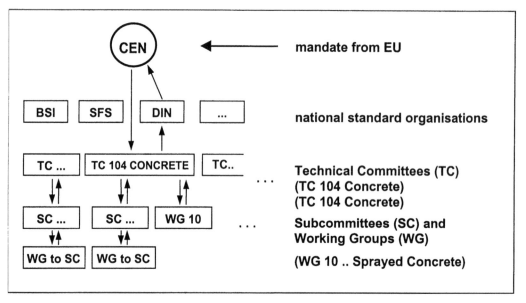

Figure 2. EN standard production, schematically. Sprayed Concrete (WG 10, TC 104, DIN) is an example.

TC 104 for a formal vote on acceptance by the delegates of all countries. At the present time an identification for the standard is not known.

2 CLASSIFICATION SYSTEMS

In the classification systems of the sprayed concrete standard, environmental classes as well as compressive strength classes are taken from the concrete standard EN 206 (2001). Those systems have been used successfully worldwide for a long time. But these regulations are not made for steel fibre reinforced concrete, therefore the European sprayed concrete standard must regulate this separately. For many purposes the typical deformation behaviour of steel fibre reinforced concrete is needed, especially in the production of a first shotcreted shell in tunneling. This typical deformation behavior is shown in Figure 3. The beam has been tested up to a large deformation. The crack bridging effect of the fibres transfers load over the crack, the beam exhibits large deformations due to increasing crack width. Using a very precise testing method allows the determination of energy dissipation.

Even in cases of beams without any fibre content the energy dissipated by the crack opening process in concrete can be measured, as represented by the curve *A* in Figure 3. The area under curve *A* represents the energy. Using commonly available fibres and a normal fibre dosage rate, a curve with a slope like *B* will be measured. The strength increases by a small amount. The load deformation curve *B* shows the energy dissipation effect of the crack bridging fibres, represented by the area between *B* and *A*. Lower fibre contents generate curves intermediate in capacity between *B* and *A*. Sometimes reference is made in the literature to curve like *C*, shown by the dotted line. Theoretically those curves may exist, but the required fibre content and aspect ratio (length/diameter) cannot be produced under the conditions of practice.

The quasi-ductility shown in the typical curve *B* in Figure 3 arises as a result of energy dissipation due to fibre deformation and fibre friction in the crack opening process. A simple model in Figure 4 may clarify this. A single fibre in the matrix has an angle θ to the load direction. The notches are only made to initiate the crack in the middle of the fibre length. With increasing crack width two effects of energy dissipation occur. First, the fibre deforms; the energy needed is the energy for plastic deformation of the crack bridging steel fibre. Second, the friction between matrix and fibre surface. The frictional surface decreases with increasing crack width, the energy dissipation according to fibre deformation must be independent of crack width. That explains the slope of the typical load deformation curve *B* in Figure 3. The decrease is only determined by decreasing fibre friction. According to Figure 4, an increase is physically impossible. If an increase is not possible, no "post-crack strength" can exist

Not all fibres act like this. For example, milled fibres have a very high tensile strength and an excellent bond to the matrix. However, the ductility of the fibre material itself is low. Those fibres are not suf-

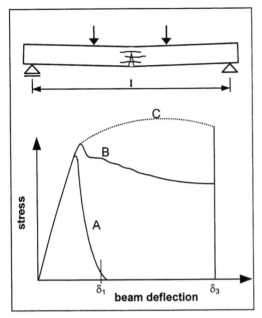

Figure 3. Bending stress-deformation curvs for steel fibre reinforced concrete compared to plain concrete.

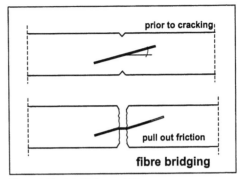

Figure 4. Crack bridging steel fibre.

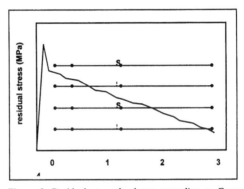

Figure 5. Residual strength classes according to European
Standard on Sprayed Concrete (2001).

ficient for quasi ductility of steel fibre reinforced
concrete as required for shotcreted shells in tunneling, but they are very useful in industrial floor application. The different types of fibres may be classified according to residual strength but at different
deformations, e.g. δ_1 or δ_3 in Figure 3. Brittle fibres
are classified according to a low deformation range.
Ductile fibres can be classified according to higher
deformation ranges.

All types of fibre reinforced concretes can be
classified by residual strength classes according to
deformation classes. According to the European
Standard for Sprayed Concrete (2001): "The purpose
of the deformation classes is to give flexibility to the
designers in the choice of deformation required of
the sprayed concrete under severe conditions. For
the purpose of design the deflection limit for each
deformation level can be considered in terms of the
equivalent angular rotation (e.g. for a beam of
450mm x 125mm x 75mm). Correspondingly, four
residual strength levels S1 – S4 are defined which in
combination with applicable deformation range D
can be specified in terms of residual strength class".

Table 2. Deformation classes.

Deformation Class	Corresponding Angular Deformation
D1	1/250
D2	1/125
D3	1/56

An illustrative example is given in Figure 5 for a
typical fibre reinforced sprayed concrete beam, this
beam fulfils the requirement for residual strength
class D1S3 (as well as D2S2 and D2S1).

The bending test allows a classification independent of fibre type, content, aspect ratio, relation between maximum aggregate size and fibre length, and
other parameters. If fibres are effective in steel fibre
reinforced concrete, a residual strength class according to a deformation range can be determined. If
fibres are not effective the class is zero in deformation class as well as in residual strength class. The
idea was to search for a criteria for performance of
steel fibre reinforced concrete independent of compressive or tensile strength, or of concrete mix parameters found in other technical papers. Different
points of view are evident, for example, in German
guidelines (DBV 2001) and in the state of discussion
at a recent international RILEM committee (RILEM
2000). Because no regulations for steel fibre reinforced concrete exist in national German standards,
German guidelines (DBV 1996, 2001) were developed which cover all types of steel fibre reinforced
concrete, not only sprayed concrete.

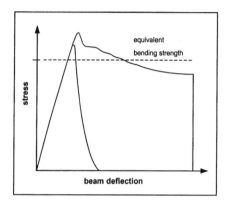

Figure 6. Equivalent bending strength

For particular structural purposes design methods will be described in this German Guideline. That is the reason for the use of a material charactcristic which seems more suitable for those methods. Figure 6 shows the same stress deformation diagram as in Figure 3.

The area under the curve in Figure 6 represents the dissipated energy related to deformation, the grey area represents that part of energy dissipation which is caused by crack bridging fibre action. Contrary to the European sprayed concrete standard, the equivalent bending strength is the required criertion, not the residual strength. This difference between standard and guidelines is unfortunate, but on the other hand, only the proposed German guidelines for design methods need this value. This value of equivalent bending strength is principally determined as the average of the measured stress values in that part of the curve that is affected by fibre bridging action. In other words, the equivalent bending strength is determined by transformation of the area of energy dissipation into a shape with a horizontal upper line.

The advantages of this approach are clear. Equivalent bending strength is less variable than residual strength estimates. As a consequence, the classification can be made in smaller ranges than classifications based on residual strength. Table 1 shows a comparison of these two types of classifications.

The deformation class depends on the maximum beam deflection in the test. The European Sprayed Concrete Standard uses three levels of deformation, compared to the German guidelines which use two "deformation classes". All levels of steel fibre concrete classifications depend strongly on deformation levels. That must be taken in account in each application for this material.

Another material property of concrete was proposed as a basis for assessment by RILEM TC 89 (1990). In this test, the specimen is a beam with a notch in the tensile zone. The evaluation of performance is made according to fracture mechanics by fracture parameters K^S_{lc} and $CTOD_c$. Generally, the quality of crack bridging capacity of fibres in concrete can be determined by these parameters. This proposal was not preferred in practice, because a structural design method for steel fibre reinforced concrete based on these values is not yet existent in guidelines or standards.

Presently in discussion is the use of a notched beam again, but not for determining fracture behavior according to fracture mechanics theory. The aim is to produce very low variability as a consequence of notching the beam. The evaluation of equivalent bending strength results in lower variability compared to residual strength. But notching makes it impossible to find a true value of strength of the material in bending test because the main crack must start from the notch, independent from the weakest zone in the body.

3 SAMPLING

Samples for testing shotcrete are usually taken from a shotcreted layer, either by drilling cores or by cutting a beam out of the layer. In the beginning of a shotcreting process fibres must be orientated parallel to the shotcreted support. With increasing thickness

Table 1. Comparison of residual strength classes and equivalent bending strength classes.

Residual strength class by European Standard (2001)	Residual stress (MPa)	Performance class according to guidelines (DBV 2001)	Equivalent bending strength (MPa)
		0	< 1.0
S1	> 1	1	> 1.0
		1.5	> 1.5
S2	> 2	2	> 2.0
		2.5	> 2.5
S3	> 3	3	> 3.0
S4	> 4		

of the layer produced in a sprayed concrete process the degree of fibre orientation will decrease.

In compression tests a very small effect due to fibre orientation occurs. In bending tests, however, the parallel orientation of fibres leads to an increase in bending strength as well as to an increase of crack bridging effect. The residual strength class as well as the performance class evaluated from equivalent bending strength is therefore sensitive to fibre orientation.

In the future it may become apparent that the equivalent bending strength is a better approach to material classification, or the use of elements of fracture mechanics may be a better classification system. But at present time it is necessary to get more experience in testing and in quality control of steel fibre reinforced sprayed concrete to be certain of this.

4 CONCLUSIONS

Shotcrete and Steel Fibre Reinforced Shotcrete (SFRS) are classified according to compressive strength and environmental classes in the EN 206 "Concrete". For SFRS, additional classification criteria are needed, these being based on deformation classes and residual strength classes.

Residual strength is measured using the well known bending test. The post-crack behaviour affects the residual strength according to the deformation determined by the deformation class. The present classification classes seems to fit all applications of SFRS, these being tunneling and strengthening of ground, repair as well as upgrading or new building constructions.

Presently under discussion is the substitution of residual strength class by the concept of equivalent bending strength class (performance class) according to guidelines promoted by the German Concrete Association. This procedure has some important advantages. In contrast to residual strength measurement, the post-crack slope of the load deformation curve is evaluated by this method. The effectiveness of steel fibres in shotcrete can therefore be seen more clearly.

In future we can expect the parallel use of both residual as well as equivalent bending strength classes in the different international standards. The present European Standard has found a sufficient compromise between experiences in practice and new methods for classifying and quality control of steel fibre reinforced sprayed concrete.

European Standard EN 206: *Concrete; Performance, production, placing and compliance criteria.* (publication expected in 2001)
Deutscher Beton-Verein (DBV) 1996. *Technologie des Stahlfaserbetons und des Stahlfaserspritzbetons.* (German Concrete Association: *Technology of steel fibre reinforced concrete and steel fibre reinforced shotcrete.* In German)
Deutscher Beton- und Bautechnikverein (DBV): *Merkblatt Stahlfaserbeton* (publication expected in 2001). (German Concrete Association: *Guidelines on fibre reinforced concrete and fibre reinforced shotcrete*).
RILEM (Technical committee TC 162-TDF): 2000 *Test design and methods for steel fibre reinforced concrete.* Materials and Structures, 33 (225):3-5.
RILEM Draft Recommendations 1990. Technical committee TC 89-FMT: *Determination of fracture parameters (K^S_{lc} and $CTOD_c$) of plain concrete using three-point bend tests.* Materials and Structures, 23 (138): 457–460.

REFERENCES

European Standard *Sprayed Concrete* (publication expected in 2001)

Shotcrete: Engineering Developments, Bernard (ed.) © 2001 Swets & Zeitlinger, Lisse, ISBN 90 5809 176 7

Developments and applications of high performance polymer fibres in shotcrete

P.C.Tatnall
SI® Concrete Systems, Marietta, Georgia, USA

J.Brooks
Master Builders, Inc., Cleveland, Ohio, USA

ABSTRACT: High Performance Polymer fibres for reinforcing shotcrete are a relatively new development in concrete reinforcing fibre technology. This paper traces the development of these fibres types, and presents a review of fibre properties. The testing methods developed to characterize their behavior in shotcrete are discussed. The performance of various High Performance Polymer fibre types, shapes and lengths in shotcrete are discussed relative to the testing methodology employed to describe them. Recent case studies of applications of wet-process shotcrete for ground support are presented and application and specification issues, based on the performance of these fibres, are presented.

1 INTRODUCTION

High Performance Polymer (HPP) fibres for reinforcing concrete are a relatively new development in fibre reinforced concrete and shotcrete (FRS) technology. During the early 1960's in the United States, the first major investigation was made to evaluate the potential of discrete steel fibres as a reinforcement for concrete (Romualdi & Batson, 1963). In the early 1980's, large scale development activities began with fine denier synthetic fibres in the 6 to 60 denier range (ACI 544.1R-96). In mid 1990 new generations of synthetic fibres were developed with a much higher denier in the 3000 to 6000 range, or a much lower aspect ratio (length/equivalent diameter) than the previously used synthetic fibres. These newer synthetic fibres have the look, feel, and sometimes the performance of steel fibres for reinforcing concrete.

In the US the first practical application of steel fibre reinforced shotcrete (SFRS) was in 1972 for slope stabilization and tunnel lining at Ririe Dam for the US Army Corps of Engineers using the dry-process. In Canada the first project was in 1977 for a slope stabilization project in Burnaby, British Columbia (Morgan & Heere, 2000). Since these early trials the use of SFRS for ground support applications has grown into a mature technology with many hundreds of thousands of cubic metres of SFRS being used around the world in ground control applications.

Low denier synthetic fibres started to be used in wet-process shotcrete in the early 1980s. At lower volumes (0.1 to 0.2 volume percent, 0.9 to 1.8 kg/m^3) for early age shrinkage crack control, and at higher dosages (0.4 to 0.7 volume percent, 4 to 6 kg/m^3) for rock slope stabilization, channel linings, and small tunnel linings.

A more detailed review of the development, properties, testing procedures and applications of the newer HPP-type fibres as used in shotcrete for ground support follows.

2 DEVELOPMENT

As the application of steel fibre reinforced concrete and shotcrete grew throughout the world in the early 1990s, investigators began looking into developing fibres of similar size and shape to steel fibres in use, but made from polymer materials. Companies such as SI® Concrete Systems, 3M Corporation, and Dow Corning Company in the US, and investigators at Dalhousie University in Nova Scotia, Canada, developed stiff polymer fibres using basic polyolefin technology, i.e., fibres made from polypropylene and co-polymers of polypropylene and polyester. These fibres were generally extruded into circular cross-sectional shapes, and are produced with diameters of about 0.35 mm to 1 mm, and lengths of about 30 to 50 mm. Various deformations and shapes are still being tried to optimize the fibre pullout resistance and the post-crack behavior of the FRS.

This paper will focus on the properties, performance and applications of one of these fibres, know as S-152 HPP, hereinafter denoted as HPP, being developed by SI Concrete Systems.

Table 1. Selected properties of S-152 HPP fibres

Property	Value
Material	Virgin Polypropylene
Specific Gravity	0.91
Tensile Strength	320 MPa
Young's Modulus	3500 MPa
Ultimate Elongation	15%
Water Absorption	Nil
Melting Temperature	175°C
Ignition Temperature	360°C
Thermal Conductivity	0.2 W/mK @ 20°C
Electrical Conductivity	Very Low
Acid/Alkali Resistance	Very High
Denier	6000
Nominal Diameter	0.9 mm
Lengths	30, 40, and 50 mm
Aspect Ratios	33, 44, and 55

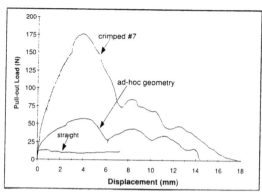

Figure 1. Single HPP fibre pullout response for three different geometries (Dubey, 1997).

3 PROPERTIES

Some of the material properties of HPP fibres for shotcrete are shown in Table 1. These properties are typical for extruded polypropylene fibres.

Note that the elongation under load for these fibres is quite large compared to steel fibres, with a modulus of elasticity about one-tenth that of hardened shotcrete. This implies that the HPP-type fibres must "stretch" a little before picking up tensile loads in a concrete composite.

An important characteristic of polypropylene is its very high resistance to acid and alkali environments. This feature provides high resistance to corrosion of the reinforcing as compared to convention steel reinforcing elements.

Another important fibre property is its bond to the concrete matrix. For these relatively large polypropylene fibres, bond is determined primarily by mechanical deformations. Figure 1 shows single fibre pullout loads versus displacement for three different fibre configurations tested at the University of British Columbia in the early development stage. Note the much higher energy required to pull out the "crimped #7" fibre versus the straight fibre of similar length and diameter.

4 TESTING PROCEDURES

The next step in the development of the HPP fibre was to demonstrate their ability to perform in concrete and shotcrete. Other fibre types, primarily steel, for shotcrete ground support applications, have used flexural beam tests such as ASTM C 1018 (ASTM 1997), JSCE SF-4 (JSCE 1984), or EF-NARC beams (EFNARC 1996) to characterize the ability of the fibres to provide post-cracking load carrying capacity or energy absorption of the composite materials. More recently panels have been used to demonstrate the ability of FRS structures to carry load and absorb energy after cracking (EF-NARC 1996, Bernard 2000).

4.1 Beam Tests

As noted above, HPP fibres must stretch before contributing to the post-crack load capability. This means that cracks will open wider than cracks in similar shotcrete with steel fibres. The use of ASTM C 1018 or JSCE SF-4 flexural beams, then, cannot adequately demonstrate the full capability of HPP-type fibres in shotcrete because the end-point deflection in these two test methods is 2 mm, and the crack opening is too small to fully characterize the post-cracking load capacity.

Figure 2 shows a typical load-deflection diagram using the ASTM C 1018 test method for shotcrete containing 1 volume percent (9 kg/m³) HPP fibres. Note how the load drops dramatically after the peak, corresponding to cracking of the matrix – to about 28 percent of the peak load, and then starts to pick up load, showing strain hardening behavior after a total central deflection of the specimen of 0.3 mm.

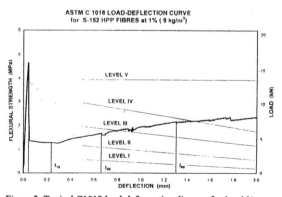

Figure 2. Typical C1018 load-deformation diagram for 1 vol.% HPP fibre reinforced shotcrete (Morgan, 1998)

The C 1018 toughness parameters for this specimen show fairly low values, $I_{60} = 20.1$ and $R_{30,60} = 37$. This means that between deflections of 0.6 and 1.2 mm the beam is carrying 37 percent of the peak load. Likewise the JSCE SF-4 Equivalent Strength is 1.91 MPa, 40 percent of the peak strength over the full range of deflection from 0 to 2 mm.

It is obvious from Figure 2 that the composite material is increasing its load capacity as the beam deflection is increased, or as the crack opens. The small end-point limits for these two test methods cannot fully describe the capability of this material when used to support ground exhibiting large deformations, and other test methods must be used to characterize this composite.

4.2 Panel Tests

In many parts of the world panel test setups are used to ensure that FRS materials meet specifications for ground support functions. The square, fully supported EFNARC panel (EFNARC 1996) has been used extensively in Europe, and although the reliability of this method versus beam testing is greater, it still has a fairly high coefficient of variation (Bernard 2000). This fact has led to the development of the Round Determinate Panel (RDP) test method (Bernard & Pircher 2000) with coefficients of variation of about 6%, exceeding the reliability of any other published procedure for the characterization of post-crack performance for fibre reinforced concrete. This test method is in the process of standardization by ASTM Committee C9 on Concrete and Aggregates.

Because these two methods allow for much larger specimens and much larger deformations (EFNARC panels to 25 mm, RDP panels to 40 mm) development of HPP fibres for shotcrete has utilized these procedures.

5 PERFORMANCE

Figure 3 shows a typical load-deflection curve for an EFNARC panel specimen tested at 28 days maturity made with shotcrete containing 1 volume percent (9 kg/m³) HPP fibres, 50 mm long. This test procedure normally requires an end-point deflection of 25 mm.

Integrating the data in Figure 3, as shown in Figure 4, we can show the cumulative energy absorption of the panel with increasing deflection. The EFNARC specification requires reporting of the total energy absorption of the panel at 25 mm deflection. In this instance the panel absorbed approximately 1100 Joules. Many performance specifications for FRS tunnel lining projects require EFNARC panel results between 500 and 1000 Joules.

Figure 5 illustrates both EFNARC panel results and RDP results for increasing HPP fibre length and dosage. The 50 mm long HPP fibre substantially out-performs the 30 mm HPP as can be expected. Also shown are panel results for two combinations of HPP and Novotex steel fibres. These blends show promise for combining the performance features of both types of fibres, but these features are beyond the scope of this paper. Figure 5 also shows the relative values attained between the EFNARC procedure and the newer RDP method.

One advantage of the panel-type tests versus beam tests is that one can also determine the energy absorbed using conventional reinforcing, for example welded wire fabric or mesh, as is used in shotcrete for ground support applications in mining. Figure 6 illustrates two different mesh configurations typically used in shotcrete linings as compared to 30 mm and 50 mm HPP tested using the EFNARC panel. This data shows that 9 kg/m³ 50mm long HPP provides the same performance in terms of total energy absorbed at 25 mm deflection as the

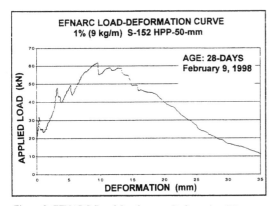

Figure 3. EFNARC Panel Load versus Deformation (Morgan, 1998)

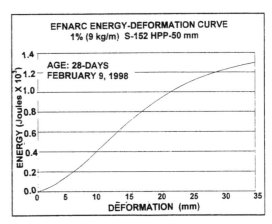

Figure 4. EFNARC Panel test result showing energy versus deformation (Morgan, 1998)

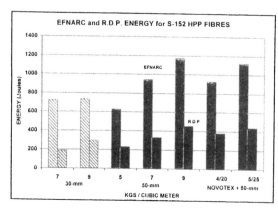

Figure 5. EFNARC and Round Determinate Panel Energy for HPP Fibres (Bernard 1998).

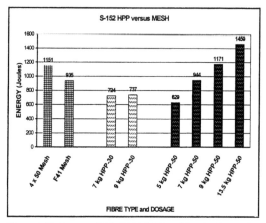

Figure 6. Mesh versus HPP, EFNARC Panel Results (Bernard, 1999, 2000)

heaviest mesh reinforcement, although the 4 by 50 mesh failed by delamination.

6 APPLICATIONS

6.1 *Lake Mead No. 2 Intake Shafts and Tunnel*

This project, completed in 1999 in Western USA, consisted of drill and blast intake and access shafts, intake tunnel, forebay and surge tunnels. Ground Support was with rock bolts and 100 mm of HPP fibre reinforced shotcrete for all tunnels. The HPP 30 mm long fibres were added to volumetrically batched concrete with an automatic dispensing system tied to the batching equipment. Shotcrete was dropped from the surface through a slick line and pumped up to 470 m. The mixture included 476 kg/m³ Type II/IV cement with 50 kg/m³ silica fume and 3.6 kg/m³ HPP fibres. Aggregates were blended

to meet Grading No. 2 of ASTM C 1436, "Specification for Materials for Shotcrete" (ASTM 1999). Master Builders' Polyheed SG was used as a water-reducing admixture, and Meyco SA 430 was used as required for acceleration.

Testing results showed an average of 35 MPa compressive strength, and 500 J on EFNARC panels. Shotcrete was dropped 115 m through the 150 mm diameter slick line to a hopper. Shotcrete was then discharged to the pump, which moved the shotcrete through 75 mm diameter lines up to 470 m to a robotic nozzle. A total of 1500 cubic metres of FRS were used on this project (Kimball & Galinat 1999).

6.2 *West Frankfort Storm Drainage Tunnel*

This combined sewer overflow project, completed in 2001, is a 1300 m long, 3.65 m diameter TBM driven tunnel in limestone in Kentucky, USA. A 125 mm shotcrete lining containing 9 kg/m³ HPP 30 mm fibre was constructed to provide impact and abrasion resistance from sand and large debris in storm run-off.

Shotcrete was supplied by ready mix trucks to the access shafts where the fibres were added and mixed at mixer manufacturer's recommended mixing speed of 10 to 12 RPM for a minimum of 40 revolutions. The mixture proportions were as shown in Table 2. Shotcrete was then pumped down 12 m and horizontally up to 220 m to handheld nozzles.

The available aggregates had a deficiency of fine materials as recommended in ASTM C 1436 (ASTM 1999), as shown in Table 3. The contractor had difficulty pumping the shotcrete when more than 3.5 kg/m³ HPP fibres were added. The fibres tended to cause plugs in the lines as the paste was pushed away from the fibres.

Table 2. Frankfort mixture proportions.

Ingredient	kg/m³
T-I Cement	390
Class F Fly Ash	89
Silica fume	30
9-mm aggregate	475
Sand	1278
Water	292
S-152 HPP-30 mm	9
Rheobuild 1000	As required

Table 3. Frankfort Aggregate Grading, percentage passing.

Sieve Size	Grading (%)	ASTM C 1436 (%)
12.5 mm	100	100
9.5 mm	94.5	90-100
4.75 mm	71.2	70-85
2.36 mm	62.5	50-70
1.18 mm	51.3	35-55
600 um	28.7	20-35
300 um	6.0	8-20
150 um	0.9	2-10
75 um	0.6	n/a

An air-entraining admixture was added to the mixture to achieve an 8 to 10% air content at the pump. Adding these air bubbles provided the "fines" that were lacking in the sand and eliminated all pumping problems for HPP fibre dosage of 9 kg/m³ as specified. It has been shown that the high air content is forced out of the shotcrete as it is compacted on the receiving surface (Morgan, 1999). The resulting air content in the shotcrete is seldom more than 4 percent, and thus only slightly effects the compressive or flexural strength of the shotcrete, while providing improved freezing and thawing durability. This project used over 2000 cubic metres of FRS containing the S-152 HPP fibre.

6.3 Nevada Mines

HPP fibres have been used for the past three years for ground support in gold mines in NE Nevada in the USA. These deep mines are using this type of fibre to accommodate the large ground movements encountered, in some cases together with conventional mesh reinforcement. Typically fibre dosage has been approximately 0.5 volume percent (4.5 kg/m³).

7 CONCLUSIONS

The development and use of new High Performance Polymer fibres in shotcrete for ground support applications has moved from the experimental stage to full application. New test methods have been developed to fully characterize their performance in shotcrete. In many applications where large deformations of the shotcrete is expected, HPP fibres can provide load capacity after cracking occurs that matches or sometimes exceeds that of conventional steel reinforcement, or steel fibre reinforcement. HPP have the additional advantage of being very user-friendly and highly resistant to aggressive environments.

REFERENCES

ACI 544.1R-96, 1996, Committee Report on Fiber Reinforced Concrete, American Concrete Institute, Farmington Hills, Michigan, 66 pp.
ASTM, 1997, C 1018-97, Standard Test Method for Flexural Toughness and First-Crack Strength of Fiber-Reinforced Concrete (Using Beam with Third-Point Loading), 1999 Annual Book of Standards, Vol. 04.02, ASTM, West Conshohocken, PA.
ASTM, 1999, C 1436-99, Standard Specification for Materials for Shotcrete, 2000 Annual Book of Standards, Vol. 04.02, ASTM, West Conshohocken, PA.
Bernard, E. S., 2000. Round Determinate Panel Testing in Australia, Shotcrete, Amercian Shotcrete Association. 2 (2): 12-15.
Bernard, E.S. & Pircher, M. 2000. Influence of Geometry on the Performance of Round Determinate Panels made with Fiber-reinforced Concrete, Civil Engineering Report CE10, School of Civic Engineering and Environment, University of Western Sydney, Nepean.
Dubey, A. 1997, Some Theoretical and Experimental Advances in the Field of Deformed-Fiber Reinforced Concrete Engineering, University of British Columbia, January.
EFNARC, 1996. European Specification for Sprayed Concrete, European Federation of National Associations of Specialist Contractors and Material Suppliers for the Construction Industry (EFNARC), 30 pp.
JSCE 1984, Japanese Society of Civil Engineers, Method of Test for Flexural Strength and Flexural Toughness of SFRS, Standard JSCE SF-4.
Kimball, R.W. and Galinat, M.A., 1999. A Synthetic Spray Solution, Concrete Engineering International, 3 (5): 14-16.
Morgan, D.R. and Heere, R., 2000. Evolution of Fiber Reinforced Shotcrete, Shotcrete, Amercian Shotcrete Association. 2 (2): 8-10.
Morgan, D.R., Heere, R, McAskill, N. and Chan, C., 1999. System Ductility of Mesh and Synthetic Fibre Reinforced Shotcretes, Proceedings, Third International Symposium on Sprayed Concrete, Gol, Norway, September 26-29
Morgan, D.R. and Heere, R, 1998. Evaluation of Synthetic Industries S-152 HPP 50 mm Polypropylene Fiber in Wet-Mix Shotcrete, Report VA04243, March 16, 25 pp.
Romualdi, J. P., & Batson, G. B. 1963. Mechanics of Crack Arrest in Concrete, Proceedings, ASCE. 89 (3): 147-168.

Shotcrete: Engineering Developments, Bernard (ed.) © 2001 Swets & Zeitlinger, Lisse, ISBN 90 5809 176 7

The role of constitutive models in the analysis of shotcrete-based ground support systems

A.H.Thomas & C.R.I.Clayton
University of Southampton, Southampton, UK

P.Norris
Mott MacDonald Ltd, Croydon, UK

ABSTRACT: Modelling of the shotcrete in a tunnel lining is critical to the assessment of the safety of an advancing tunnel face. However, this area still presents engineers with considerable difficulty, often because of the limitations of modelling of the shotcrete. Given the characteristics of sprayed concrete and its behaviour, it is clear that simple linear elastic constitutive models are not adequate for the accurate prediction of the stress and strain distribution in a tunnel lining. A programme of 2D and 3D numerical analyses is underway to investigate how this modelling could be improved.

1 INTRODUCTION

Compared with a segmental lining, which is built in a single construction stage from preformed units of mature concrete, a sprayed concrete lined (SCL) tunnel has a more complex early life, in that it is loaded while it is still young. This complicates the already difficult task of predicting stresses and strains in tunnel linings. Having outlined the main differences in material properties between sprayed concrete and conventionally placed concrete, this paper will review the current practice in numerical analysis, in the context of SCL tunnels in soft ground. It is recognised that the role of shotcrete is different in rock tunnels, where questions of block stability or large tectonic stresses predominate (Rabcewicz 1969).

As part of a recent BRITE-EURAM research project, a set of large-scale laboratory tests on shotcrete rings was performed. Using these laboratory tests as a case study, a series of 2D and 3D numerical analyses has been performed to investigate the role of constitutive models in the analysis of SCL tunnels. This paper discusses the initial results.

The 4½ year long BRITE-EURAM project - BRE-CT92-0231 "New Materials, Design and Construction Techniques for Underground Structures in Soft Rock and Clay Media" – was part funded by the Commission of the European Communities and led by Mott MacDonald. The wide-ranging project sought to understand the practical and design-related issues, which need to be addressed if confidence in sprayed concrete as a permanent structural material is to be developed (Norris & Powell 1999).

2 INFLUENCES ON SPRAYED CONCRETE

The methods of construction of SCL tunnels and of placement of sprayed concrete require a different composition of the concrete and impart different characteristics to the lining as a whole, compared to conventionally placed concrete.

The composition of the concrete is tailored so that: it can be conveyed to the nozzle and sprayed with a minimum of effort; it will adhere to the excavated surface, support its own weight and the ground loading as it develops; it attains the strength and durability requirements for its purpose in the medium to long-term. In general this leads to a mix (in both the wet or dry process), which has more sand, a higher cement content, smaller sized aggregate and more additives compared with conventionally cast concrete (see Table 1). Furthermore the water-cement ratio is relatively high. In general, the consequences of this are: a faster development of strength and other properties with age and more pronounced creep and shrinkage (see Table 2). The higher water-cement ratios also lead to lower ultimate strengths. In addition to inelastic strains due to creep or shrinkage, the increase in stiffness with age leads to anelastic strains, where the strain recovered on unloading is less than the strain induced when first loading because the stiffness of the material has increased (see Figure 1). More detailed discussions of the composition of sprayed concrete can be found in Malmberg (1993) and Brooks (1999).

The influence of the spraying process depends on whether the dry or wet mix process is used, since increasingly more automated equipment is being used with the wet mix process and therefore the nozzle-man has less influence. However, in both methods, there is evidence to suggest that, because the lining is formed by a series of layers, it is anisotropic. The addition of reinforcement (either in the form of mesh or fibres) increases this anisotropy.

The sequential construction method for SCL tunnels suggests that the mode of action of the lining changes with time (from a cantilever/arch in the top-heading to a completed ring, mainly in compression) and that the lining, which is usually considered as a monolithic shell, actually consists of a series of sections. If one considers the top-heading, bench and invert of a typical tunnel, one can see that there may be a considerable difference in age between adjacent sections. This may lead to differential strains due to the difference in shrinkage and stiffness.

Traditionally, sprayed concrete has been regarded as being of inferior quality to conventionally cast concrete, because, inter alia, of the lack of curing, possible poor workmanship at the numerous construction joints, the influence of the action of spraying, "shadowing" around bar reinforcement, and the variation in lining thickness and shape. While variation in shape is routinely taken into account in the design of segmental linings (eg: through checks on ovalisation), this and the variation in lining thickness are not normally considered in the case of SCL tunnels. Typically, the standard deviation in compressive strengths at 28 days may be 5 MPa for a 25 MPa primary lining wet mix (Bonapace 1997) or 10 MPa for a permanent lining SFRS wet mix with a mean of 61 MPa (Ripley et al. 1998). Typical values for conventional concrete are about 4 MPa (Neville 1995). That said, recent technological advances in spraying equipment and mix design are continuing to reduce the effect of these detrimental influences.

3 CONSTITUTIVE MODELS

In recent years numerous researchers have investigated the role of the constitutive model of the ground in numerical analysis of tunnels (e.g. Van der Berg 1999). Taking clays as an example, the effects of spatial variation of properties, the variation in stiffness from small to large strains (strain-softening behaviour) and plasticity, are now incorporated in the geotechnical model for the analysis as the norm. In contrast, relatively crude models are normally used for sprayed concrete. Consideration of the behaviour of sprayed concrete would suggest that a comprehensive model would incorporate age-dependent parameters, the non-linear stress-strain curve, creep, and possibly shrinkage.

However, the most commonly used model is a linear elastic one, because of its simplicity and computational efficiency. Typically, elastic models predict axial forces and bending moments in linings,

Table 1. Typical mix

Material	High quality wet-mix shotcrete *	Cast in situ concrete (Neville 1995)
Grade	C40	C40
Water/cement ratio	0.43	0.40
Cement inc. PFA, etc.	430 kg/m^3	375 kg/m^3
Accelerator	4 %	-
Plasticiser	1.6 %	1.5%
Stabiliser	0.7 %	-
Micro-silica	60 kg/m^3	-
Max aggregate size	10 mm	30 mm
Aggregate < 0.6 mm	30 - 55 %	32 %

* Data from the Heathrow Express project (Darby & Leggett 1997).

Table 2. Typical properties

Property	High quality sprayed concrete*	Cast insitu concrete (Neville 1995)
Uniaxial compressive strength after 1 day	20 MPa	6 MPa (est.)
Uniaxial compressive strength after 28 days	59 MPa	44 MPa
Elastic modulus after 28 days	34 GPa	31 GPa (est.)
Shrinkage after 100 days	0.1 – 0.12 %	0.03 – 0.08 %
Specific creep after 160 days	0.01 - 0.06 %/MPa	0.008 %/MPa
Density (kg/m^3)	2140 – 2235	2200 – 2600
Total porosity	15 – 20 %	15 – 19 %
Permeability (m/s)	10^{-12} to 10^{-14}	10^{-11} to 10^{-12}

Data from the Heathrow Express project (Darby & Leggett 1997).

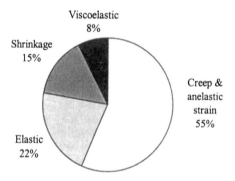

Figure 1. Composition of strains after 240 hours for SFRS (Ding 1998).

which are generally believed to be unrealistically high. The development of the elastic modulus with the age of the sprayed concrete has been incorporated into linear elastic models, resulting in considerable reductions in the predicted bending moments (of up to 50%) along with smaller increases in deformation and reductions in axial forces in the lining (Soliman et al. 1994). Pottler (1985) proposed the use of a Hypothetical Effective Modulus (HME), which is smaller than the actual elastic modulus to allow for creep and shrinkage. While this has been a widely used "short-cut", there is essentially no theoretical basis behind the choice of the values of the HME.

Non-linear stress-strain behaviour under both compression and tension starts at low stress levels due to micro-cracking (Figure 2). Fundamentally this is a plastic process but the irreversibility of the deformation is only of concern if unloading occurs. In tunnels this only occurs to any large extent in special cases, such as when junctions are formed. A relatively simple mathematical formulation of a non-linear elastic model (and other models) can be used because of the largely biaxial stress state in a tunnel lining (Meschke 1996).

However, to account for the non-linearity of the stress-strain behaviour, most researchers have tended to use a strain-hardening Drucker-Prager plasticity model with age-dependent parameters (e.g. Meschke 1996, Hellmich et al. 1999). One potential drawback of this is that the yield surface is circular in the deviatoric plane whereas the shape of the Mohr-Coulomb surface reflects better the behaviour of concrete at low mean stresses (see Figure 8 later). Typically such plasticity models predict an increase of 15 - 30% in the magnitude of deformations and a reduction of 10 - 25% in the magnitude of bending moments, compared to an age-dependent elastic model (Hafez 1995).

Creep of sprayed concrete has long been regarded as a likely mechanism for easing stress concentrations in linings (Rabcewicz 1969). Considerable efforts have been made to investigate this aspect of the behaviour of sprayed concrete. The models proposed include various rheological models, power laws and a modified Rate of Flow method (Aldrian 1991). The parameters for these models are determined by curve-fitting the model to data from uniaxial creep tests. However, further work on the relative importance of creep compared to other aspects of material behaviour remains to be done.

Differential strains due to thermal expansion during initial hydration and shrinkage can lead to cracking, which is of considerable concern if the sprayed concrete forms part of the permanent lining. With the exception of the thermo-chemomechanical model of Hellmich et al. (1999), and the Rate of Flow method, current models ignore this aspect of behaviour, on the grounds that these strains are much smaller than those induced by the ground loading. More sophisticated models of tensile behaviour usually adopt a smeared crack approach with exponential or linear softening for unreinforced shotcrete (e.g. Moussa 1993). The tension-stiffening effect of reinforcement is rarely incorporated.

Considering the importance of other factors, anisotropy of the shotcrete itself can probably be neglected since the variation in stiffness (measured perpendicular and parallel to the direction of spraying) is only about 25% (Aldrian 1991). Damage due to early loading appears to occur only at high stresses (greater than 70% of peak strength) (Moussa 1993). The questions of variation in shotcrete quality, thickness, and cross-sectional shape have not been addressed adequately and may have more influence on lining stresses and strains than some of the nuances of material behaviour.

Finally, this review has assumed that, with increases in computing power, tunnels will be analysed with 3D models as the norm in the near future. At the moment this is rarely done. Hence, the empirical methods to account for 3D effects in 2D analyses have not been discussed.

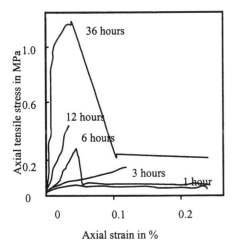

Figure 2. Uniaxial tensile stress-strain behaviour at different ages for SFRS (Mott MacDonald 1998)

4 NUMERICAL MODELLING

In the final part of a recent BRITE-EURAM research project on sprayed concrete, large-scale load tests were performed on three 2.5 m internal diameter steel fibre reinforced shotcrete rings, with a thickness to diameter ratio of 1:17. The ring was surrounded by a 280 mm thick annulus of remoulded clay, similar to London clay, and loaded by a circumferential arrangement of jacks. The loads were applied incrementally in 4 steps during the first 260 hours after spraying the concrete and then unloaded in one step (Figure 3). The loading was intended to replicate a typical ground load at a depth of about 30 m, which increased with time and in which the ratio of "vertical" to "horizontal" stress was 0.85. Typically, SCL tunnels in London have thickness to diameter ratios of between 1:20 and 1:27 and are situated 15 to 30 m below ground level. Further details

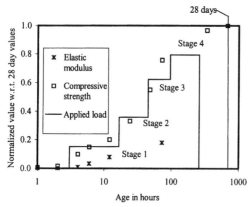

Figure 3. Variation of compressive strength and modulus (normalised w.r.t 28 day values) and the applied load (normalised w.r.t. full overburden pressure) with time.

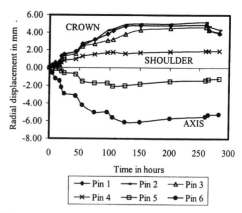

Figure 5. Radial displacements of pins 1 to 6, measured during the BRITE EURAM load test – inward movement is positive.

on the BRITE EURAM project's tests, including the details of SFRS mix , can be found in Norris & Powell (1999).

One quadrant of this ring has been analysed using the FLAC 2D and 3D finite difference programs at the University of Southampton. These analyses are the first stage of research into the modelling of SCL tunnels in soft ground such as London clay. Because of the difficulties in performing the tests on sprayed concrete at early ages, a close agreement between the analysis and recorded data was not expected but it was hoped that a qualitative assessment would provide some insight into the material behaviour during this period.

Having trialled meshes of various numbers of zones, the analyses were performed using a fine

mesh with the cross-sectional area (in the X-Z plane) divided into 180 zones, each with aspect ratios of between 1:1.4 to 1:2.0. Figure 4 shows the mesh, along with the positions of the six displacement pins on the intrados of this quadrant of the sprayed concrete ring. The shotcrete lining was modelled by a layer of 3 zones. The plane strain models were supported with rollers at both ends and restrained from moving in the Y-direction on the top and bottom surfaces.

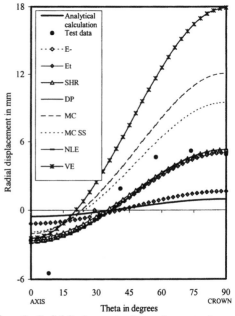

Figure 6. Radial displacements of sprayed concrete after Load Stage 4, inward movement is positive.

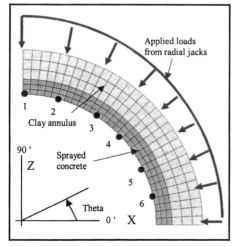

Figure 4. Schematic showing a quadrant of the FLAC 3D mesh in the X-Z plane with the applied load

Figure 5 shows the measured displacements of the six pins in the laboratory test. One can see that time-dependent behaviour accounts for a large proportion of the increment of displacement after each load increment, the instantaneous elastic response accounts for only about one third of the displacement increment and the amount of strain recovered on unloading is small.

Table 3 lists the constitutive models used in the numerical analyses, along with the identification code for each of the sprayed concrete constitutive models. This code will be used to identify the models in the rest of the paper for the sake of brevity. Where possible, the parameters for the models were taken from other tests on same SFRS mix (e.g. Figures 2 and 3). Otherwise, they were based on a review of published data. The simplifications of the test (such as the construction of the ring in one pass) sought to minimise the influence of the other factors present in a real tunnel, leaving the material behaviour of the shotcrete as the major un-quantified influence.

5 DISCUSSION

Considering the situation in the "crown" of this model of a tunnel, one can see that the numerical model with a constant modulus of 25 GPa (E-) agrees reasonably well with the prediction using the analytical method of Einstein & Schwartz (1979) (Figure 6). The agreement is not perfect because the laboratory test differs slightly from the analytical case. The measured displacements in the test differ in pattern from the analytical prediction. This is probably due to variations in the thickness of the lining. Generally the lining was 20 mm thicker than the original design thickness of 125 mm, but the overspray varied considerable around the ring, from –13 to +65 mm. Given that this ring was sprayed under laboratory conditions, this underlines the potential variability in the construction of sprayed concrete linings.

The radial deformation predicted by this constant stiffness model is much smaller compared to the other models. From Figure 7 it can be seen that this assumption results in significantly larger bending moments and axial forces. The introduction of an age-dependent modulus increases the deformation by a factor of three and reduces the bending moment by 22%. The axial force is only reduced by 5%.

Compared to the measured displacements in Figure 5, the time-dependent stiffness model (Et) appears to over-predict the elastic deformation.

A common indicator of the performance of sprayed concrete linings is the "utilization factor", α, which is the current stress/strength ratio. In this case, the utilization factor is the deviatoric stress, r, divided by current yield strength. Figure 8 shows the deviatoric stresses in the crown of the ring after Load Stages 1 to 3, along with the Drucker-Prager and Mohr-Coulomb yield surfaces at those ages. Considering the utilization factors, it can be seen that yielding is mostly confined to the initial load steps. α falls from about 1.1 in Load Stage 1 to about 0.8 at Load Stage 4. Since the yield strength was chosen as 40 % of the ultimate strength of the concrete, this means that the stresses are only about 32 to 44% of the ultimate strength. These results are for an applied loading, which was consistent with field measurements of the ground loading, increasing from 15 and 80% of the full overburden pressure over Load Stages 1 to 4 (see Figure 3). The stresses from the numerical analyses agree well with field

Table 3. FLAC constitutive models

Clay	Parameters	
Elastic model	$E = 10^5$ kPa, $\nu = 0.15$	

Sprayed concrete		
ID code	Description	Parameters
E-	Elastic constant modulus	$E = 25$ GPa, $\nu = 0.2$
Et	Elastic age-dependent modulus	$E = 27$ GPa after 28 days, $\nu = 0.2$
Shr	Et model with shrinkage	Calculated according to ACI method with ultimate shrinkage strain, $\varepsilon_{shr} = 0.001$
NLE	Non-linear elastic Et model	According to Saenz's equation, using a biaxial stress state formulation (Chen 1982) and constant Poisson's ratio
DP	Et model with age-dependent perfect plasticity	Drucker Prager yield criterion; yield strength is 0.40 times ultimate strength; strengths vary with age; ultimate compressive strength = 16 MPa at 340 hours
MC	Et model with age-dependent perfect plasticity	Mohr Coulomb yield criterion; yield strength is 0.40 times ultimate strength; strengths vary with age
MC SS	Et model with age-dependent plasticity	Strain hardening post-yield behaviour with strain at peak strength estimated from published data
VE	Et model with visco-elasticity	Maxwell model with a constant but different viscosity in each load stage – i.e. pseudo-age-dependent, $\eta = 1.25 \times 10^9$ kPa.s to 5.33×10^{12} kPa.s

data from pressure cells in SCL tunnels in London clay. This reinforces the suggestion that the factor of safety for such linings is about 3 (Bonapace 1997), although it was not possible to check this directly since the sprayed concrete rings in the laboratory tests were too thin for tangential pressure cells to be installed in them.

At such low utilization factors it would seem reasonable to conclude that the variation with age of the elastic modulus would be more important than plasticity or the non-linearity of the stress-strain behaviour. A comparison of the results from the age-dependent elastic model with the Drucker-Prager and non-linear elastic models tend to support this (Figures 7 and 9).

However, the analyses performed using a Mohr-Coulomb yield criterion resulted in significant additional deformation (twice as much as most of the other models, see Figure 6). Given that the concrete

is generally under triaxial compression, one would expect the discrepancy between the Drucker-Prager and Mohr-Coulomb perfectly plastic models to be small (since the parameters for the former are set to match the Mohr-Coulomb model on the compressive meridians - see Figure 8). The fact that the concrete is on the borderline between elasticity and yielding may have exaggerated the effect of the slightly higher yield strength, which was 16% larger for the Drucker-Prager model. As one might expect, the use of a perfectly plastic (Mohr-Coulomb) model as opposed to a strain-hardening model results in an additional deformation, in this case an additional 25%.

Considering time-dependent behaviour, if one examines the measured displacements in Figure 5, one can see that most of the displacement after each load increment is due to time-dependent behaviour. In line with this, the visco-elastic model shows large additional deformations due to creep and the elastic component of the deformations is only about one third of the total (see Figure 6). However, the magnitudes of the displacements are much greater than those recorded in the tests. The effect on the axial force is a considerable reduction of 25%, and the bending moment is reduced by 45%, compared to the age-dependent elastic model (see Figure 7).

Shrinkage has a much smaller effect on deformations but increases the tensile stresses in the ring due to the restraints at the boundaries and increases the bending moment slightly.

Similar results were obtained for the rest of the ring. The reasons for the discrepancy in magnitude of deformations may lie in the input parameters for the models, or other factors such as the exact thickness of the lining. This programme of work is continuing with an investigation of the effects of modelling the exact lining thickness and different creep models.

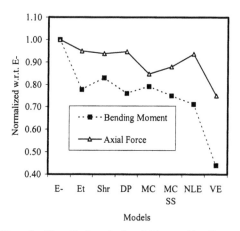

Figure 7. Normalised results for axial force and bending moment in the "crown"

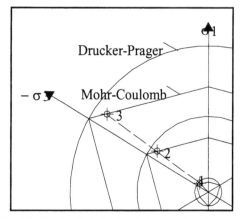

Figure 8. Stress paths in the deviatoric stress plane for results from MC SS model, after Load Stages 1,2 & 3.

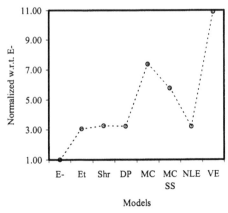

Figure 9. Normalised results for radial displacement in the "crown".

Guided by this work, 3D analyses of an advancing tunnel face will be performed to investigate the influence of constitutive models and the modelling of the lining on the distribution of stress and strain within the lining. It is recognised that while these factors may have a considerable influence on the stress and strain in the lining, they may have relatively little influence on the situation in the ground around the tunnel and the surface settlement. Therefore, there may be scope to adjust the complexity of the constitutive models for either the ground or the sprayed concrete depending on which is of most concern.

6 CONCLUSIONS

At present the determination of the distribution of stress and strain in SCL tunnel linings, either analytically or numerically, remains a difficult task. One reason for this is the complex behaviour of sprayed concrete at an early age. The most commonly used constitutive models tend to be very simplistic. If more sophisticated models are applied, the type of model required will depend on various factors, such as the utilization factor of the lining, whether significant unloading occurs and the time-scale in question. This is because the importance of plasticity, creep and shrinkage will vary depending on these factors.

Introducing age-dependency of the stiffness into an elastic model results in a considerable increase in deformation, a small decrease in the axial force and a considerable reduction in the bending moment (more than 20%). Creep behaviour may reduce both axial forces and bending moments to an even greater extent. In the case study presented here, the effects of plasticity, non-linearity of the stress-strain curve, and creep, were small but generally led to a further reduction in the bending moment. Additional work is required to examine the role of the constitutive model at higher utilization values and the effect of construction defects, such as poor shape control, weak joints, and thickness variation, to obtain better agreement between predicted values and measurements from laboratory tests and real tunnels.

ACKNOWLEDGEMENTS

The authors would like to acknowledge the contribution of all the partners in the BRITE EURAM project to that research. The first author would like to thank the Royal Commission for the Exhibition of 1851 and Mott MacDonald for their generous support for the current work at the University of Southampton.

SYMBOLS AND ABBREVIATIONS

E	elastic modulus
est.	estimated
HME	Hypothetical Modulus of Elasticity
PFA	pulverised flue ash
SCL	sprayed concrete lined
SFRS	steel fibre reinforced shotcrete
v	Poisson's ratio

REFERENCES

Aldrian, W. 1991. *Beitrag zum Materialverhalten von fruh belastetem Spritzbeton*, Diplomarbeit Montanuniversitat Leoben.

Bonapace, P. 1997. Evaluation of stress measurements in NATM tunnels at the Jubilee Line Extension Project. In Hinkel, Golser & Schubert (eds), *Tunnels for People*: 325 - 330.

Brooks, J. 1999. Shotcrete for ground support as used in the Asia Pacific region, *Rapid Excavation and Tunnelling Conference Proceedings*: 473 – 524.

Chen, W.F. 1982. *Plasticity in reinforced concrete*, New York: McGraw-Hill.

Darby, A. & Leggett, M. 1997. *Use of shotcrete as the permanent lining of tunnels in soft ground*, Mott MacDonald internal project report.

Ding, Y. 1998. *Technologische Eigenschaften von jungem Stahlfaserbeton und Stahlfaserspritzbeton*. PhD Thesis, University of Innsbruck.

Einstein, H.H. & Schwartz, C.W. 1979. Simplified Analysis for Tunnel Supports. *ASCE Journal of Geotechnical Engineering Division*: 499 – 517.

Hafez, N. M. 1995. *Post-failure modelling of three-dimensional shotcrete lining for tunnelling*, PhD Thesis, University of Innsbruck.

Hellmich, C., Ulm, F-J. & Mang, H.A. 1999. Multisurface Chemoplasticity. I: Material Model for Shotcrete. *ASCE Journal of Engineering Mechanics*, June: 692 – 701.

Malmberg, B. 1993. Shotcrete for Rock Support: a Summary Report on the State of the Art in 15 Countries, ITA report. *Tunnelling and Underground Space Technology*, 8 (4): 441-270.

Meschke, G. 1996. *Elasto-viskoplastische Stoffmodelle fur numerische Simulationen mittels der Methode der Finiten Elemente*. Habilitationsschrift, TU Wien.

Mott MacDonald 1998. *Final Technical Report, BRITE EURAM BRE2-CT92-0231 New Materials, Design and Construction Techniques for Underground Structures in Soft Rock and Clay Media*.

Moussa, A. M. 1993. *Finite Element Modelling Of Shotcrete In Tunnelling*. PhD thesis, University of Innsbruck.

Neville, A. M. 1995. *Properties of concrete*. Addison Wesley Longman Ltd, Harlow.

Norris, P. & Powell, D. 1999. Towards quantification of the engineering properties of steel fibre reinforced sprayed concrete. *3rd Int. Symp. on Sprayed Concrete*, Gol, Norway.

Pottler R. 1985. Evaluating the stresses acting on the shotcrete in rock cavity constructions with the Hypothetical Modulus of Elasticity. *Felsbau*, 3 (3): 136-139.

Rabcewicz, L. v. 1969. Stability of tunnels under rock load Part 2. *Water Power*, July: 266-273.

Ripley, B.D., Rapp, P.A. & Morgan, D.R. 1998. Shotcrete design, construction and quality assurance for the Stave Falls tunnels. *Canadian Tunnelling*: 141 - 156.

Soliman, E., Duddeck, H. & Ahrens, H. 1994. Effects of development of stiffness on stresses and displacements of single and double tunnels. *Tunnelling and Ground Conditions*, Abdel Salam (ed.): 549 - 556.

Van der Berg, J.P. 1999. *Measurement and prediction of ground movements around three NATM tunnels*. PhD Thesis, University of Surrey.

Shotcrete: Engineering Developments, Bernard (ed.) © 2001 Swets & Zeitlinger, Lisse, ISBN 90 5809 176 7

Application of Yield Line Theory to Round Determinate Panels

V.N.G.Tran & A.J.Beasley
University of Tasmania, Hobart, Australia

E.S.Bemard
University of Western Sydney, Nepean, Australia

ABSTRACT: The Round Determinate panel test has been found to provide reliable and economical post-crack performance assessment for Fibre Reinforced Concrete (FRC) and Shotcrete (FRS). However, it suffers the problem that performance parameters obtained from this test are difficult to relate to the in situ behaviour of FRS. This investigation has sought to improve understanding about the behaviour of FRS in structures by examining the relationship between post-crack behaviour in beams and corresponding performance in Round Determinate panels. Several types of FRS exhibiting post-crack strain softening have been studied using Yield Line theory to predict the load-deflection response of round determinate panels based on moment-crack rotation relationships developed from tests on beams.

1 INTRODUCTION

Post-crack performance assessment for Fibre Reinforced Concrete (FRC) and Shotcrete (FRS) has been conducted using a variety of tests in recent years. The majority of these have involved beams (eg. JSCE 1984, EFNARC 1996, ASTM 1997), although panel-based procedures such as the EFNARC panel test (EFNARC 1996) and Round Determinate panel test (Bernard and Pircher 2000) also exist. Issues that require consideration when selecting the most appropriate test to use for performance assessment purposes include the reliability of results, the cost of testing, and whether the sample is truly representative of the *in situ* concrete. In all of these respects, the Round Determinate panel has been shown to be highly effective (Bernard 1998a).

Despite these advantages, the Round Determinate panel test suffers the disadvantage that the results are difficult to relate directly to the behaviour of FRS and FRC in structures such as tunnel linings and floors. While it is intuitively obvious that the load to cause first crack of a panel is related to the Modulus of Rupture, the mechanism by which the latter can be deduced from the former is not immediately obvious. The situation with respect to post-crack performance is even less clear. An investigation was therefore instigated to develop a theoretical basis for interpreting the results of Round Determinate panel tests so that they can be used to calculate behaviour in structures such as FRS tunnel linings. The study has focussed on the relationship between the per-formance of FRS beams and the corresponding behaviour of Round Determinate panels.

1.1 *Yield Line Theory*

In the design of concrete structures, engineers consider structural behaviour both prior to and after cracking of the concrete matrix. The load to cause cracking is therefore important. If conventional steel reinforcement is employed, the load to cause first yield of the steel also plays an important role in behaviour. Yield Line theory (Johansen 1972) has proven to be a simple and effective means of calculating the load to cause yielding of steel bars in conventionally reinforced concrete structures (Jones and Wood 1967) and first crack in fibre reinforced concrete floors (Concrete Society 1994). However, rational application of this theory to materials that display post-crack strain softening has been limited to date, and only a few examples exist in which post-crack behaviour in FRC has been modelled using Yield Line theory (Holmgren 1993).

Yield Line theory is widely used for moment redistribution and for the determination of collapse loads in suspended concrete slabs (Warner et al. 1998). However, design rules incorporating this method of analysis are qualified by the requirement that slabs be under-reinforced (eg. AS3600 1997). This is because collapse loads calculated on the basis of Yield Line theory are not valid at large deflections unless the moment capacity of elements within the slab display quasi-elastic perfectly plastic behaviour (Johansen 1972). Slabs that are over-reinforced exhibit strain softening at low to moderate levels of deformation. Despite this, the determination of load capacity using Yield Line theory is not strictly limited to elastic perfectly plastic materials since this theory is based on the absorption of energy by deforming components of a chosen collapse mechanism. If the moment capacity of a component within a mechanism is altered, the work done in resisting external load changes and the load capacity will similarly change. This feature is the key to applying Yield Line theory in a step-wise analysis of strain softening materials.

In the present investigation, the Round Determinate panel has been considered a simple structure for which it is required to determine the load to cause first crack and post-crack behaviour. When a point load is introduced at the centre of a laboratory specimen, flexural stresses are developed throughout the panel. Based on elastic plate theory (Timoshenko and Woinowsky-Krieger 1959), the maximum tensile stress in an uncracked panel is predicted to occur on the opposite face along three radial lines between the supporting pivots (see Figure 1, from Bernard and Pircher 2000). As the load is increased, the tensile strength of the concrete matrix is eventually exceeded and a crack forms at the centre. This bifurcates and runs to the edges along the lines of maximum tensile stress to form a symmetric arrangement of three radial cracks. If the out-of-plane moment capacity of the material comprising the panel is truly elastic-perfectly plastic, the load resistance will be maintained as the deflection is increased. However, as the cracks widen they ultimately separate at the centre as a result of geometric constraints. No material will continue to support a moment across a discontinuity, so the load resistance will ultimately drop at severe deflections.

Consideration of experimentally observed collapse behaviour in Round Determinate panels suggests many similarities to the premises upon which Yield Line theory is based. To understand the similarities, it is necessary to examine the assumptions

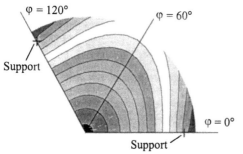

Figure 1. Radial stress distribution in a Round Determinate panel under an arbitrary central point load according to elastic plate theory (Bernard and Pircher 2000).

made in undertaking a Yield Line analysis. These include, that:
1. Each yield line (ie. crack) is a straight line. In reality, cracks are jagged and seldom straight.
2. The individual parts of a panel between the yield lines are regarded as plane. In real structures, quasi-elastic deformations arise from flexural, membrane, shear, and torsional stresses.
3. The deformation that occurs at each yield line consists of a relative rotation of the two adjoining parts of the panel about axes whose location depends upon the placement of supports.
4. Yield lines are taken to occur simultaneously. In reality, cracks propagate from the centre of the opposite face to the edges in succession.

To analyse a new structure for which the collapse mechanism is unknown, the following steps are undertaken as part of a Yield Line analysis:
1. A possible yield line pattern is adopted.
2. The ultimate moment capacity m per unit length is determined for the various yield lines.
3. The collapse load P corresponding to the assumed yield line pattern is calculated by consideration of structural equilibrium. This calculation can be achieved by the use of static or virtual work principles.
4. If necessary, the dimensions of the particular failure pattern are adjusted to minimise P.
5. If a different yield line pattern is possible, this pattern is assumed and steps 2 to 4 are repeated until a minimum value of P is found.

Central to this procedure is the virtual work theorem which states that the external work U_{ext} and the internal work U_{int} in a mechanism must be equal to maintain structural equilibrium. The external work is the summation of the products of applied (external) forces and their conjugate displacements that arise within the virtual displacement system. The internal work is the summation of the products of the internal stress resultants and their conjugate strains. In a Round Determinate panel test, the external force is the point load, P, applied at the centre of the panel, and its conjugate displacement is the deflection at the centre, δ. The internal stress resultants are the moments of resistance at each yield line, m, and their conjugate strains are the corresponding crack rotation angles.

The predicted load capacity of a structure depends on the pattern of yield lines chosen for analysis. According to Johansen (1972), the pattern that results in the lowest estimate of the equilibrium load will govern behaviour. Unfortunately, no method exists for predicting this pattern. Instead, it must be determined by trial and error, or by educated guesses. Not all patterns of yield lines are admissible. In the selection of a yield line pattern the following conditions must be satisfied:

1. A yield line between two parts of a panel must pass through the point of intersection of their axes of rotation (fold lines).
2. Each yield line pattern is determined by the axes of rotation of the various parts of the panel and the ratios between the rotations.
3. A line support must be an axis of rotation for a panel segment.
4. A point support must be on an axis of rotation.
5. A negative yield line must form at a fixed support.

For the case of a Round Determinate panel, several patterns of yield lines are possible and have been observed in laboratory tests. The two most common patterns are analysed as follows. The less common consists of a single diametral crack through the centre resulting in a beam-like failure of the panel. The more common consists of three radial cracks running from the centre to the free edges of the panel between the three pivot supports.

For the diametral pattern of yield lines (shown in Figure 2), the external and internal energies are expressed (Bernard and Pircher 2000)

$$U_{ext} = P\delta \tag{1}$$

and

$$U_{int} = 2R\theta m \tag{2}$$

where R is the radius of the panel, θ is the rotation at the yield line, and m is the moment of resistance per unit length of yield line. The rotation at the yield line is found as

$$\theta = \theta' + \theta'' \tag{3}$$

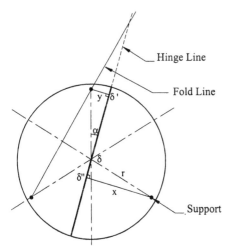

Figure 2. Diametral yield line pattern for Round Determinate panels.

where

$$\theta' = \frac{\delta'}{x} = \delta\frac{(1 - 2\cos\alpha\sin(\pi/6 - \alpha))}{r\sin\alpha} \tag{4}$$

and

$$\theta'' = \frac{\delta''}{y} = \delta\frac{(1 + 2\sin^2(\pi/6 - \alpha))}{r\cos(\pi/6 - \alpha)} \tag{5}$$

The failure load for this pattern is therefore

$$P = \frac{2Rm}{r}\left[\frac{1 - 2\cos\alpha\sin(\pi/6 - \alpha)}{\sin\alpha} + \frac{1 + 2\sin^2(\pi/6 - \alpha)}{\cos(\pi/6 - \alpha)}\right] \tag{6}$$

which is equal to a minimum value of

$$P = \frac{6mR}{r} \tag{7}$$

for $\alpha = \pi/6$. The pattern of three radial yield lines is analysed for the general case of three unequal angles between yield lines (see Figure 3) as follows. The external energy is expressed by Eqn. 1, but the internal energy is

$$U_{int} = R(m_1\theta_1 + m_2\theta_2 + m_3\theta_3) \tag{8}$$

where R is the radius of the panel, m_1, m_2, and m_3 are the moments of resistance per unit length along the three yield lines, and θ_1, θ_2, and θ_3 are the angles of rotation between the sets of planes. By the virtual work theorem, $U_{ext} = U_{int}$, hence

$$P = R(m_1\theta_1 + m_2\theta_2 + m_3\theta_3)/\delta \tag{9}$$

Since the uncracked portions of the panel are assumed to remain plane, the rotation angles at the yield lines are determined by their location and the geometry of the panel. The three angles of rotation are calculated below with reference to Figure 4, which shows a Round Determinate panel with three radial cracks arranged at arbitrary angles $\gamma_1, \gamma_2, \gamma_3$ with respect to the bisectors of the unsupported sides. Considering the yield line radiating to the lower-most corner (E), the distance from pivot K to the closest point on the yield line, B, is found as

$$h_{13} = r.\sin(\pi/3 + \gamma_1) \tag{10}$$

and the distance from pivot I to the closest point on the yield line, C, is found as

$$h_{12} = r.\sin(\pi/3 - \gamma_1) \tag{11}$$

where r is the radius to the pivoted supports. The deflections at points B and C are found as

$$\delta_B = (z - x).\,\delta/z \tag{12}$$

$$\delta_C = (z - y).\,\delta/z \tag{13}$$

where x is the distance from the centre, A, to B,

$$x = r.\cos(\pi/3 + \gamma_1) \tag{14}$$

y is the distance from A to C,

$$y = r.\cos(\pi/3 - \gamma_1) \tag{15}$$

and z is the distance from A to E,

$$z = r.\cos(\pi/3 - \gamma_1) + h_{12}/\tan\alpha_2$$

$$= r.\sin(\alpha_2 + \pi/3 - \gamma_1)/\sin\alpha_2 \tag{16}$$

To find δ_B and δ_C it is firstly necessary to determine the relationship between the corner angles α_1, α_2, α_3, α_4, α_5, and α_6. Consideration of triangles *JUZ*, *AU3*, *AVZ* and *KV3* gives:

$$\xi - \alpha_5 = \gamma_3 - \pi/6 \tag{17}$$

$$\phi + \alpha_6 = \gamma_3 + \pi/6 \tag{18}$$

Similarly,

$$\phi - \alpha_1 = \gamma_1 - \pi/6 \tag{19}$$

$$\nu + \alpha_2 = \gamma_1 + \pi/6 \tag{20}$$

$$\nu - \alpha_3 = \gamma_2 - \pi/6 \tag{21}$$

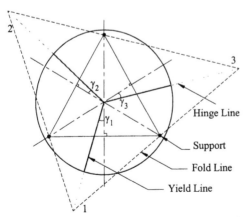

Figure 3. General pattern of three radial yield lines at unequal angles for a Round Determinate panel.

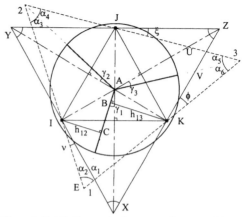

Figure 4. Detail of geometry for a general pattern of three radial yield lines at unequal angles for a Round Determinate panel. Note: ξ, ϕ, ν are the angles between the sides of the general triangle 123 and symmetric triangle XYZ.

$$\xi + \alpha_4 = \gamma_2 + \pi/6 \tag{22}$$

From (17) to (22),

$$\alpha_1 + \alpha_6 - \gamma_3 + \gamma_1 = \pi/3 \tag{23}$$

$$\alpha_2 + \alpha_3 - \gamma_1 + \gamma_2 = \pi/3 \tag{24}$$

$$\alpha_4 + \alpha_5 - \gamma_2 + \gamma_3 = \pi/3 \tag{25}$$

Based on the expressions listed above, it is possible to develop a system of equations to solve for the angle α_2, and subsequently find the remaining angles. Using this approach,

$$\alpha_2 = \operatorname{atan}\left[\frac{\cos(\gamma_1 + \frac{\pi}{6})}{-\sqrt{3}.\sin\gamma_1 + \dfrac{\cos(\gamma_1 - \frac{\pi}{6})}{A_1}}\right] \tag{26}$$

where $A_1 = \tan(\frac{\pi}{3} - \gamma_1 + \gamma_3 - A_2)$

$$A_2 = \operatorname{atan}\left[\frac{\cos(\gamma_3 + \frac{\pi}{6})}{-\sqrt{3}.\sin\gamma_3 + \dfrac{\cos(\gamma_3 - \frac{\pi}{6})}{A_3}}\right]$$

$A_3 = \tan(\frac{\pi}{3} - \gamma_3 + \gamma_2 - A_4)$ and

$$A_4 = \operatorname{atan}\left[\dfrac{\cos(\gamma_2 + \frac{\pi}{6})}{-\sqrt{3}.\sin\gamma_2 + \dfrac{\cos(\gamma_2 - \frac{\pi}{6})}{\tan(\frac{\pi}{3} - \gamma_2 + \gamma_1 - \alpha_2)}}\right]$$

Similarly,

$$\alpha_4 = \operatorname{atan}\left[\dfrac{\cos(\gamma_2 + \frac{\pi}{6})}{-\sqrt{3}.\sin\gamma_2 + \dfrac{\cos(\gamma_2 - \frac{\pi}{6})}{B_1}}\right] \qquad (27)$$

where $B_1 = \tan(\frac{\pi}{3} - \gamma_2 + \gamma_1 - B_2)$

$$B_2 = \operatorname{atan}\left[\dfrac{\cos(\gamma_1 + \frac{\pi}{6})}{-\sqrt{3}.\sin\gamma_1 + \dfrac{\cos(\gamma_1 - \frac{\pi}{6})}{B_3}}\right]$$

$$B_3 = \tan(\frac{\pi}{3} - \gamma_1 + \gamma_3 - B_4)$$

$$B_4 = \operatorname{atan}\left[\dfrac{\cos(\gamma_3 + \frac{\pi}{6})}{-\sqrt{3}.\sin\gamma_3 + \dfrac{\cos(\gamma_3 - \frac{\pi}{6})}{\tan(\frac{\pi}{3} - \gamma_3 + \gamma_2 - \alpha_4)}}\right]$$

and

$$\alpha_6 = \operatorname{atan}\left[\dfrac{\cos(\gamma_3 + \frac{\pi}{6})}{-\sqrt{3}.\sin\gamma_3 + \dfrac{\cos(\gamma_3 - \frac{\pi}{6})}{C_1}}\right] \qquad (28)$$

where $C_1 = \tan(\frac{\pi}{3} - \gamma_3 + \gamma_2 - C_2)$

$$C_4 = \operatorname{atan}\left[\dfrac{\cos(\gamma_2 + \frac{\pi}{6})}{-\sqrt{3}.\sin\gamma_2 + \dfrac{\cos(\gamma_2 - \frac{\pi}{6})}{C_3}}\right]$$

$$C_3 = \tan(\frac{\pi}{3} - \gamma_2 + \gamma_1 - C_4)$$

$$C_4 = \operatorname{atan}\left[\dfrac{\cos(\gamma_1 + \frac{\pi}{6})}{-\sqrt{3}.\sin\gamma_1 + \dfrac{\cos(\gamma_1 - \frac{\pi}{6})}{\tan(\frac{\pi}{3} - \gamma_1 + \gamma_3 - \alpha_6)}}\right]$$

The corner angles can be found for any γ_1, γ_2, and γ_3 by solving equations (26) to (28) through iteration. The rotation of yield line AE is then expressed as

$$\theta_1 = \operatorname{atan}(\delta_C/h_{12}) + \operatorname{atan}(\delta_B/h_{13}) \qquad (29)$$

which can be re-arranged as

$$\theta_1 = \operatorname{atan}(A\delta) + \operatorname{atan}(B\delta) \qquad (30)$$

where

$$A = \left[\dfrac{\sin(\alpha_2 + \frac{\pi}{3} - \gamma_1) - \cos(\frac{\pi}{3} - \gamma_1).\sin\alpha_2}{r.\sin(\frac{\pi}{3} - \gamma_1).\sin(\alpha_2 + \frac{\pi}{3} - \gamma_1)}\right] \qquad (31)$$

$$B = \left[\dfrac{\sin(\alpha_2 + \frac{\pi}{3} - \gamma_1) - \cos(\frac{\pi}{3} + \gamma_1).\sin\alpha_2}{r.\sin(\frac{\pi}{3} + \gamma_1).\sin(\alpha_2 + \frac{\pi}{3} - \gamma_1)}\right] \qquad (32)$$

Similarly, the rotations of the other two yield lines can be expressed as

$$\theta_2 = \operatorname{atan}(C\delta) + \operatorname{atan}(D\delta) \qquad (33)$$

$$\theta_3 = \operatorname{atan}(E\delta) + \operatorname{atan}(F\delta) \qquad (34)$$

where

$$C = \left[\dfrac{\sin(\alpha_4 + \frac{\pi}{3} - \gamma_2) - \cos(\frac{\pi}{3} - \gamma_2).\sin\alpha_4}{r.\sin(\frac{\pi}{3} - \gamma_2).\sin(\alpha_4 + \frac{\pi}{3} - \gamma_2)}\right] \qquad (35)$$

$$D = \left[\frac{\sin(\alpha_4 + \frac{\pi}{3} - \gamma_2) - \cos(\frac{\pi}{3} + \gamma_2).\sin \alpha_4}{r.\sin(\frac{\pi}{3} + \gamma_2).\sin(\alpha_4 + \frac{\pi}{3} - \gamma_2)} \right] \quad (36)$$

$$E = \left[\frac{\sin(\alpha_6 + \frac{\pi}{3} - \gamma_3) - \cos(\frac{\pi}{3} - \gamma_3).\sin \alpha_6}{r.\sin(\frac{\pi}{3} - \gamma_3).\sin(\alpha_6 + \frac{\pi}{3} - \gamma_3)} \right] \quad (37)$$

$$F = \left[\frac{\sin(\alpha_6 + \frac{\pi}{3} - \gamma_3) - \cos(\frac{\pi}{3} + \gamma_3).\sin \alpha_6}{r.\sin(\frac{\pi}{3} + \gamma_3).\sin(\alpha_6 + \frac{\pi}{3} - \gamma_3)} \right] \quad (38)$$

The load to cause first crack of the concrete matrix can be found when the deflection tends to zero because elastic deformation prior to cracking is ignored. Therefore, from Eqn. 9,

$$P_{Crack} = \lim_{\delta \to 0} P$$
$$= R \left[m_1 \lim_{\delta \to 0} \frac{\theta_1}{\delta} + m_2 \lim_{\delta \to 0} \frac{\theta_2}{\delta} + m_3 \lim_{\delta \to 0} \frac{\theta_3}{\delta} \right] \quad (39)$$

Applying L'Hopital's Rule,

$$\lim_{\delta \to 0} \frac{\theta_1}{\delta} = \lim_{\delta \to 0} \frac{[a \tan(A\delta) + a \tan(B\delta)]}{\delta} = A + B \, (40)$$

Similarly,

$$\lim_{\delta \to 0} \frac{\theta_2}{\delta} = C + D \quad \text{and} \quad \lim_{\delta \to 0} \frac{\theta_3}{\delta} = E + F \quad (41)$$

Thus,

$$P_{Crack} = R(m_1(A+B) + m_2(C+D) + m_3(E+F)) \quad (42)$$

For the symmetric case in which all included angles between yield lines equal 120° (ie. all midpoint angles γ equal zero), $m_1 = m_2 = m_3 = m$ and

$$A = B = C = D = E = F = \frac{\sqrt{3}}{2r} \quad (43)$$

Thus,

$$P_{Crack} = 3\sqrt{3}m\frac{R}{r} \quad (44)$$

This expression is the same as Eqn 12 in Bernard (1998b) which was obtained by a simplified analysis of the symmetric case. The magnitude of this estimate of P_{crack} is 13 percent lower than the value

given in Eqn. 7 for the diametral mode of failure, so the symmetric mode of failure will theoretically govern behaviour.

The analysis described above is applicable to any collapse mechanism for which the moment resistance at each yield line is known. At the point of first crack of the concrete matrix, the moment resistance is the moment to cause first crack in beam elements representing the one-way bending capacity of the panel. Post-crack capacity can be determined by increasing the displacement at the centre of the panel and using the moment of resistance offered by beams at each corresponding crack rotation angle to find the load at equilibrium with these moments. The post-crack analysis must be performed in a step-wise manner to model the changing moment of resistance offered by each of the yield lines as the rotation angles increase.

1.2 Energy Calculation

The theory described above outlines the approach required to solve for geometric constraints upon yield line formation in a round determinate panel. However, there are two alternative approaches to the solution of the virtual work theorem. The difference between them is related to the way the internal energy is calculated, as shown below.

1.2.1 Standard Model

In this model the moment of resistance at each yield line is taken to be constant between the onset of loading and the crack rotation angle under consideration, but the degree of resistance to deformation changes as the crack rotation angle increases. The internal energy is calculated for each yield line as

$$U_{int} = ml\theta \quad (45)$$

where m is taken to be the instantaneous moment capacity offered by a particular yield line of length l at a crack rotation angle of θ. This is represented graphically by the shaded area in Figure 5.

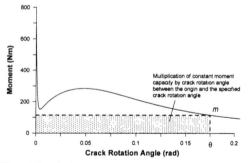

Figure 5. Standard approach to determination of moment-crack rotation relation between onset of cracking and a crack rotation angle of θ.

250

1.2.2 Integration Model

An alternative approach is to consider the variation in moment capacity at each yield line up to the level of deformation under consideration and incorporate this into the energy calculation expression. This is equivalent to

$$U_{int} = \int m(\theta)l d\theta \qquad (46)$$

in which m is taken to vary as a function of θ as shown in Figure 6. The two approaches produce the same result for elastic perfectly plastic materials, but produce different results for other types of post-cracking behaviour.

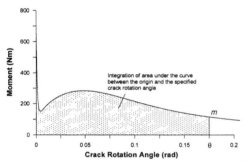

Figure 6. Integration of moment-crack rotation relation between onset of cracking and a crack rotation angle of θ.

1.3 Numerical Analysis

The theory outlined above was developed into a numerical program to carry out a step-wise analysis of post-crack behaviour in Round Determinate panels. For each increment of displacement at the centre, the rotation at each crack was calculated and used to find the moment of resistance based on a moment-crack rotation relationship derived from beams. The virtual work theorem was then used to solve for the load resistance using both the standard and integration approaches to internal energy calculation. The structure and operation of the program are described in Figure 7.

2 EXPERIMENTAL VALIDATION

Validation of the theory and numerical methods described above was carried out using data obtained from an experimental study by Bernard et al. (2000). This study involved tests on large numbers of Round Determinate panels and centrally loaded beams produced using identical materials. Four sets of specimens were produced using shotcrete reinforced with seven different types of fibre. Each set consisted of 20 beams and 20 panels in order to develop very reliable estimates of the characteristic behaviour. The mix design used by Bernard et al (2000) for the concrete is listed in Table 1, and the fibre types and dosages are listed in Table 2. Note that more than one type of fibre was used in some of the mixes to achieve certain post-crack characteristics.

2.1 Beam testing

Moment-crack rotation relationships were measured using the centrally loaded beam test, developed by Bernard (1999). This test involves the imposition of a central point load under displacement control on a saw-cut FRS beam and measurement of rotation at

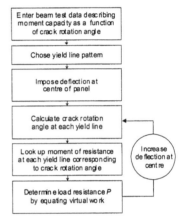

Figure 7. Structure and operation of numerical code for the estimation of post-crack behaviour using Yield Line theory.

Table 1. Mix design for FRC, all quantities in kg/m³ unless otherwise indicated.

Ingredient	Sets 1-3	Set 4
Coarse agg. (5/7 mm)	640	640
Coarse sand (5 mm)	620	560
Fine sand (2 mm)	420	380
Cement (ASTM Type 1)	360	380
Fly ash	-	40
Silica Fume	40	40
Water reducer	1900 mL	1900 mL
Slump	65 mm	65 mm

Table 2. Fibre dosages (and source) used for each specimen set.

Set	Fibre type	Dosage (kg/m³)
1	Novotex 0730 (Synthetic Ind.)	34
	256 EE (BHP Fibresteel)	27
2	50 mm HPP (Synthetic Ind.)	12
3	52 mm polyolefin (Dalhousie)	7.5
4	Dramix RC65/35 (Bekaert)	20
	Dramix BP80/35 (Bekaert)	15
	50 mm HPP (Synthetic Ind.)	3

251

the crack as a function of the applied moment. In contrast to data produced using conventional third-point loaded beams (ASTM 1997, EFNARC 1996) this test results in data of direct structural relevance. The size of the specimen used is the same as that used in the EFNARC third-point beam test (75×125×550 mm, on a 450 mm span). The method used to measure and calculate the relationship between moment and crack rotation in these specimens is described in detail in Bernard et al. (2000).

2.2 *Panel testing*

In the Round Determinate panel test (Bernard and Pircher 2000) a central point load is imposed on a specimen measuring 75×800 mm diameter, supported on three radial points located on a 750 mm diameter. Specimens tested by Bernard et al (2000) were placed in a test fixture located within an Instron 8506 servo-hydraulic test machine and loaded in displacement-control up to 100 mm total central deflection.

3 RESULTS

The results of the experiments by Bernard et al. (2000) consisted of data representing the moment-crack rotation relationships for four sets of Centrally Loaded beams, and load-deflection histories for four corresponding sets of Round Determinate panels. An example of the results for specimen Set 3 are shown in Figures 8 and 9. The results for each set of nominally identical specimens have been super-imposed to illustrate the level of variability typical for the beams and panels. Note that the beams generally suffered a very abrupt drop in moment capacity immediately after cracking. This was particularly pronounced in the sets reinforced with polymer fibres.

A proprietary curve-fitting program called *Tablecurve 2-D* was used to perform a least-squares adjusted curve fit of 3667 linear and non-linear two dimensional expressions to each set of experimental data. This was carried out so that the beam results could be used as input to the numerical analysis. The panel test results were curve-fitted so that they could be compared to the results of the numerical analysis. The curve-fitted equations for specimen Set 3 are superimposed as dark lines in Figures 8 and 9. For

each set of specimens, all 20 results were analysed simultaneously to arrive at the most suitable expression. The panel data used in these analyses was not corrected for thickness or diameter.

The expressions that resulted from curve-fitting the beam test data were used as input to the Yield Line analysis to produce estimates of post-crack behaviour in the panels. These have been compared to the experimental results from the panel tests in Figures 10 to 13 and in Table 3. In each of these figures, the dark line represents the curve-fitted expression for the results of 20 panel tests. The other two lines represent the results of numerical analysis based on moment-crack rotation data obtained from the beam tests. All the numerical analyses were performed for a symmetric arrangement of 3 radial yield lines and the pre-crack displacement has been subtracted from the record.

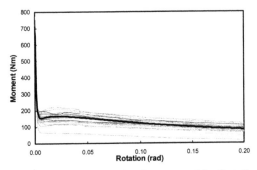

Figure 8. Moment-crack rotation data generated for Centrally Loaded beams from Set 3.

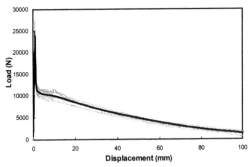

Figure 9. Load-displacement data generated for Round Determinate panels from Set 3.

Table 3. Comparison of experimental and theoretical results obtained by Yield Line analysis

Panel Set	Load to Cause First Crack (N)			Residual Load at 40 mm (N)				
	Experiment	Theory	Test/Theory	Experiment	Standard	Test/Standard	Integration	Test/Integration
1	33160	37263	0.890	1316	1358	0.969	7940	0.1710
2	23371	32165	0.727	4128	4745	0.870	10459	0.3947
3	28179	31828	0.885	5696	3860	1.476	7088	0.8036
4	43557	45144	0.965	7713	7219	1.068	14627	0.5273

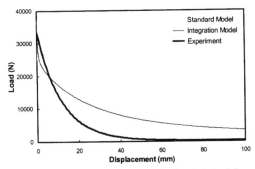

Figure 10. Load-displacement curves for Set 1 derived from experiments and Yield Line analysis.

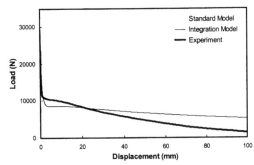

Figure 12. Load-displacement curves for Set 3 derived from experiments and Yield Line analysis.

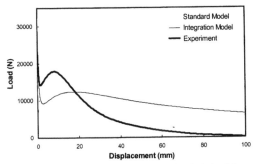

Figure 11. Load-displacement curves for Set 2 derived from experiments and Yield Line analysis.

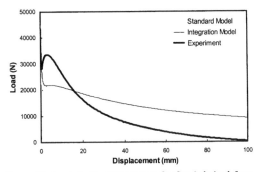

Figure 13. Load-displacement curves for Set 4 derived from experiments and Yield Line analysis.

4 DISCUSSION

On inspection of the results it is apparent that the numerical estimate of the load to cause first crack of the panels is always greater than that found experimentally. However, in the immediate post-crack range, the numerical estimates of residual load capacity were lower than the experimental result. It is also apparent that the integration model results in a higher estimate of post-crack capacity than the standard model, especially at large deflections. There are a number of possible reasons for this.

Firstly, experimental factors may have caused differences in material behaviour between the beams and panels. While the specimens examined in this investigation were all cured under the same conditions, the panels were tested in a surface dry state and the beams were tested in a surface saturated state. Bernard and Clements (2001) have shown that panels tested in a surface dry state exhibit a 16 per cent reduction in the load to cause first crack compared to panels tested in a surface saturated state. In the present investigation, the numerical estimates of the load to cause first crack in the panels were 14 per cent greater on average than the experimental results for the panels. Furthermore, the beams were cut on all faces, but the panels had a cast tensile face and a trowelled upper face. The condition of the tensile and compression faces therefore differed between the two types of specimen.

The difference between the numerical and experimental estimates of residual load capacity in the immediate post-crack range may have arisen out of inherent problems with beam testing. Beams are known to exhibit unstable behaviour immediately after cracking if residual load capacity is low. All the present beam tests were undertaken in displacement control and several of the mixes exhibited very low residual load capacity immediately after cracking which make them prone to instability. The consequences of unstable post-crack behaviour have been widely debated (Mindess et al 1995), but the present results suggest that unstable beam behaviour may under-estimate capacity immediately after cracking.

The discrepancies between the experimental and numerical results may also be due to shortcomings in the numerical analyses. Firstly, compressive arch action between the load and supports was ignored. Secondly, the yield lines were assumed to be symmetrically arranged, but in the laboratory panel tests the angles between each yield line were close to but not equal to 120°. As the discrepancy in magnitude between the three angles is increased, the numerical

253

analysis predicts greater load resistance both at first crack and in the post-crack range.

The numerical estimates of residual load capacity at high levels of deformation in the panels were also greater than found in the experiments, especially when the integration model was used to calculate internal energy. The most likely explanation for this is that the beam data upon which the numerical results were based did not experience tensile axial loads equivalent to the membrane stresses suffered by the panels at large deflections. The uncracked parts of each panel gradually separated at the centre as the deflection was increased, hence the moment resistance offered would have been lower than exhibited by simply supported beams at similar crack rotation angles. This is a phenomenon that is very difficult to incorporate into a yield line analysis because each yield line is assumed to experience a constant moment of resistance along its length.

Perhaps the most surprising result of the numerical analyses is that the standard model of internal energy calculation produced better overall estimates of post-crack performance than the integration model, despite the inconsistency of this approach. It appears that the errors in internal energy calculation inherent in this approach were cancelled out by neglect of the effect of axial tension across cracks.

5 CONCLUSION

A theoretical relationship between the behaviour of Round Determinate panels and Centrally Loaded beams made of FRS was developed on the basis of Yield Line theory. This was validated using experimental data obtained from a large number of tests on FRS specimens incorporating different fibre types and dosages. The theoretical analysis was found to predict behaviour with reasonable accuracy, but agreement was limited by differences between the methods of preparing and testing the beams and panels.

The results of this investigation have indicated that Yield Line theory is capable of modelling post-crack behaviour in strain softening FRS, but the accuracy of the prediction depends on the method used to calculate equilibrium within the failure mechanism. There is also a requirement to develop procedures that account for membrane tension across yield lines as this appears to exert a significant influence on behaviour.

6 ACKNOWLEDGEMENTS

The authors gratefully acknowledge the support of the following organisations and individuals in this investigation: CSR Readymix Concrete, through their representative Dr. Dak Baweja, for providing concrete; Jetcrete Australia P/L, for assistance in producing specimens, and Synthetic Industries for support in funding the investigation. The respective fibre suppliers are also thanked for fibres.

REFERENCES

American Society for Testing and Materials, Standard C-1018, 1997, "Standard Test Method for Flexural Toughness and First-Crack Strength of Fiber-Reinforced Concrete (Using Beam With Third-point Loading)", ASTM, West Conshohocken.

Australian Standard AS3600 1995, *Concrete Structures*, Standards Australia, Sydney.

Bernard, E.S., 1998a, "Measurement of Post-cracking Performance in Fibre Reinforced Shotcrete", *Australian Shotcrete Conference 1998*, Sydney, October 8-9.

Bernard, E.S., 1998b, "The Behaviour of Round Steel Fibre Reinforced Concrete Panels under Point Loads", *Engineering Report CE8*, Department of Civil and Environmental Engineering, University of Western Sydney.

Bernard, E.S., 1999, "Correlations in the Performance of Fibre Reinforced Shotcrete Beams and Panels", *Engineering Report CE9*, Civic Engineering and Environment, University of Western Sydney.

Bernard, E.S. and Clements, M.J.K., 2001, "The Influence of Curing on the Performance of Fibre Reinforced Shotcrete Panels", *International Conference on Engineering Developments in Shotcrete*, April 2-4, Hobart, Tasmania.

Bernard, E.S., Fagerberg, K. M. S., and Overmo, E. A., 2000, "Moment-Crack Rotation Relationships for Fibre Reinforced Shotcrete Beams and Panels", *Engineering Report CE13*, Civic Engineering and Environment, University of Western Sydney.

Bernard, E.S. and Pircher, M., 2000, "Influence of Geometry on Performance of Round Determinate Panels made with Fibre Reinforced Concrete", *Civil Engineering Report CE10*, School of Civic Engineering and Environment, UWS Nepean.

Concrete Society, 1994, *Concrete Industrial Ground Floors – A Guide to their Design and Construction*, Technical Report No. 34, 2nd ed., London.

European Specification for Sprayed Concrete, 1996, European Federation of National Associations of Specialist Contractors and Material Suppliers for the Construction Industry (EFNARC), Aldershot.

Holmgren, J., 1993, "The Use of Yield-Line Theory in the Design of Steel Fibre Reinforced Concrete Slabs", *Shotcrete for Underground Support VI*, Proceedings of the Engineering Foundation Conference, Niagara-on-the-Lake, Canada, May 2-6.

Japanese Society of Civil Engineers, 1984, "Method of Test for Flexural Strength and Flexural Toughness of SFRS", Standard JSCE-SF4.

Johansen, K.W., 1972, "Yield Line Theory", Cement and Concrete Association, London.

Jones, L.L. & Wood, R.H., 1967, *"Yield-line Analysis of Slabs"*, Thames and Hudson, London.

Mindess, S. Taerwe, L. Lin, Y-Z., Ansari, F. and Batson, G. "Standard Testing", Chpt. 10, *High Performance Fiber Reinforced Cement Composites 2*, ed. A.E. Naaman and H.W. Reinhart, E&FN Spon, London, 1995.

Timoshenko S.P., and Woinowsky-Krieger S., 1959, "Theory of Plates and Shells", McGraw-Hill, New York.

Warner, R.F., Rangan, B.V., Hall, A.S. and Faulkes, K.A., 1998, "Concrete Structures", Addison Wesley Longman, Sydney.

Shotcrete: Engineering Developments, Bernard (ed.) © 2001 Swets & Zeitlinger, Lisse, ISBN 90 5809 176 7

Author index

9 789058 091765